Ohio State University Mathematical Research Institute Publications 7

Editors: Gregory R. Baker, Walter D. Neumann, Karl Rubin

Ohio State University
Mathematical Research Institute Publications

1 Topology '90, *B. Apanasov, W. D. Neumann, A. W. Reid,
   L. Siebenmann* (Eds.)
2 The Arithmetic of Function Fields, *D. Goss, D. R. Hayes,
   M. I. Rosen* (Eds.)
3 Geometric Group Theory, *R. Charney, M. Davis, M. Shapiro* (Eds.)
4 Groups, Difference Sets, and the Monster, *K. T. Arasu, J. F. Dillon,
   K. Harada, S. Sehgal, R. Solomon* (Eds.)
5 Convergence in Ergodic Theory and Probability, *V. Bergelson,
   P. March, J. Rosenblatt* (Eds.)
6 Representation Theory of Finite Groups, *R. Solomon* (Ed.)

The Monster and Lie Algebras

Proceedings of a Special Research Quarter
at The Ohio State University, May 1996

Editors

J. Ferrar
K. Harada

Walter de Gruyter · Berlin · New York 1998

Editors
JOSEPH FERRAR, KOICHIRO HARADA
Department of Mathematics, The Ohio State University
231 West 18th Avenue, Columbus, OH 43210, USA

Series Editors:
Gregory R. Baker
Department of Mathematics, The Ohio State University, Columbus, Ohio 43210-1174, USA
Karl Rubin
Department of Mathematics, Stanford University, Stanford, CA 94305-2125, USA
Walter D. Neumann
Department of Mathematics, The University of Melbourne, Parkville, VIC 3052, Australia

1991 Mathematics Subject Classification: 20-06, 17-06; 17Bxx

⊚ Printed on acid-free paper which falls within the guidelines of the ANSI to ensure permanence and durability.

Library of Congress — Cataloging-in-Publication Data

The Monster and Lie algebras : proceedings of a special research
  quarter held at the Ohio State University, Spring 1996 /
  editors, J. Ferrar, K. Harada.
        p.   cm.   — (Ohio State University Mathematical
  Research Institute publications ; 7)
     ISBN 3-11-016184-2 (alk. paper)
     1. Lie algebras — Congresses.   2. Vertex operator alge-
  bras — Congresses.   3. Group theory — Congresses.   4. Mathematical physics — Congresses.
  I. Ferrar, J. (Joseph)   II. Harada, Koichiro, 1941— .   III. Series
  QC20.7.L54M66   1998
  512'.55—dc21   98-29397
                                                          CIP

Die Deutsche Bibliothek — Cataloging-in-Publication Data

The Monster and Lie algebras : proceedings of a special research
  quarter at The Ohio State University, May 1996 / ed. J. Ferrar ; K. Harada. — Berlin ; New York : de Gruyter, 1998
  (Ohio State University Mathematical Research Institute
  publications ; 7)
  ISBN 3-11-016184-2

Printed in Germany.
Typeset using the authors' T$_E$X files: I. Zimmermann, Freiburg.
Printing: WB-Druck GmbH & Co., Rieden/Allgäu. Binding: Lüderitz & Bauer GmbH, Berlin.
Cover design: Thomas Bonnie, Hamburg.

Preface

These Proceedings are an outgrowth of a conference culminating a special research quarter held at the Ohio State University in Spring 1996 and supported by the O.S.U. Mathematical Research Institute and the National Science Foundation. The focus of the conference was groups, Lie algebras and the Monster, and considerable emphasis was placed on presenting the various aspects of group theory and Lie algebra theory in a modern perspective. The Monster served nobly as a central theme with deep roots in both Lie algebra and group theory. The papers in this volume represent only a portion of the material presented at the conference but they give a fascinating picture, in both expository presentations and state-of-the-art research reports, of the breadth and vitality of current activity in these areas and of the extraordinary interconnections that have repeatedly surfaced in their study.

It is a pleasure to acknowledge the efforts of a number of people who helped make the special quarter and conference successful. The Ohio State University Mathematics Department staff—particularly Marilyn Radcliffe—provided much needed organizational support. Terry England played an important role in the preparation of manuscripts for publication. Professor Tom Gregory shared much of the responsibility for planning the special quarter and the conference, and for hosting long-term visitors.

The Monster.   Three years after the first, the second Columbus Monster conference was held in May, 1996. In the preface of the proceedings of the first Columbus Monster Conference (de Gruyter, Berlin–New York, 1996), I wrote:

'The *moonshine* has not yet turned into the *sunshine*, far from it actually. Our progress, however, is slow and steady. The year 1993, when the conference was held, is the 20th year since the Monster first appeared in the world. The Monster is now an adult at least physically if not mentally.'

The Monster now is 24 years old and the *moonshine* is still a *moonshine*. The Monster, however, has not exhausted its supply of surprises for us. Given below is a brief description of each individual work submitted for these proceedings.

A. Baker contributed a paper which is to give algebraic topologists an introduction to some of the algebraic ideas associated with vertex operator algebras and to show to algebraists that the vertex operator algebras have topological significance, especially in Conformal Field Theory. Vertex operator algebras have appeared in connection with elliptic genera and elliptic cohomology also.

Dong–Li–Mason–Montague investigates the radical of a vertex operator algebra V. If v is an element of the homogeneous component V_k of V, then the endomorphism v_n maps V_m to $V_{m+k-n-1}$. In particular, the *zero mode* $o(v) = v_{k-1}$ acts on V_n for each $n \in \mathbb{Z}$. The radical $J(V)$ of the vertex operator algebra V is defined by

$$J(V) = \{v \in V \mid o(v) = 0\}.$$

They show that under a certain condition for V, $J(V)$ can be determined.

Dong–Li–Mason–Norton constructs many interesting (maximal) associative subalgebras in the weight 2 subspace V_2 of the moonshine module $V^\natural$. The key idea of Dong–Li–Mason–Norton is to relate this question to the theory of root systems and the Niemeier lattices. For example, the $L(\frac{1}{2}, 0)^{\otimes 24}$ corresponds to the lattice A_1^{24}.

R. L. Griess constructs a vertex operator algebra with automorphism group $O^+(10, 2)$, one of the 3-transposition groups.

T. Hsu deals with quilts (developed by Norton, Parker, Conway, and Hsu). Some basic definitions and results are first given. He shows how quilts may be used to study an action of $\mathbf{B}_3$, the 3-string braid group, on pairs of elements of a finite group. The theory of quilts may be relevant to Norton's generalized moonshine conjectures though no strong evidence is yet to been seen. Some examples are worked out.

M. Miyamoto has defined automorphisms for a vertex operator algebra which contains an element v whose components v_n of the vertex operator $Y(v, z)$ generate a Virasoro algebra of central charge $\frac{1}{2}$. If such an element exists, then using the fusion rule of the rational vertex operator algebra $L(\frac{1}{2}, 0)$, one can construct an involutive automorphism on V. Miyamoto's initial idea to produce involutive automorphisms appears to be able to give rise to automophisms of order larger than 2.

S. Norton has been involved in the Monster/moonshine since its birth some 24 year ago. His name is associated with such notions as the generalized moonshine conjecture, quilts, and of course, the epoch-making paper of Conway–Norton. In the paper submitted for these proceedings, Norton discusses his new game called net, which is a finite geometry associated with the triple (a, b, c) where a, b, c are three involutions of a group G. The subgroup $\langle a, b, c \rangle$ is called the net group.

C. Simons's paper deals the Y-diagram whose group of automorphisms is isomorphic to the Bimonster, the wreathed product of $\mathbb{M}$ by a cyclic group of order 2, the fact first being conjectured and then proved, by a combined effort of J. H. Conway, A. D. Pritchard, S. P. Norton, A. A. Ivonov. Let τ be the wreathing involution of the Bimonster. τ is called the reflection element of the Bimonster. The generators of the Y-diagram correspond to reflections called fundamental Monster roots. Simons shows that the naming relationship will provide an easy way to compute with the Monster.

The conference was stimulating for all participants. The speakers and their topics were the following (in the order of presentation):

John McKay, Concordia University (Canada), *Some indiscrete thoughts about the Monster*;

Alex Ryba, University of Minnesota, *Modular moonshine*;

Ching Hung Lam, Ohio State University, *Vertex operator algebras with $V_1 = 0$*;

Robert L. Griess, Jr., University of Michigan, *Automorphisms of VOAs*;

Simon Norton, University of Cambridge (U.K.), *Footballs and the Monster*;

Geoffrey Mason, University of California at Santa Cruz, (Colloquium talk) *Modular functions of genus zero and the generalized moonshine conjectures*;

Elizabeth Jurisich, University of Chicago, *On integrable modules for generalized Kac–Moody algebras*;

Masahiko Miyamoto, University of Tsukuba (Japan), *Tensor products of Ising models and the moonshine module*;

Steven Smith, University of Illinois at Chicago, *Finite-geometry structures associated with the Monster*;

Gerald Hoehn, University of California at Santa Cruz, *Self-dual codes, lattices and vertex operator algebras — a fourfold analogy*;

Chongying Dong, University of California at Santa Cruz, *Compact automorphism groups of vertex operator algebras*;

Haisheng Li, University of California at Santa Cruz, *Certain associative algebras similar to $U(sl_2)$ and Zhu's algebra $A(V_L)$*;

Chris Cummins, Concordia University (Canada), *Modular equations and moonshine functions*;

Michael P. Tuite, University College, Galway (Ireland) and Dublin Institute for Advanced Studies, Dublin (Ireland), *Moonshine for Siegel forms?*;

Paul S. Montague, University of California at Santa Cruz, *On the self-duality of the Monster module and other lattice orbifold theories*;

Timothy Hsu, Princeton University, *Quilts and braid actions on elements of finite groups: an overview*;

Chris Simons, Princeton University, $\mathbb{M}_{666}$ *and Monster roots*.

Koichiro Harada

Lie Algebras.   1996 marked the 25th anniversary of a Conference on Lie Algebras held at the Ohio State University and featuring presentations by Nathan Jacobson and A. Adrian Albert. At the time of that first conference, Kac–Moody algebras were in their infancy, Lie superalgebras were yet to appear on the scene, the classification of simple modular Lie algebras was but a wishful thought, and quantum groups were not yet conceived. By the date of this most recent conference, much of this had changed, and with it the thrust of research efforts directed towards the study of Lie algebras and related topics. The goal of the conference organizers was to represent as broad a view as possible of the directions in which recent advances have pushed the study of Lie algebras. Presentations covered topics from geometry, representation theory, modular Lie algebras of low characteristic, (extended) affine Lie algebras, quantum groups, groups associated with Lie algebras, and applications of Hopf algebras in the study of Lie algebras. The papers contributed to these proceedings are a representative cross section of the invited presentations. They are well-suited for a reader interested in surveying current research trends associated with Lie algebras.

A complete list of speakers, and their topics, for this portion of the conference follows:

M. Kuznetsov, Nizhny Novgorod State University (Russia), *Simple modular graded Lie algebras*;

G. Benkart, University of Wisconsin, Madison, *Towards a representation theory for Lie algebras graded by finite root systems*;

S. J. Kang, Seoul National University (Korea), *Denominator identity for graded Lie algebras and Lie superalgebras*;

K. Misra, N. Carolina State University, *Demazure modules for the affine Lie algebra $sl(2)$*;

R. Farnsteiner, University of Wisconsin, Milwaukee, *Auslander–Reiten theory for restricted Lie algebras*;

J. Feldvoss, Hamburg University (Germany), *Burnside's Theorem for restricted Lie algebras*;

D. Nakano, Utah State University, *Support varieties for induced modules over quantum groups*;

I. Kantor, Lund University (Sweden), *A generalization of a Jordan approach to Riemannian symmetric spaces*;

J. Faulkner, University of Virginia, *Elementary groups for Kantor pairs*;

R. Griess, University of Michigan, *Recent progress in finite subgroups of Lie groups*;

B. Allison, University of Alberta (Canada), *Extended affine Lie algebras*;

I. Kryliouk, Gainesville, Florida, *The automorphism groups of some quasi-simple Lie algebras*;

R. Block, University of California, Riverside, *A Hopf algebra generalization of the elimination theory for free Lie algebras*;

N. Jing, N. Carolina State University, *On twisted quantum affine algebras*.

Joe Ferrar

Table of Contents

PART I

THE MONSTER

Vertex operators in algebraic topology

Andrew Baker

Introduction

This paper is intended for two rather different audiences. First we aim to provide algebraic topologists with a timely introduction to some of the algebraic ideas associated with vertex operator algebras. Second we try to demonstrate to algebraists that many of the constructions involved in some of the most familiar vertex operator algebras have topological (and indeed geometric) significance. We hope that both of these mathematical groups will benefit from recognition of their links in this area. Rather than simply attempting to survey the area, we have reworked some aspects to emphasise integrality and other algebraic features that are less well documented in the literature on vertex operator algebras, but probably well understood by experts.

The notion of a *vertex operator algebra* is due to R. Borcherds and arose in the algebraicization of structures first uncovered in the context of Conformal Field Theory and representations of infinite dimensional Lie algebras and groups. A spectacular example is provide by the *Monster vertex operator algebra, $V^\sharp$*, whose automorphism group is the Monster simple group $\mathbb{M}$. As well as the book of Frenkel, Lepowsky and Meurman [5], the paper of Dong [2] and the memoir of Frenkel, Huang and Lepowsky [4] provide algebraic details on vertex operator algebras, and we take these as basic references.

The work of [6] already gives a hint that there is an 'integral' structure underlying some of the algebraic aspects of Conformal Field Theory. In this paper we will show that there are integral (at least after inverting 2) structures within some of the most basic examples of vertex operator algebras associated to positive definite even lattices. We will also interpret such algebras in terms of the (co)homology of spaces related to the classifying space of K-theory. In future work we will further clarify the topological connections by explaining their origins in the geometry of certain free loop spaces as described in the work of Pressley and Segal [9], [8].

Although our topological interpretation of vertex operator algebras involves *homology*, it could just as easily (and perhaps more naturally) be given in terms of *cohomology*. We could even describe such structures in generalized (co)homology theories, particularly complex oriented theories. There is some evidence that vertex operators may usefully be viewed as giving rise to families of unstable operations in such theories, perhaps leading to algebraic generalizations of vertex operator algebras appropriate to the study of some important examples, and we intend to consider these issues in future work.

For the benefit of topologists we note that vertex operator algebras have appeared in work of H. Tamanoi and others in connection with elliptic genera and elliptic cohomology as well the study of loop spaces and particularly loop groups. Indeed it is possible that a geometric model of elliptic cohomology will involve vertex operator algebras and their modules.

My understanding of the material in this paper owes much to the encouragement and advice of Jack Morava and Hirotaka Tamanoi, as well as the many preprints supplied by Geoff Mason and Chongying Dong. I also wish to acknowledge financial support from the EU, University of Glasgow, IHES, NSF, Ohio State University and Osaka Prefecture. Finally I would like to thank Koichiro Harada for organising the timely Ohio meeting for the Friends of the Monster.

§1 Vertex operator algebras and their modules

Let $\Bbbk$ be a field of characteristic 0. Let $\mathcal{V} = \mathcal{V}_\bullet$ denote a $\mathbb{Z}$-graded vector space over $\Bbbk$; following [5], we denote the n th grading by $\mathcal{V}_{(n)}$ and for $v \in \mathcal{V}_{(n)}$ we refer to n as the *weight* of v and write $\operatorname{wt} v = n$. Whenever we refer to elements of $\mathcal{V}$, we always assume that they are homogeneous. Suppose that there is a $\Bbbk$-linear map

$$Y(\,,z): \mathcal{V} \longrightarrow \operatorname{End}_\Bbbk(\mathcal{V})[[z, z^{-1}]],$$

where for any abelian group M,

$$M[[z, z^{-1}]] = \left\{ \sum_{n \in \mathbb{Z}} m_n z^n : m_n \in M \right\}.$$

We write

$$Y(v, z) = \sum_{n \in \mathbb{Z}} v_n z^{-n-1},$$

where $v_n \in \operatorname{End}_\Bbbk(\mathcal{V})$, $v_n u = v_n(u)$ and

$$Y(v, z)u = \sum_{n \in \mathbb{Z}} (v_n u) z^{-n-1}.$$

The pair $(\mathcal{V}, Y)$ gives rise to a *vertex operator algebra* if the following axioms are satisfied.

VOA-1   For each $n \in \mathbb{Z}$, $\dim_\Bbbk \mathcal{V}_{(n)} < \infty$.

VOA-2   For $n \ll 0$, $\dim_\Bbbk \mathcal{V}_{(n)} = 0$.

VOA-3   Given elements $u, v \in \mathcal{V}$, $u_n v = 0$ for $0 \ll n$.

VOA-4   There are two distinguished elements $\mathbf{1}, \omega \in \mathcal{V}$ and a rational number rank $\mathcal{V}$. We set $\omega_{n+1} = \mathrm{L}_n$.

VOA-5   For any $v \in \mathcal{V}$, we have the identities

$$Y(\mathbf{1}, z) = \operatorname{Id}_\mathcal{V};$$
$$Y(v, z)\mathbf{1} \in \mathcal{V}[[z]];$$
$$Y(v, 0)\mathbf{1} = \lim_{z \to 0} Y(v, z)\mathbf{1} = v.$$

VOA-6   The following identity amongst operator valued Laurent series in the variables z_0, z_1, z_2 holds:

$$z_0^{-1}\delta\left(\frac{z_1 - z_2}{z_0}\right) Y(u, z_1)Y(v, z_2) - z_0^{-1}\delta\left(\frac{z_2 - z_1}{-z_0}\right) Y(v, z_2)Y(u, z_1)$$

$$= z_2^{-1}\delta\left(\frac{z_1 - z_0}{z_2}\right) Y(Y(u, z_0)v, z_2),$$

where the expansion of the Dirac function δ will be discussed below.

VOA-7   The elements L_n (as operators on $\mathcal{V}$) satisfy

$$[L_m, L_n] = (m - n)L_{m+n} + \frac{(m^3 - m)}{12}(\text{rank } \mathcal{V})\delta_{m+n,0}.$$

VOA-8   For $v \in \mathcal{V}_{(n)}$,

$$L_0 v = nv = (\text{wt } v)v,$$
$$L_n \mathbf{1} = 0 \quad \text{if } n \geqslant -1,$$
$$L_{-2}\mathbf{1} = \omega,$$
$$L_0\omega = 2\omega.$$

VOA-9   As formal series in z,

$$\frac{d}{dz}Y(v, z) = Y(L_{-1}v, z),$$
$$[L_{-1}, Y(v, z)] = Y(L_{-1}v, z),$$
$$[L_0, Y(v, z)] = Y(L_0 v, z) + zY(L_{-1}v, z).$$

These axioms are essentially those of [3], [5]. We use the notation

$$(\mathcal{V}_\bullet, Y, \mathbf{1}, \omega, \text{rank } \mathcal{V})$$

to denote such a vertex operator algebra, often just writing $(\mathcal{V}_\bullet, Y, \mathbf{1}, \omega)$ or even $(\mathcal{V}_\bullet, Y)$. For each element $v \in \mathcal{V}$, the series $Y(v, z)$ is called the *vertex operator* corresponding to v. The operators L_n are called the *Virasoro operators* and generate an action of the so-called *Virasoro algebra* on the vertex operator algebra. Provided rank $\mathcal{V} \neq 0$, this action implies that $\dim_\Bbbk \mathcal{V}_{(n)} \neq 0$ infinitely often (or equivalently that $\mathcal{V}_\bullet$ is infinite dimensional); this is enough to ensure that a vertex operator algebra is non-trivial, and indeed all examples are complicated to construct.

To expand the Dirac function δ referred to in VOA-6, we define

$$\delta(z) = \sum_{n\in\mathbb{Z}} z^n, \tag{1.1}$$

and for three variables z_0, z_1, z_2,

$$\delta\left(\frac{z_1 - z_2}{z_0}\right) = \sum_{n\in\mathbb{Z}} z_0^{-n} z_1^n \left(1 - \frac{z_2}{z_1}\right)^n$$

$$= \sum_{n\in\mathbb{Z}} \sum_{0\leqslant k} (-1)^k \binom{n}{k} z_0^{-n} z_1^{n-k} z_2^k. \tag{1.2}$$

In other words, we expand in terms of the *second* variable in the numerator of the argument.

From [2] and [5], we also record the definition of a *module* $(\mathcal{M}, Y_{\mathcal{M}})$ *over a vertex operator algebra* $(\mathcal{V}, Y, \mathbf{1}, \omega, \mathrm{rank}\ \mathcal{V})$. This consists of a $\mathbb{Q}$-graded $\Bbbk$-module $\mathcal{M} = \mathcal{M}_\bullet$ together with a $\Bbbk$-linear map

$$Y_{\mathcal{M}}: \mathcal{V} \longrightarrow \mathrm{End}(\mathcal{M})[[z, z^{-1}]];$$

$$v \longmapsto Y_{\mathcal{M}}(v, z) = \sum_{n\in\mathbb{Z}} v_n z^{-n-1},$$

satisfying the following conditions.

VOM-1   For each $n \in \mathbb{Q}$, $\dim_\Bbbk \mathcal{M}_{(n)} < \infty$.
VOM-2   For $n \ll 0$, $\dim_\Bbbk \mathcal{M}_{(n)} = 0$.
VOM-3   Given elements $u, v \in \mathcal{V}$, $u_n v = 0$ for $0 \ll n$.
VOM-4   We have the identity

$$Y_{\mathcal{M}}(\mathbf{1}, z) = \mathrm{Id}_{\mathcal{M}}.$$

VOM-5   The following identity amongst Laurent series in z_0, z_1, z_2 holds:

$$z_0^{-1}\delta\left(\frac{z_1 - z_2}{z_0}\right) Y_{\mathcal{M}}(u, z_1) Y_{\mathcal{M}}(v, z_2) - z_0^{-1}\delta\left(\frac{z_2 - z_1}{-z_0}\right) Y_{\mathcal{M}}(v, z_2) Y_{\mathcal{M}}(u, z_1)$$

$$= z_2^{-1}\delta\left(\frac{z_1 - z_0}{z_2}\right) Y_{\mathcal{M}}(Y_{\mathcal{V}}(u, z_0)v, z_2).$$

VOM-6   The elements $L_n = \omega_{n+1}$ (as operators on $\mathcal{M}$) satisfy

$$[L_m, L_n] = (m - n)L_{m+n} + \frac{(m^3 - m)}{12}(\mathrm{rank}\ \mathcal{V})\delta_{m+n,0}.$$

VOM-7   For $w \in \mathcal{V}_{(n)}$,

$$L_0 w = nw = (\mathrm{wt}\ w)w.$$

VOM-8   As formal series in z,

$$\frac{\mathrm{d}}{\mathrm{d}z} Y(v, z) = Y(L_{-1}v, z),$$

$$[L_{-1}, Y(v, z)] = Y(L_{-1}v, z),$$

$$[L_0, Y(v, z)] = Y(L_0 v, z) + z Y(L_{-1}v, z).$$

Actually, this 'naive' notion of module is insufficient for many purposes and there is a more general one of a *twisted module,* which seems to occur in particular in connection with bundles over loop spaces and the elliptic cohomology of non-simply connected spaces.

§2 The homology of the classifying space for *K*-theory

We take as a general reference for the topology of this section the classic work of Adams [1], which contains full details on many of the points we mention.

We begin with the simplest case, which provides a starting point for several generalizations. We assume throughout that $\Bbbk$ is a commutative, unital ring and $H_*(\ )$ denotes the ordinary homology functor $H_*(\ ;\Bbbk)$.

We recall the space $BU \times \mathbb{Z}$ classifying the *K*-theory functor

$$KU^0(\ ) \cong [\ \ ; BU \times \mathbb{Z}].$$

In homology we have

$$H_*(BU \times \mathbb{Z}) = H_*(BU)[[1], [-1]],$$

where $[n]$ denotes the component of $BU \times \mathbb{Z} = \coprod_{n \in \mathbb{Z}} BU \times \{n\}$, and we identify BU with $BU \times \{0\}$. We follow the notation of Ravenel and Wilson [7] by writing $[n] = [1]^n$. This homology ring is a Laurent polynomial ring over the ring $H_*(BU)$ on the generator $[1]$.

We assign the following new grading: an element $x \in H_{2n}(BU)$ is assigned weight n and the element $x[m]$ is assigned weight $m^2 + n$. With this grading, $H_*(BU \times \mathbb{Z})$ is no longer a graded ring. We will write $H_\bullet(BU \times \mathbb{Z})$ to emphasise this regrading.

Now we recall the structure of the bicommutative Hopf algebra $H_*(BU)$. There are standard algebra generators $b_n \in H_{2n}(BU)$ (we set $b_0 = 1$), coming from complex projective space under the natural embedding in homology. These span a binomial coalgebra, thus their generating function $b(T) = \sum_{n \geq 0} b_n T^n$ is a grouplike element of $H_*(BU)[[T]]$. There are also the primitive elements $p_n \in H_{2n}(BU)$, satisfying the Newton recurrence relation

$$p_1 = b_1,$$
$$p_n = b_1 p_{n-1} - b_2 p_{n-2} + b_3 p_{n-3} - \cdots + (-1)^{n-2} b_{n-1} p_1 + (-1)^{n-1} n b_n. \quad (2.1)$$

When $\Bbbk$ is a $\mathbb{Q}$-algebra, this is equivalent to the generating function identity

$$\ln b(T) = \sum_{n \geq 1} \frac{(-1)^{n-1} p_n}{n} T^n,$$

where $\ln(1 + Z) = \sum_{n \geq 1} (-1)^{n-1} Z^n / n$ is the formal logarithmic series. The primitive submodule of $H_{2n}(BU)$ is generated by p_n, but these elements are not algebra generators of $H_*(BU)$ over a general ring $\Bbbk$, in particular over integers $\mathbb{Z}$. They do however generate over the rationals $\mathbb{Q}$, which accounts for the fact that in [5] they are used in giving explicit formulæ for vertex operators.

Dually, in $H^*(BU)$ we have the universal Chern classes $c_n \in H^{2n}(BU)$, which are also polynomials generators and generate a binomial coalgebra by the Cartan formula. The duality is given by the formula

$$\langle c_n, b_1^{r_1} \cdots b_k^{r_k} \rangle = \begin{cases} 1 & \text{if } b_1^{r_1} \cdots b_k^{r_k} = b_1^n, \\ 0 & \text{else.} \end{cases}$$

The primitive submodule in $H^{2n}(BU)$ has generator s_n satisfying the recurrence relation

$$\begin{aligned} s_1 &= c_1, \\ s_n &= c_1 s_{n-1} - c_2 s_{n-2} + c_3 s_{n-3} - \cdots + (-1)^{n-2} c_{n-1} s_1 + (-1)^{n-1} n c_n. \end{aligned} \tag{2.2}$$

We also have the duality formula

$$\langle s_n, b_1^{r_1} \cdots b_k^{r_k} \rangle = \begin{cases} 1 & \text{if } b_1^{r_1} \cdots b_k^{r_k} = b_n, \\ 0 & \text{else.} \end{cases}$$

We will find it convenient to use the *total symmetric functions* h_n, recursively defined by

$$\sum_{0 \leqslant k \leqslant n} (-1)^k c_{n-k} h_k = 0,$$

and also satisfying the recurrence relation

$$\begin{aligned} s_1 &= h_1, \\ s_n &= n h_n - (h_1 s_{n-1} + h_2 s_{n-2} + h_3 s_{n-3} + \cdots + h_{n-1} s_1). \end{aligned} \tag{2.3}$$

These formulæ combine to give the following remarkable result.

Theorem 2.1. *There is an isomorphism of (graded) Hopf algebras*

$$\Phi \colon H_*(BU) \longrightarrow H^*(BU);$$
$$\Phi(b_n) = h_n$$

under which $\Phi(p_n) = (-1)^{n-1} s_n$.

Let A be a Hopf algebra over a ring $\Bbbk$ and let $A^* = \operatorname{Hom}_{\Bbbk}(A, \Bbbk)$ be its dual. Then there is a $\Bbbk$-algebra homomorphism

$$A^* \longrightarrow \operatorname{End}_{\Bbbk}(A);$$
$$\alpha \longmapsto \alpha \cdot,$$

where

$$\alpha \cdot a = \sum \alpha(a') a'',$$

with $\sum a' \otimes a''$ denoting the coproduct on a. This gives a canonical action of A^* on A.

In the case where $A = H_*(BU)$, and $A^* = H^*(BU)$, this action agrees with the *cap product* action of cohomology on homology. Thus, for $u \in H^{2m}(BU)$ and $x \in H_{2n}(BU)$ we have $u \cdot x = u \cap x$.

Now given a $\Bbbk$-algebra homomorphism $\varphi\colon A \longrightarrow A^*$, we have an induced action of A on itself given by

$$a \cdot x = \varphi(a) \cdot x, \quad \text{for } a, x \in A.$$

Taking the case of $A = H_*(BU)$, we obtain the action

$$u \cdot x = \Phi(u) \cap x.$$

Proposition 2.2. *We have the following formulæ.*

1) *For $a, b, x \in H_*(BU)$, $(ab) \cdot x = a \cdot (b \cdot x)$. Hence $\cdot$ is a left action of $H_*(BU)$ on itself.*

2) *For the primitives p_n $(n > 0)$,*

$$p_m \cdot p_n = n\delta_{m,n}.$$

3) *For the standard generators b_k,*

$$b_m \cdot b_n = h_m \cap b_n = \begin{cases} b_0 = 1 & \text{if } m = n, \\ b_{n-m} & \text{if } 0 \leqslant m < n, \\ 0 & \text{otherwise.} \end{cases}$$

In terms of the generating function $b(T)$, this formula is equivalent to

$$b(X) \cdot b(Y) = (1 - XY)^{-1}b(Y).$$

In the statement of part (2), the Kronecker symbol $\delta_{m,n}$ is determined by

$$\delta_{m,n} = \begin{cases} 1 & \text{if } m = n, \\ 0 & \text{otherwise.} \end{cases}$$

We can now describe a vertex operator algebra associated to $H_\bullet(BU \times \mathbb{Z})$ which we take as the underlying $\Bbbk$-module. We will define a linear map

$$Y(\ , z)\colon H_\bullet(BU \times \mathbb{Z}) \longrightarrow \mathrm{End}(H_\bullet(BU \times \mathbb{Z}))[[z, z^{-1}]]$$

with the relevant properties. The indeterminate z may be interpreted as indexing copies of $BU \times \mathbb{Z}$ corresponding to the spaces classifying the functors $KU^{-2k}(\ )$. Thus we can consider the *graded space*

$$BU \times \mathbb{Z}[z, z^{-1}] = \{BU \times \mathbb{Z}\{z^k\}\}_{k \in \mathbb{Z}},$$

where

$$KU^{-2k}(\ ) \cong [\ \ ; BU \times \mathbb{Z}\{z^k\}] \xrightarrow[\cong]{t^k} KU^0(\ ),$$

with the latter being multiplication by the kth power of the Bott element. We can therefore view z as t^{-1} and each vertex operator $Y(u, z)$ as an element of the graded tensor product

$$H_\bullet(BU \times \mathbb{Z}[z, z^{-1}]) \widehat{\underset{\Bbbk}{\otimes}} H^\bullet(BU \times \mathbb{Z}) = H_\bullet(BU \times \mathbb{Z}) \widehat{\underset{\Bbbk}{\otimes}} H^\bullet(BU \times \mathbb{Z})[[z, z^{-1}]],$$

which has an action on $H_\bullet(BU \times \mathbb{Z})$ induced by

$$(a \otimes c) \cdot x = a(c \cap x).$$

Using this interpretation, the *normal ordering* convention found in [5] becomes the usual commutation rule in the tensor product.

We begin by defining the operators $Y([a], z)$ for $a \in \mathbb{Z}$. Set

$$Y([a], z) = [a]b(z)^a \otimes h\,(-1/z)^{2a}\,z^{[a]},$$

where $z^{[a]}$ is given by linearly extending the formula

$$z^{[a]}(x[n]) = x[n]z^{2an} \quad \text{for } x \in H_*(BU) \text{ and } n \in \mathbb{Z}.$$

This is essentially the formula found in [5, section 7.1]. Next we must explain how to define the more general vertex operators $Y(x[n], z)$ for arbitrary elements $x \in H_*(BU)$ and $n \in \mathbb{Z}$. Again this is done in [5, section 8.5], at least on the assumption that the ground ring $\Bbbk$ is a rational algebra. Our description is valid for any ground ring $\Bbbk$.

By linearity, it suffices to describe the elements $Y(b_{r_1} \cdots b_{r_k}[a], z)$. This is done using the generating function

$$b(w_1) \cdots b(w_k) = \sum_{r_j \geqslant 0} b_{r_1} \cdots b_{r_k} w_1^{r_1} \cdots w_k^{r_k},$$

and the formula

$$
\begin{aligned}
&Y(b(w_1) \cdots b(w_k)[a], z) \\
&= b(z + w_1) \cdots b(z + w_k) \otimes h(-1/(z + w_1)) \cdots h(-1/(z + w_k))z^{[k]}Y([a], z) \\
&= [a]b(z + w_1) \cdots b(z + w_k)b(z)^a \otimes h(-1/(z + w_1)) \cdots \\
&\qquad\qquad\qquad\qquad \cdots h(-1/(z + w_k))h(-1/z)^{2a}z^{[a+k]}. \quad (2.4)
\end{aligned}
$$

To determine $Y(b_{r_1} \cdots b_{r_k}[a], z)$, we read off the term in the monomial $w_1^{r_1} \cdots w_k^{r_k}$ appearing in the expansion of $Y(b(w_1) \cdots b(w_k)[a], z)$ in terms of the variable z, using Equation 2.4. Such 'preferential' treatment of certain variables lies at the heart of the formal variable calculations of [5] and is related to normal ordering and time ordering in Quantum Field Theory.

Given the structure we have described we obtain the following result. The proof is essentially given in [5], where the case of a $\mathbb{Q}$-algebra $\Bbbk$ is covered; for the general case we need to verify that our integral formulæ agree with their rational results. Set

$$\mathcal{V}(\mathbb{Z})_\bullet = H_\bullet(BU \times \mathbb{Z}),$$
$$Y_\mathbb{Z} = Y,$$
$$\mathbf{1} = [0],$$
$$\omega = \frac{1}{2}b_1^2.$$

Theorem 2.3. *The quintuple $(\mathcal{V}(\mathbb{Z})_\bullet, Y_\mathbb{Z}, \mathbf{1}, \omega, 1)$ is a vertex operator algebra over any $\mathbb{Z}[1/2]$-algebra $\Bbbk$.*

Remark 2.4. The attentive reader will have noticed the appearance of factors of 2 in some of the above formulæ. This is not an accident and corresponds to the fact that we are viewing the integers as the root lattice $\mathbb{A}_1$ of the simple Lie algebra $\mathfrak{sl}_2$. In Section 3 we will make this more explicit. It is perhaps worth remarking that we could use the more 'natural' inner product $\langle a, b \rangle = ab$ to define an algebraic object which is an example of a somewhat weaker notion, that of a *quasi-vertex operator algebra*, as defined in [4] and possessing most of the properties of a vertex operator algebra except for those involving the Virasoro operators L_n, which are only required to occur for $n = 0, \pm 1$.

§3 Vertex operator algebras based on positive, even lattices

In this section we generalize the vertex operator algebra construction of Section 2 starting with a positive definite, even lattice $(\mathbb{L}, \langle \, , \, \rangle)$ in place of the integers $\mathbb{Z}$. Thus, $\mathbb{L}$ is a finite rank free abelian group equipped with a positive definite inner product

$$\langle \, , \, \rangle : \mathbb{L} \times \mathbb{L} \longrightarrow \mathbb{Z}$$

which is *even* in the sense that for $\ell \in \mathbb{L}$,

$$\langle \ell, \ell \rangle \in 2\mathbb{Z}.$$

Following [5], we may construct a vertex operator algebra $(\mathcal{V}(\mathbb{L}), Y_{\mathbb{L}})$ for which the underlying $\Bbbk$-module is

$$\mathcal{V}(\mathbb{L}) = \Bbbk\{\mathbb{L}\} \underset{\Bbbk}{\otimes} \mathcal{S}(\mathbb{L}).$$

Here $\Bbbk\{\mathbb{L}\}$ is the free $\Bbbk$-module on the elements of $\mathbb{L}$, given the weight grading for which wt $\ell = \langle \ell, \ell \rangle / 2$. Also the factor $\mathcal{S}(\mathbb{L})$ is a infinite tensor product

$$\mathcal{S}(\mathbb{L}) = \bigotimes_{k \geqslant 1} \Bbbk[b(\ell)_k : \ell \in \mathbb{L}]/(\text{relations})$$

where the symbols $b(\ell)_k$ stand for elements of weight k satisfying relations

$$b(\ell_1 + \ell_2)_k = \sum_{0 \leqslant j \leqslant k} b(\ell_1)_j b(\ell_2)_{k-j}.$$

Thus, if $\mathbb{L}$ has a basis $\ell_1, \ldots, \ell_{\operatorname{rank} \mathbb{L}}$, we have

$$\mathcal{S}(\mathbb{L}) = \Bbbk[b(\ell_1)_k, \ldots, b(\ell_{\operatorname{rank} \mathbb{L}})_k : k \geqslant 1].$$

The recursion of Equation 2.1 can be used to define elements $p(\ell)_k$ for which

$$\begin{aligned}
p(\ell)_1 &= b(\ell)_1, \\
p(\ell)_n &= b(\ell)_1 p(\ell)_{n-1} - b(\ell)_2 p(\ell)_{n-2} + b(\ell)_3 p(\ell)_{n-3} - \cdots \\
&\quad + (-1)^{n-2} b(\ell)_{n-1} p(\ell)_1 + (-1)^{n-1} n b(\ell)_n.
\end{aligned} \tag{3.1}$$

Our $p(\ell)_k$ is essentially equivalent to the element denoted $\ell(-k)$ in [5]. We will use the notation

$$b(T)^\ell = \sum_{k \geqslant 0} b(\ell)_k T^k.$$

We define elements $c(\ell)_k$, $h(\ell)_k$ in the dual of $\mathcal{S}(\mathbb{L})$ by requiring that they satisfy the formulæ

$$c(X)^\ell = \sum_{k \geqslant 0} c(\ell)_k X^k,$$

$$\langle c(X)^\ell, b(Y)^{\ell'} \rangle = (1 - XY)^{\langle \ell, \ell' \rangle},$$

and

$$h(X)^\ell = \sum_{k \geqslant 0} h(\ell)_k X^k,$$

$$\langle h(X)^\ell, b(Y)^{\ell'} \rangle = (1 - XY)^{-\langle \ell, \ell' \rangle}.$$

and form divided power sequences under the coproduct, i.e., for any $x, y \in \mathcal{S}(\mathbb{L})$,

$$\langle c(X)^\ell, xy \rangle = \langle c(X)^\ell, x \rangle \langle c(X)^\ell, y \rangle$$

and

$$\langle h(X)^\ell, xy \rangle = \langle h(X)^\ell, x \rangle \langle h(X)^\ell, y \rangle.$$

Using these together with the approach of Section 2, we have an action of $\mathcal{S}(\mathbb{L})$ on itself by

$$b(X)^\ell \cdot b(Y)^{\ell'} = (1 - XY)^{-\langle \ell, \ell' \rangle} b(Y)^{\ell'}. \tag{3.2}$$

We can inflict K-theory with coefficients in $\mathbb{L}$ by forming the even degree K-theory functors

$$KU\mathbb{L}^{-2k}(\ ) = KU^{-2k}(\ ) \underset{\mathbb{Z}}{\otimes} \mathbb{L}.$$

These are all represented by a space

$$BU\mathbb{L} \times \mathbb{L},$$

where $BU\mathbb{L}$ is connected. Of course, given a basis $\{\ell_1, \ldots, \ell_{\text{rank }\mathbb{L}}\}$ for $\mathbb{L}$, we obtain a decomposition (of infinite loop spaces)

$$BU\mathbb{L} \cong BU \times \cdots \times BU,$$

with one factor for each ℓ_j. The homology of this classifying space is of form

$$H_*(BU\mathbb{L} \times \mathbb{L}, \Bbbk) \cong H_*(BU\mathbb{L})[[\ell_j], [\ell_j]^{-1} : 1 \leqslant j \leqslant \text{rank }\mathbb{L}],$$

where

$$H_*(BU\mathbb{L}) = H_*(BU) \circ [\ell_1] \underset{\Bbbk}{\otimes} \cdots \underset{\Bbbk}{\otimes} H_*(BU) \circ [\ell_{\text{rank }\mathbb{L}}].$$

Here the notation $H_*(BU) \circ [\ell]$ denotes the homology of the space

$$BU \times \{[\ell]\} \subset BU\mathbb{L} \times \{[\ell]\}.$$

This notation is suggested by the Hopf ring notation of [7]; we are working with a 'Hopf module' $H_*(BU\mathbb{L} \times \mathbb{L})$ over the Hopf ring $H_*(BU \times \mathbb{Z})$. We will frequently write ℓ in place of $[\ell]$ when there is no chance of confusion.

We now proceed to describe the vertex operators for this example. For $\ell \in \mathbb{L}$, let z^ℓ denote the operator for which

$$z^\ell \cdot (x[\ell']) = x[\ell']z^{\langle \ell, \ell' \rangle},$$

where $\ell' \in \mathbb{L}$ and $x \in H_*(BU\mathbb{L})$. We make the following definitions:

$$Y([\ell], z) = [\ell]b(z)^\ell \otimes h\,(-1/z)^\ell\, z^\ell,$$

and

$$
\begin{aligned}
Y(b&(w_1)^{\ell_1} \cdots b(w_k)^{\ell_k}[\ell], z) \\
&= b(z + w_1)^{\ell_1} \cdots b(z + w_k)^{\ell_k} \otimes \\
&\quad h(-1/(z + w_1))^{\ell_1} \cdots h(-1/(z + w_k))^{\ell_k} z^{\ell_1 + \cdots + \ell_k} Y([\ell], z) \\
&= [\ell]b(w_1 + z)^{\ell_1} \cdots b(w_k + z)^{\ell_k} b(z)^\ell \otimes \\
&\quad h(-1/(w_1 + z))^{\ell_1} \cdots h(-1/(w_k + z))^{\ell_k} h\,(-1/z)^\ell\, z^{\ell_1 + \cdots + \ell_k + \ell}. \qquad (3.3)
\end{aligned}
$$

Also set

$$
\begin{aligned}
\mathcal{V}(\mathbb{L})_\bullet &= H_\bullet(BU\mathbb{L} \times \mathbb{L}), \\
Y_\mathbb{L} &= Y, \\
\mathbf{1} &= [0], \\
\omega &= \sum_{1 \leqslant j \leqslant \text{rank }\mathbb{L}} \frac{1}{2\langle \ell_j, \ell_j \rangle} b(\ell_j)_1^2,
\end{aligned}
$$

where $\{\ell_1, \ldots, \ell_{\text{rank }\mathbb{L}}\}$ is a basis for $\mathbb{L}$.

Theorem 3.1. *The quintuple* $(\mathcal{V}(\mathbb{L})_\bullet, Y_\mathbb{L}, \mathbf{1}, \omega, \text{rank }\mathbb{L})$ *is a vertex operator algebra over any* $\mathbb{Z}[1/2]$*-algebra* $\Bbbk$.

Again, the proof is essentially given in [5].

The case discussed in Section 2 amounts to taking the lattice $\mathbb{L} = \mathbb{Z}\sqrt{2}$, rather than $\mathbb{Z}$ itself. This is the root lattice $\mathbb{A}_1$, and root lattices of semi-simple Lie algebras provide many interesting examples.

§4 Some modules over vertex operator algebras

In this section we briefly describe some modules over the vertex operator algebras constructed earlier, including their irreducibles.

Again we assume the data of Section 3. In general, the linear map $\mathbb{L} \longrightarrow \mathrm{Hom}_{\mathbb{Z}}(\mathbb{L}, \mathbb{Z})$ given by

$$\ell \longmapsto \ell^* = \langle \ell, \ \rangle$$

is injective but not surjective; if it *is* an isomorphism then the lattice $\mathbb{L}$ is said to be *unimodular.* We also set

$$\mathbb{L}^0 = \{\ell \in \mathbb{Q} \otimes \mathbb{L} : \ \forall \ell' \in \mathbb{L}, \ \langle \ell, \ell' \rangle \in \mathbb{Z}\} \supseteq \mathbb{L}.$$

The index $|\mathbb{L}^0/\mathbb{L}|$ is of course finite and equal to 1 if and only if $\mathbb{L}$ is unimodular.

We may generalize the definition of the vertex operator algebra

$$(\mathcal{V}(\mathbb{L})_\bullet, \mathrm{Y}_\mathbb{L}, \mathbf{1}, \boldsymbol{\omega}, \mathrm{rank}\ \mathbb{L})$$

as follows. We replace the set of components $\mathbb{L}$ of $BU\mathbb{L} \times \mathbb{L}$ by a coset $\mathbb{L} + \ell$ of $\mathbb{L}$ in $\mathbb{L}^0$. Thus we take

$$\mathcal{V}(\mathbb{L} + \ell) = \Bbbk\{\mathbb{L} + \ell\} \underset{\Bbbk}{\otimes} \mathcal{S}(\mathbb{L}),$$

where the $\Bbbk$-module is free on the elements of the coset $\mathbb{L} + \ell$. We grade $\mathcal{V}(\mathbb{L} + \ell)$ by decreeing that wt $\ell = \langle \ell, \ell \rangle/2$ for $\ell \in \mathbb{L}+\ell$ and extending this to the whole of $\mathcal{V}(\mathbb{L}+\ell)$. Of course, this is now a grading over $\mathbb{Q}$ rather than just $\mathbb{Z}$.

Theorem 4.1. *The graded $\Bbbk$-module admits the structure of a module $(\mathcal{V}(\mathbb{L} + \ell), \mathrm{Y})$ over the vertex operator algebra $(\mathcal{V}(\mathbb{L})_\bullet, \mathrm{Y}_\mathbb{L}, \mathbf{1}, \boldsymbol{\omega}, \mathrm{rank}\ \mathbb{L})$. In the case of $\mathbb{L} + \ell = \mathbb{L}$, this agrees with the adjoint module (i.e., the natural module structure of a vertex operator algebra over itself).*

These modules may be combined into a single $\mathcal{V}(\mathbb{L})$-module

$$\mathcal{V}(\mathbb{L}^0) = \bigoplus_{\mathbb{L}+\ell \in \mathbb{L}^0/\mathbb{L}} \mathcal{V}(\mathbb{L} + \ell)$$

graded by $\mathbb{L}^0/\mathbb{L}$.

As a particular example, we may consider the situation of Section 2, in which we are taking $\mathbb{L} = \mathbb{Z}\sqrt{2}$. Then $\mathbb{L}^0 = \mathbb{Z}(1/\sqrt{2})$ and

$$\mathbb{L}^0/\mathbb{L} = \{\mathbb{L}, \mathbb{L} + (1/\sqrt{2})\}.$$

The following result of Dong [2] explains the significance of these examples.

Theorem 4.2. *Over the vertex operator algebra $(\mathcal{V}(\mathbb{L})_\bullet, \mathrm{Y}_\mathbb{L}, \mathbf{1}, \boldsymbol{\omega}, \mathrm{rank}\ \mathbb{L})$, each of the modules $(\mathcal{V}(\mathbb{L} + \ell), \mathrm{Y})$ $(\mathbb{L} + \ell \in \mathbb{L}^0/\mathbb{L})$ is irreducible; moreover, every irreducible module is isomorphic to precisely one of these.*

For the root lattice $\mathbb{A}_n$ of $\mathfrak{sl}_{n+1}$, $\mathbb{A}_n^0/\mathbb{A}_n \cong \mathbb{Z}/(n+1)$; hence there are exactly $n+1$ irreducibles. These are indexed by the fundamental weights associated to the $n+1$ nodes in the extended Dynkin diagram for $\mathbb{A}_n$ which has a rotational symmetry of order $n+1$; this symmetry also appears in the vertex operator algebra $\mathcal{V}(\mathbb{A}_n)$ and has the effect of twisting these modules by shifting the indexing element of $\mathbb{Z}/(n+1)$ by 1 modulo $(n+1)$.

In fact, it is possible to incorporate all such modules into a single $\mathbb{Z}/(n+1)$-graded object which possesses the algebraic structure of what has been called an *Abelian intertwining algebra* by Dong and Lepowsky [3]. In terms of the underlying geometry of [8] this is probably the most natural algebraic object to study.

References

[1] J. F. Adams, Stable homotopy and generalised homology, The University of Chicago Press, Chicago–London 1974.

[2] C. Dong, Vertex operator algebras associated with to even lattices, J. Algebra **161** (1993), 245–265.

[3] C. Dong, J. Lepowsky, Abelian intertwining algebras – a generalization of vertex operator algebras, in: Algebraic Groups and their generalizations: quantum and infinite-dimensional methods (University Park, PA, 1991), Proc. Sympos. Pure Math. **56**, Part 2 (1994), 261–293.

[4] I. B. Frenkel , Y.-Z. Huang, J. Lepowsky, On axiomatic approaches to vertex operator algebras and modules, Mem. Amer. Math. Soc. **494**, Amer. Math. Soc., Providence 1993.

[5] I. B. Frenkel, J. Lepowsky, A. Meurman, Vertex operator algebras and the monster, Pure Appl. Math. **134**, Academic Press, Boston 1988.

[6] T. Katsura, Y. Shimizu, K. Ueno, Complex cobordism ring and conformal field theory over $\mathbb{Z}$, Math. Ann. **291** (1991), 551–71.

[7] D. C. Ravenel, W. S. Wilson, The Hopf ring for complex cobordism, J. Pure Appl. Algebra **9** (1977), 241–280.

[8] A. Pressley, G. Segal, Loop Groups, Oxford University Press, Oxford 1986.

[9] G. Segal, Unitary representations of some infinite dimensional groups, Comm. Math. Phys. **80** (1981), 301–42.

Department of Mathematics
University of Glasgow
Glasgow G12 8QW
Scotland
E-mail: a.baker@maths.gla.ac.uk

The radical of a vertex operator algebra

C. Dong, H. Li, G. Mason, and P. S. Montague

1. Introduction

Suppose that V is a vertex operator algebra with canonical $\mathbb{Z}$-grading

$$V = \coprod_{n \in \mathbb{Z}} V_n.$$

Each $v \in V$ has a vertex operator $Y(v, z) = \sum_{n \in \mathbb{Z}} v_n z^{-n-1}$ attached to it, where $v_n \in \operatorname{End} V$. For the conformal vector ω we write $Y(\omega, z) = \sum_{n \in \mathbb{Z}} L(n) z^{-n-2}$. If v is *homogeneous of weight* k, that is $v \in V_k$, then one knows that

$$v_n : V_m \to V_{m+k-n-1}$$

and in particular the *zero mode* $o(v) = v_{\operatorname{wt} v - 1}$ induces a linear operator on each V_m. We extend the "o" notation linearly to V, so that in general $o(v)$ is the sum of the zero modes of the homogeneous components of v. Then we define the *radical* of V to be

$$J(V) = \{v \in V \mid o(v) = 0\}. \tag{1.1}$$

The problem arose in some work of the first three authors in [DLiM] and in work of the fourth author in [M] of describing $J(V)$ precisely. We will essentially solve this problem in the present paper in an important special case, namely that V is a vertex operator algebra of *CFT type*. This means that the $\mathbb{Z}$-grading on V has the shape

$$V = \coprod_{n=0}^{\infty} V_n \tag{1.2}$$

and moreover that $V_0 = \mathbb{C}\mathbf{1}$ is spanned by the vacuum vector $\mathbf{1}$.

V is said to be a vertex operator algebra of *hermitian* CFT type if, in addition, V has the structure of a Hilbert space and further we have an involution $v \mapsto \bar{v}$ on V such that

$$(\bar{v})_n{}^2 = \left(e^{L(1)} v\right)_{-n},$$

and $\bar{\omega} = \omega$.

In [M], the concept of a *deterministic* conformal field theory (a vertex operator algebra of hermitian CFT type in our current notation) was introduced as one for which $o(v) = 0$ and $L(1)v = 0$ imply that $v \in V_1$. Equivalently, the definition may be restated in terms of the modes $p(v) \equiv v_{\operatorname{wt} v - 2}$, i.e., if $L(1)v = 0$ and $p(v) = 0$ then $v \in V_0$. The motivation is that if we have a state $v \in \coprod_{n=2}^{\infty} V_n$ such that $p(v) = 0$ then $o\left(\frac{1}{L(0)-1} v\right)$ is a generator of a (possibly trivial) continuous symmetry of the conformal field theory. We show in this paper that all conformal field theories are deterministic (restricting attention only to states which are finite sums of components coming from distinct V_n's).

To describe our result, let

$$J_1(V) = J(V) \cap V_1. \tag{1.3}$$

We observe that $J(V)$ is in general *not* a $\mathbb{Z}$-graded subspace of V, which accounts for some of the difficulty that its study offers, nevertheless the radical elements of weight one will turn out to play a special rôle. We will prove

Theorem 1. *Suppose that V is a vertex operator algebra of CFT type. Then*

$$J(V) = J_1(V) + (L(0) + L(-1))V.$$

Related to Theorem 1 are the next two theorems.

Theorem 2. *Suppose that V is a vertex operator algebra of CFT type. Let v be homogeneous of weight at least one. Then there is an integer t such that $0 \le t \le$ wt v and*

$$Y(v, z) = \sum_{n<0} v_n z^{-n-1} + \sum_{n \ge t} v_n z^{-n-1} \tag{1.4}$$

where each operator v_n in (1.4) is non-zero.

We define the *degree* of v to be t and write $\deg v = t$ if v satisfies the hypotheses and conditions of Theorem 2. For general $v \in V$ define $\deg v$ to be the least of the degrees of the homogeneous components of v, and $\deg v = -1$ if $v \in V_0$. For $d \ge 0$ set

$$V^d = \{v \in V \mid \deg v \ge d\}.$$

Then $V^0 = \coprod_{n \ge 1} V_n$ and $V^0 \supset V^1 \supset V^2 \supset \cdots$ defines a filtration on V^0. We prove

Theorem 3. *Suppose that V is a vertex operator algebra of CFT type. If $d \ge 1$ then*

$$V^d = L(-1)^{d-1} J_1(V) + L(-1)^d V.$$

In favorable situations we will have $J_1(V) = 0$, in which case Theorems 1 and 3 take a simpler form. Although we will say something about this situation below, we do not yet have a complete description of those vertex operator algebras of CFT type for which $J_1(V)$ is not zero. We note here only that the vertex operator algebra associated with the Heisenberg algebra has $J_1(V) \ne 0$; indeed $J_1(V) = V_1$ in this case.

We refer the reader to [FHL], [FLM], [DGM] and [DHL] for background definitions and elementary results about vertex operator algebras.

We gratefully acknowledge the following partial support: C.D.: faculty research funds granted by the University of California, Santa Cruz and DMS-9303374; G.M.: faculty research funds granted by the University of California, Santa Cruz and NSF grant DMS-9401272.

2. Quasi-primary elements

Recall that $v \in V$ is called *quasi-primary* in case $L(1)v = 0$. It is convenient to introduce the term *semi-primary* for an element $v \in V$ if v is either quasi-primary or of weight 1. Note that v is quasi-primary or semi-primary if, and only if, each homogeneous component of v has the same property. Also note that, for a vertex operator algebra of hermitian CFT type, all weight one states are quasi-primary and so the terms quasi- and semi-primary are synonomous.

In this section we prove the following results under the assumption that V is a vertex operator algebra of CFT type.

Theorem 2.1. *If $v \in V$ is homogeneous and satisfies $v_{\mathrm{wt}\,v} = 0$, then $v \in V_0$.*

Theorem 2.2. *Suppose that $v \in V$ is semi-primary and satisfies $o(v) = 0$. Then $v \in V_1$.*

Theorem 2.3. *Suppose that $v \in V$ is quasi-primary and homogeneous of weight at least 2. If $v_0 = 0$ then $v = 0$.*

The following is well-known.

Lemma 2.4. *The following are equivalent for a homogeneous element $v \in V$:*
(a) $v \in V_0$.
(b) $[L(-1), v_n] = 0$ *for all* $n \in \mathbb{Z}$.

Proof. This follows from the identity

$$[L(-1), Y(v, z)] = Y(L(-1)v, z) = \frac{d}{dz} Y(v, z) \tag{2.1}$$

together with the fact that $L(-1)$ is injective on V_n for $n \neq 0$ (cf. Corollary 2.4 of [DLM]). $\qquad\square$

Turning to the proof of Theorem 2.1, the component version of (2.1) is

$$[L(-1), v_n] = (L(-1)v)_n = -nv_{n-1}. \tag{2.2}$$

So as $v_{\mathrm{wt}\,v} = 0$ then we see inductively that

$$v_n = 0 \tag{2.3}$$

for $0 \leq n \leq \mathrm{wt}\,v$.

Now in general one has [FLM]

$$[a_m, b_n] = \sum_{t \geq 0} \binom{m}{t} (a_t b)_{m+n-t}. \tag{2.4}$$

Taking $b_n = L(-1) = \omega_0$ in (2.4) and taking (2.3) into account, we see that

$$[v_m, L(-1)] = \sum_{t > \mathrm{wt}\,v} \binom{m}{t} (v_t \omega)_{m-t}. \tag{2.5}$$

Now $\operatorname{wt} v_t \omega = \operatorname{wt} v - t + 1$, so $v_t \omega = 0$ if $t > \operatorname{wt} v + 1$. So (2.5) reduces to

$$[v_m, L(-1)] = \binom{m}{\operatorname{wt} v + 1}(v_{\operatorname{wt} v + 1}\omega)_{m - \operatorname{wt} v - 1}. \tag{2.6}$$

However, $\operatorname{wt} v_{\operatorname{wt} v + 1}\omega \in V_0$, so $(v_{\operatorname{wt} v + 1}\omega)_{m - \operatorname{wt} v - 1} = 0$ unless $m - \operatorname{wt} v - 1 = -1$, that is $m = \operatorname{wt} v$. In this case $\binom{m}{\operatorname{wt} v + 1} = 0$. So (2.6) yields

$$[v_m, L(-1)] = 0 \tag{2.7}$$

for all $m \in \mathbb{Z}$.

Theorem 2.1 is now a consequence of (2.7) and Lemma 2.4.

We next present the proof of Theorem 2.2. Denote the homogeneous components of v by v^i, that is

$$v = \sum_{i \geq 0} v^i, \quad v^i \in V_i. \tag{2.8}$$

Note that we do not necessarily have $o(v^i) = 0$, although each v^i is certainly semi-primary.

Lemma 2.5. *The following hold:*

(i)   *If u is a homogeneous and quasi-primary then*

$$[L(1), u_t] = 2(\operatorname{wt} u - t/2 - 1)u_{t+1}. \tag{2.9}$$

(ii) *If u has weight 1 then*

$$[L(1), u_0] = 0. \tag{2.10}$$

Proof. Use (2.2) and (2.4) with $a_m = L(1) = \omega_2$ to see that

$$[L(1), u_t] = (2 \operatorname{wt} u - t - 2)u_{t+1} + (L(1)u)_t.$$

So if u is quasi-primary then (2.9) holds. If $\operatorname{wt} u = 1$ and $t = 0$ then $[L(1), u_0] = (L(1)u)_0$. But $L(1)u \in V_0 = \mathbb{C}\mathbf{1}$, so that $(L(1)u)_0 = 0$. □

Turning to the proof of Theorem 2.2, since $o(v) = 0$ we see from (2.8)–(2.10) that

$$0 = [L(1), o(v)] = \sum_{i \geq 0}[L(1), o(v^i)] = \sum_{i \geq 2} 2(\operatorname{wt} v^i - (\operatorname{wt} v^i - 1)/2 - 1)v^i_{\operatorname{wt} v^i}.$$

So

$$\sum_{i \geq 2}(i - 1)v^i_i = 0. \tag{2.11}$$

We now apply the operator $[L(1), [L(-1), *]]$ to (2.11). We have

$$[L(1), [L(-1), u_t]] = [L(1), -tu_{t-1}] = -t(2 \operatorname{wt} u - t - 1)u_t$$

by (2.9). So (2.11) implies that

$$\sum_{i \geq 2}(i - 1)i(i - 1)v^i_i = 0.$$

Continuing in this fashion, we get

$$\sum_{i\geq 2}(i-1)^k i^{k-1} v_i^i = 0$$

for all $k \geq 1$. It follows that each $v_i^i = 0$ for all $i \geq 2$, and therefore $v^i = 0$ for $i \geq 2$ by Theorem 2.1. Hence

$$v = v^0 + v^1.$$

To finish the theorem we need to also show that $v^0 = 0$. In fact we have $o(v) = v_{-1}^0 + v_0^1 = 0$, so that $0 = v_{-1}^0 \mathbf{1} + v_0^1 \mathbf{1}$. But $v_0^1 \mathbf{1} = 0$, so $v^0 = v_{-1}^0 \mathbf{1} = 0$, and the proof of Theorem 2.2 is complete.

To prove Theorem 2.3, let v be as in the statement of the theorem. Then $v_0 = 0$, and we can apply (2.9) to see inductively that $v_n = 0$ for $0 \leq n \leq 2(\mathrm{wt}\, v - 1)$. As $\mathrm{wt}\, v \geq 2$ then in particular we get $v_{\mathrm{wt}\, v} = 0$ and so we can apply Theorem 2.1 to complete the proof.

3. The shape of a vertex operator

We prove Theorems 2 and 3 in this section. Let $v \in V$ be a non-zero homogeneous vector of weight at least one and with vertex operator

$$Y(v, z) = \sum_{n\in\mathbb{Z}} v_n z^{-n-1}.$$

We have $Y(v, z)\mathbf{1} = e^{zL(-1)}v$, in particular if $n < 0$ then $v_n \mathbf{1} = \frac{L(-1)^{-n-1}}{(-n-1)!}v$. As $L(-1)$ is injective on V_n for $n \neq 0$ we deduce that $v_n \mathbf{1} \neq 0$, so that $v_n \neq 0$ for $n < 0$.
Next, if $v_n = 0$ with $n \geq 0$ then (2.2) yields $0 = [L(-1), v_n] = -nv_{n-1}$. This then shows that $v_m = 0$ for $0 \leq m \leq n$. By Theorem 2.1 we find that $v_n \neq 0$ for $n \geq \mathrm{wt}\, v$, and now Theorem 2 follows immediately.
Now we prove Theorem 3. We start with

Lemma 3.1. *If $v \in V$ is homogeneous of degree t, then $L(-1)^k v$ has degree $k+t$ for all $k \geq 0$.*

Proof. We may take $k = 1$. Now $Y(v, z)$ has the shape (1.4), whence

$$Y(L(-1)v, z) = \frac{d}{dz}Y(v, z) = -\sum_{n<0}(n+1)v_n z^{-n-2} - \sum_{n\geq t}(n+1)v_n z^{-n-2}$$

i.e.,

$$Y(L(-1)v, z) = -\sum_{n<0}nv_{n-1}z^{-n-1} - \sum_{n\geq t+1}nv_{n-1}z^{-n-1}. \tag{3.1}$$

Since each v_n is (3.1) is non-zero we see that $\deg L(-1)v = t + 1$ as required.   □

Corollary 3.2. *If $d \geq 1$ then*

$$L(-1)^{d-1}J_1(V) + L(-1)^d V \subset V^d \subset \bigoplus_{n \geq d} V_n.$$

Proof. $L(-1)^{d-1}J_1(V) + L(-1)^d V \subset V^d$ follows from Lemma 3.1. The containment $V^d \subset \oplus_{n \geq d} V_n$ reflects the fact that $\deg v \leq \mathrm{wt}\, v$. $\qquad\square$

We also need

Lemma 3.3. *If $n \neq 1$ there is a direct sum decomposition*

$$V_n = \ker(L(1) : V_n \to V_{n-1}) \bigoplus \mathrm{im}(L(-1) : V_{n-1} \to V_n).$$

Proof. See, for example, Proposition 3.4 of [DLM]. $\qquad\square$

To prove Theorem 3, suppose that there is v of degree d with $v \notin L(-1)^{d-1}J_1(V) + L(-1)^d V$. By Lemma 3.3 we can choose v to be homogeneous and such that it has an expression of the form

$$v = \sum_{n=0}^{d-1} L(-1)^n u^n$$

with each u^n homogeneous and semi-primary. Moreover, we can take $\mathrm{wt}\, v = \mathrm{wt}(L(-1)^n u^n)$, i.e., $\mathrm{wt}\, v = n + \mathrm{wt}\, u^n$ for $0 \leq d - 1$.

Now if $\mathrm{wt}\, u^n = 1$ then $\mathrm{wt}\, v = n + 1$, so as $n \leq d - 1$ then $\mathrm{wt}\, v \leq d$. So in fact $\mathrm{wt}\, v = d$ and $n = d - 1$ in this case. So if $n \leq d - 2$ then u^n is quasi-primary of weight at least 2, so that $\deg(L(-1)^n u^n) = n$ if $u^n \neq 0$ by Lemma 3.1 and Theorem 2.3. Since v has degree $d \geq n + 1$ and $\deg(L(-1)^{d-1}u^{d-1}) \geq d - 1$ we see that $L(-1)^n u^n = 0$ for $n \leq d - 2$.

So now $v = L(-1)^{d-1}u^{d-1}$ with $\mathrm{wt}\, u^{d-1} = 1$, forcing $\deg u^{d-1} = 1$ by Lemma 3.1. So $u^{d-1} \in J_1(V)$, and the theorem follows. $\qquad\square$

4. Theorem 1

First we prove

Lemma 4.1. $J_1(V) + (L(0) + L(-1))V \subset J(V)$.

Proof. It is enough to show that $(L(0) + L(-1))v$ lies in $J(V)$ for homogeneous $v \in V$. But we have $L(0)v = (\mathrm{wt}\, v)v$ and $(L(-1)v)_n = -nv_{n-1}$, so taking $n = \mathrm{wt}(L(-1)v) - 1 = \mathrm{wt}\, v$ shows that $o(L(0)v + L(-1)v) = 0$ as required. $\qquad\square$

To begin the proof of Theorem 1, pick $v \in J(V)$. As before we can write

$$v = \sum_{n=0}^{m} L(-1)^n u^n \tag{4.1}$$

where each u^n is semi-primary and where $u^m \neq 0$. We prove by induction on m that v lies in $J_1(V) + (L(0) + L(-1))V$.

Suppose first that $m = 0$. Then $v = u^m$ is semi-primary, whence the condition $o(v) = 0$ forces $v \in V_1$ by Theorem 2.2. So in fact $v \in J_1(V)$ in this case.

In general, set $x = L(-1)^{m-1}u^m$ and $y = \sum_{n=0}^{m-1} L(-1)^n u^n$. Thus $v = L(-1)x + y$. Now from Lemma 4.1 we have $(L(0) + L(-1))x \in J(V)$, that is

$$0 = o(v) = o(L(-1)x) + o(y) = -o(L(0)x) + o(y) = o(y - L(0)x).$$

We easily check that $L(0)x = (m-1)L(-1)^{m-1}u^m + L(-1)^{m-1}L(0)u^m$ so that

$$y - L(0)x = \sum_{n=0}^{m-2} L(-1)^n u^n + L(-1)^{m-1}((m-1)u^m + L(0)u^m + u^{m-1})$$

lies in $J(V)$. Since $L(0)u^m$ is semi-primary, we conclude by induction that $y - L(0)x$ lies in $J_1(V) + (L(0) + L(-1))V$. But then the same is true of $v = y - L(0)x + (L(0) + L(-1))x$. This completes the proof of the theorem.

Remark. The referee has kindly pointed out that one can also deduce Theorem 1 from Theorem 3.

We make some remarks about Heisenberg vertex operator algebras. These are constructed from a finite-dimensional abelian Lie algebra H equipped with a non-degenerate, symmetric, bilinear form $\langle \, , \, \rangle$. One then forms the $\mathbb{Z}$-graded affine Lie algebra

$$L = H \bigotimes \mathbb{C}[t, t^{-1}] \oplus \mathbb{C}c$$

where $[u \otimes t^m, v \otimes t^n] = \langle u, v \rangle m\delta_{m+n,0}c$ and $[c, L] = 0$. The corresponding vertex operator algebra has underlying Fock space

$$M(1) = S(\oplus_{n<0} H \otimes t^n)$$

with the usual action of L. In particular, c acts as 1. If $h \in H$ is identified with $h \otimes t^{-1}$ we have

$$Y(h, z) = \sum_{n \in \mathbb{Z}} (h \otimes t^n)z^{-n-1}$$

so that $h_n = h \otimes t^n$. In particular, we see that $o(h) = h_0$ acts trivially on $M(1)$, that is $o(h) = 0$. So in fact $M(1)_1 = H = J_1(M(1))$ in this case.

More generally, we see that if $M(1)$ is as above and if V is any vertex operator algebra then the tensor product $M(1) \otimes V$ (cf. [FHL]) is such that $M(1) \otimes \mathbf{1} \subset J_1(M(1) \otimes V)$.

There is a converse to this observation. To explain this, suppose that V is a vertex operator algebra of CFT type. On V_1 there is a canonical symmetric bilinear form defined by

$$\langle u, v \rangle = u_1 v.$$

Proposition 4.2. *Suppose that* $H \subset J_1(V)$ *is a subspace such that the restriction of* $\langle \, , \rangle$ *to* H *is non-degenerate. Then* V *is a tensor product*

$$V \simeq M(1) \bigotimes W$$

for some vertex operator algebra W.

Proof. Given the condition on H, it generates a sub vertex operator algebra of V (with a different Virasoro element) isomorphic to $M(1)$. Now V is a $M(1)$-module, that is an L-module (L as above). As is well-known (e.g., Theorem 1.7.3 of [FLM]) this implies that there is an isomorphism of L-modules $V \simeq M(1) \otimes W$ where W is the space of highest weight vectors for the action of L on V. That is,

$$W = \{v \in V \mid h_n v = 0, \ h \in H, \ n \geq 0\}.$$

This is the so-called commutant of $M(1)$ (cf. Theorem 5.1 of [FZ]), and is known (loc. cit.) to be also a vertex operator algebra . The proposition follows.   □

We do not know if elements of $J_1(V)$ can occur in ways other than those described above, though for a vertex operator algebra of *hermitian* CFT type, the canonical symmetric bilinear form introduced above is non-degenerate (see, for example, [DGM]), and so Proposition 4.2 provides the complete picture in this case.

We may consider $O_\infty(V)$, an object closely related to the radical of V, which we define as follows:

$$O_\infty(V) = \{v \in V \mid o(v)_M = 0 \text{ for all modules } M\},$$

where $o(v)_M$ is the action of the zero mode of the vertex operator corresponding to v on the module M. This is to be compared to Zhu's $O(V)$, which is the set of all states in V whose zero modes annihilate the states of lowest conformal weight in each module. (In fact, $O_\infty(V)$ may alternatively be defined as the intersection of objects $O_n(V)$ as described in [DLiM].) Clearly, $O_\infty(V) \subset J(V) = J_1(V) + (L(0) + L(-1)) V$. Further, if $v \in J_1(V) \cap O_\infty(V)$, then, by Proposition 4.2, V splits up into a tensor product $M(1) \otimes W$. We have modules for $M(1)$ on which the zero mode of v is non-zero, and so tensoring with the adjoint module for W gives a module M for V with $o(v)_M \neq 0$. We deduce that $O_\infty(V) = (L(0) + L(-1)) V$.

Bibliography

[DGM]   L. Dolan, P. Goddard, P. Montague, Conformal field theories, representations and lattice constructions, `hep-th/9410029`, Comm. Math. Phys. **179** (1996), 61–120.

[DHL]   C. Dong, H. Li, Y. Huang, Introduction to vertex operator algebras I–III, Moonshine and vertex operator algebra, Lecture Notes **904**, Research Institute for Mathematics Sciences, Kyoto University, 1995, 1–76.

[DLiM]   C. Dong, H. Li, G. Mason, Vertex operator algebras and associative algebras, q-alg/9612010.

[DLM]   C. Dong, Z. Lin, G. Mason, On vertex operator algebras as sl_2-modules, in: Groups, difference sets, and the monster, Proc. of a special research quarter at The Ohio State University, Spring 1993, K. T. Arasu, J. F. Dillon, K. Harada, S. Sehgal, R. Solomon (Ed.), Walter de Gruyter, Berlin–New York, 1996, 349–362.

[FHL]   I. B. Frenkel, Y. Huang, J. Lepowsky, On axiomatic approaches to vertex operator algebras and modules, Mem. Amer. Math. Soc. **494**, Providence 1993.

[FLM]   I. B. Frenkel, J. Lepowsky, A. Meurman, Vertex operator algebras and the monster, Pure Appl. Math. **134**, Academic Press, Boston 1988.

[FZ]   I. Frenkel and Y. Zhu, Vertex operator algebras associated to representations of affine and Virasoro algebras, Duke Math. J. **66** (1992), 123–168.

[M]   P. S. Montague, Continuous symmetries of lattices and their $\mathbb{Z}_2$-orbifolds, `hep-th/` `9410218`, Phys. Lett. B **343** (1995), 113–121.

Mathematics Department
University of California
Santa Cruz, CA 95064, U.S.A.
E-mail: dong@cats.ucsc.edu
hli@crab.rutgers.edu
gem@cats.ucsc.edu
pmontagu@maths.adelaide.edu.au

Associative subalgebras of the Griess algebra and related topics

C. Dong, H. Li, G. Mason, and S. P. Norton

1. Introduction

In [DMZ] it was shown that the moonshine module $V^\natural$ contains a sub vertex operator algebra isomorphic to the tensor product $L(\frac{1}{2}, 0)^{\otimes 48}$ where $L(\frac{1}{2}, 0)$ is the vertex operator algebra associated to the highest weight unitary representation for the Virasoro algebra with central charge $\frac{1}{2}$. This containment turns out to be fundamental, because it allows us to deduce properties of $V^\natural$ from those of $L(\frac{1}{2}, 0)$, which are very much easier to discern (loc. cit.). This approach has been used to prove, for example, that $V^\natural$ is holomorphic [D], and to construct twisted sectors and intertwining algebras for $V^\natural$ [DLM], [H]. One can even base a simplified approach to the existence of $V^\natural$ on $L(\frac{1}{2}, 0)^{\otimes 48}$ (cf. Miyamoto's lecture at this conference).

The reason why $L(\frac{1}{2}, 0)$ is so attractive is that it is the first non-trivial *discrete series* representation of the Virasoro algebra and all modules and fusion rules are known for this family of vertex operator algebras. It is therefore a natural question to ask if $V^\natural$ contains other similar tensor products of discrete series representations, and more generally to ask which discrete series representations can be generated by idempotents in the Griess algebra B. We do not have complete answers to these questions, but we will take an approach which allows us, for example, to exhibit in a fairly painless way a sub vertex operator algebra of $V^\natural$ of the shape

$$\left(L(\frac{1}{2}, 0) \bigotimes L(\frac{7}{10}, 0) \bigotimes L(\frac{4}{5}, 0) \right)^{\otimes 12} \tag{1.1}$$

as well as another (of central charge less than 24) which is the tensor product of the first 24 members of the discrete series. It turns out that the key idea is to relate these questions to the theory of root systems and Niemeier lattices.

These issues are naturally related to the question of describing the maximal associative subalgebras of the Griess algebra. In this form, these question were first studied in [MN]. We show that each Niemeier lattice (and its attendant root system) determines (in many ways) certain maximal associative subalgebras of B. For example, (1.1) is associated to the Niemeier lattice of type A_2^{12}, while $L(\frac{1}{2}, 0)^{\otimes 48}$ corresponds to A_1^{24}.

G. M. would like to thank Gerald Hoehn for stimulating conversations and Jeff Harvey for pointing out not only the occurrence of the parafermion in Theorem 2.7, but also that he (together with Lance Dixon) had done a number of the calculations in Section 2 independently (and much earlier). We gratefully acknowledge the following partial support: C. D.: a faculty research funds granted by the University of California, Santa Cruz and DMS-9303374; G. M.: NSF grant DMS-9401272.

2. Algebras and root systems

We consider a simple (i.e., irreducible) root system Φ of type A, D, E. Let Φ^+ be the set of positive roots and let $l \geq 1$ be the rank of Φ, $N = |\Phi^+|$, and h the Coxeter number of Φ. It is well-known that we have

$$2N = lh. \tag{2.1}$$

Any $\alpha \in \Phi^+$ determines a partition of Φ^+, namely

$$\begin{aligned}
\Delta_0(\alpha) &= \{\alpha\} \\
\Delta_1(\alpha) &= \{\beta \in \Phi^+ \mid (\alpha, \beta) \neq 0,\ \beta \neq \alpha\} \\
\Delta_2(\alpha) &= \{\beta \in \Phi^+ \mid (\alpha, \beta) = 0,\ \beta \neq \alpha\}.
\end{aligned} \tag{2.2}$$

Here, $(\cdot, \cdot)$ denotes the usual inner product associated with Φ, normalized so that $(\alpha, \alpha) = 2$ for $\alpha \in \Phi$. We often write $\alpha \sim \beta$ in case $\beta \in \Delta_1(\alpha)$; of course $\sim$ is a symmetric relation.

We will define a certain $\mathbb{Q}$-algebra $A = A(\Phi)$. Additively it is a free abelian group with a distinguished basis consisting of elements $t(\alpha), u(\alpha)$ with $\alpha \in \Phi^+$. Thus A has rank $2N$. To define multiplication, note that if $\alpha \sim \beta$ with $\alpha, \beta \in \Phi^+$ then there is a unique $\gamma \in \Phi^+ \cap \mathbb{Z}\langle \alpha, \beta \rangle$ such that $\alpha \sim \gamma \sim \beta$. We use this observation to define

$$t(\alpha)t(\beta) = \begin{cases} t(\alpha) + t(\beta) - t(\gamma), & \text{if } \alpha \sim \beta \sim \gamma \sim \alpha \\ 0, & \text{if } \beta \in \Delta_2(\alpha) \end{cases}$$

$$u(\alpha)u(\beta) = \begin{cases} u(\alpha) + u(\beta) - t(\gamma), & \text{if } \alpha \sim \beta \sim \gamma \sim \alpha \\ 0, & \text{if } \beta \in \Delta_2(\alpha) \end{cases}$$

$$u(\alpha)t(\beta) = t(\beta)u(\alpha) = \begin{cases} u(\alpha) + t(\beta) - u(\gamma), & \text{if } \alpha \sim \beta \sim \gamma \sim \alpha \\ 0, & \text{if } \beta \in \Delta_0(\alpha) \cup \Delta_2(\alpha) \end{cases}$$

$$t(\alpha)^2 = 8t(\alpha), \quad u(\alpha)^2 = 8u(\alpha).$$

Obviously, $A = A(\Phi)$ is a commutative (but in general non-associative) algebra. The motivation for introducing this algebra will be explained later.

The following is well-known (cf. Chap. IV, 1.11, Proposition 32 of [B]).

Lemma 2.1. *If* $\alpha \in \Phi^+$ *then* $|\Delta_1(\alpha)| = 2h - 4$.

We use this to prove

Proposition 2.2. $A(\Phi)$ *has an identity element, namely*

$$\delta = \frac{1}{4h} \sum_{\alpha \in \Phi^+} (t(\alpha) + u(\alpha)).$$

Proof. If $\beta \in \Phi^+$ then $t(\beta)\delta$ is equal to

$$\frac{1}{4h}\left(t(\beta)^2 + t(\beta)\sum_{\alpha \in \Delta_1(\beta)} t(\alpha) + t(\beta)\sum_{\alpha \in \Delta_1(\beta)} u(\alpha)\right) \tag{2.3}$$

$$= \frac{1}{4h}\left(8t(\beta) + \sum_{\alpha \in \Delta_1(\beta)} (t(\beta) + t(\alpha) - t(\gamma(\alpha,\beta))) + \sum_{\alpha \in \Delta_1(\beta)} (t(\beta) + u(\alpha) - u(\gamma(\alpha,\beta)))\right)$$

where $\gamma(\alpha, \beta)$ is the element of Φ^+ determined by α and β whenever $\alpha \sim \beta$. But $\gamma(\alpha, \beta)$ ranges over $\Delta_1(\beta)$ as α does, so all terms in (2.3) cancel except for the $t(\beta)$'s. Lemma 2.1 tells us that (2.3) is thus equal to $\frac{1}{4h}(8t(\beta) + (2h-4)t(\beta) + (2h-4)t(\beta)) = t(\beta)$, that is $t(\beta)\delta = t(\beta)$. Similarly $u(\beta)\delta = u(\beta)$, and the proposition is proved. $\square$

We observe that the $\mathbb{Q}$-span of the $t(\alpha)$ for $\alpha \in \Phi^+$ is a subalgebra of A, which we denote by $T(\Phi)$. The same proof shows

Lemma 2.3. $T(\Phi)$ *has an identity, namely*

$$\epsilon = \frac{1}{2h+4}\sum_{\alpha \in \Phi^+} t(\alpha).$$

Now introduce a symmetric bilinear form $\langle\ ,\ \rangle$ on $A(\Phi)$ as follows:

$$\langle t(\alpha), t(\alpha)\rangle = \langle u(\alpha), u(\alpha)\rangle = 4$$

$$\langle t(\alpha), t(\beta)\rangle = \langle u(\alpha), u(\beta)\rangle = \begin{cases} 1/2, & \alpha \sim \beta \\ 0, & \alpha \in \Delta_2(\beta) \end{cases}$$

$$\langle t(\alpha), u(\beta)\rangle = \begin{cases} 1/2, & \alpha \sim \beta \\ 0, & \alpha \in \Delta_0(\beta) \cup \Delta_2(\beta). \end{cases}$$

We remark that $\langle\ ,\ \rangle$ is not necessarily non-degenerate (see below).

For an element $a \in A$ we define

$$c(a) = 8\langle a, a\rangle. \tag{2.4}$$

We may refer to $c(a)$ as the *central charge* of a, though usually we reserve this term for the case that a is an idempotent of A such as δ or ϵ. In these cases we prove

Lemma 2.4. *We have*

$$c(\delta) = l, \quad c(\epsilon) = \frac{lh}{h+2}.$$

Proof. We have

$$c(\delta) = \frac{8}{(4h)^2} \sum_{\alpha \in \Phi^+} \left(\langle t(\alpha), t(\alpha) \rangle + \langle u(\alpha), u(\alpha) \rangle \right)$$

$$+ \frac{8}{(4h)^2} \sum_{\substack{\alpha \in \Phi^+ \\ \beta \in \Delta_1(\alpha)}} \left(\langle t(\alpha), t(\beta) \rangle + \langle u(\alpha), u(\beta) \rangle + \langle t(\alpha), u(\beta) \rangle + \langle u(\alpha), t(\beta) \rangle \right)$$

$$= \frac{1}{2h^2} (8N + N(2h-4)2)$$

$$= \frac{2N}{h}.$$

Now use (2.1) to see that $c(\delta) = l$. The calculation of $c(\epsilon)$ is similar. □

We now consider the possibility of decomposing the identity δ of A (or the identity ϵ of T) into a sum of idempotents $\delta = e_1 + e_2 + \cdots + e_k$ such that $\langle e_i, e_j \rangle = 0$ if $i \neq j$ and $e_i e_j = \delta_{i,j} e_i$. Such a decomposition corresponds to a particular kind of associative subalgebra of A isomorphic to a direct sum of k copies of $\mathbb{Q}$. One way to find such a decomposition is as follows: locate within the root system Φ a subsystem Φ', so that $A(\Phi')$ is a subalgebra of $A(\Phi)$. Then decompose δ as $\delta' + (\delta - \delta')$ where δ' is the identity of $A(\Phi')$. This has the desired properties (with $k = 2$). For larger values of k one can iterate this procedure, considering chains of root systems $\Phi \supset \Phi' \supset \Phi'' \cdots$.

We illustrate this procedure in the case that Φ is of type A_l – the case of most interest to us. Having fixed a system $\pi = \{\alpha_1, \ldots, \alpha_l\}$ of simple roots in Φ, we let Φ_i be the sub-system whose simple roots consist of $\alpha_1, \ldots, \alpha_i$, with A_i and T_i the corresponding algebras. Thus $A = A_l$ and $T = T_l$.

We begin by writing $\delta = \epsilon + (\delta - \epsilon)$ and verifying that indeed $\langle \epsilon, \delta - \epsilon \rangle = 0$, which we leave to the reader. Then we find that

$$c(\delta - \epsilon) = c(\delta) - c(\epsilon) = \frac{2l}{h+2}.$$

As Φ is of type A_l we have $h = l + 1$, so that

$$c(\delta - \epsilon) = \frac{2l}{l+3}. \tag{2.5}$$

Now consider the containment $T_{l-1} \subset T_l$ and let ϵ' be the identity of T_{l-1}, with l', N' having the obvious meaning.

Lemma 2.5. *We have* $\langle \epsilon - \epsilon', \epsilon' \rangle = 0$.

Proof. If $h' = h - 1$ is the Coxeter number of Φ_{l-1} then by Lemma 2.4 we have $\langle \epsilon', \epsilon' \rangle = \frac{(l-1)h'}{8(h'+2)}$. On the other hand we have

$$\langle \epsilon, \epsilon' \rangle = \frac{1}{(2h+4)(2h'+4)} \sum_{\alpha \in \Phi_l^+} \sum_{\beta \in \Phi_{l-1}^+} \langle t(\alpha), t(\beta) \rangle. \tag{2.6}$$

If $\alpha \in \Phi_{l-1}^+$ then $\sum_{\beta \in \Phi_{l-1}^+} \langle t(\alpha), t(\beta) \rangle = 4 + \frac{2h'-4}{2} = h' + 2$. If $\alpha \in \Phi_l^+ \backslash \Phi_{l-1}^+$ then $\sum_{\beta \in \Phi_{l-1}^+} \langle t(\alpha), t(\beta) \rangle = \frac{l-1}{2}$ since $|\Delta_1(\alpha) \cap \Phi_{l-1}^+| = l - 1$ for each α. Since $|\Phi_l^+ \backslash \Phi_{l-1}^+| = l$, we find that (2.6) yields

$$
\begin{aligned}
\langle \epsilon, \epsilon' \rangle &= \frac{1}{(2h+4)(2h'+4)} \left(N'(h'+2) + \frac{l(l-1)}{2} \right) \\
&= \frac{1}{4(h'+2)(h'+3)} \left(\frac{h'l'(h'+2)}{2} + \frac{l(l-1)}{2} \right) \\
&= \frac{h'(h'+3)(l-1)}{8(h'+2)(h'+3)} \\
&= \frac{h'(l-1)}{8(h'+2)}.
\end{aligned}
$$

So indeed $\langle \epsilon, \epsilon' \rangle = \langle \epsilon', \epsilon' \rangle$, as desired. $\qquad\square$

Lemma 2.6. *We have* $c(\epsilon - \epsilon') = 1 - \dfrac{6}{(l+2)(l+3)}$.

Proof. After Lemma 2.5 we have $c(\epsilon - \epsilon') = c(\epsilon) - c(\epsilon')$. Now use Lemma 2.4 to see that

$$
\begin{aligned}
c(\epsilon - \epsilon') &= \frac{lh}{h+2} - \frac{l'h'}{h'+2} \\
&= \frac{l(l+1)}{l+3} - \frac{l(l-1)}{l+2} \\
&= 1 - \frac{6}{(l+2)(l+3)}.
\end{aligned}
$$

If we iterate this procedure, we obtain the following result.

Theorem 2.7. *Suppose that* $A = A(\Phi)$ *corresponding to the root system* Φ *of type* A_l. *Then the identity* δ *of* A *can be decomposed into a sum of* $l + 1$ *idempotents* $e_1, e_2, \ldots, e_{l+1}$ *satisfying the following:*

(i) $\delta = e_1 + e_2 + \cdots e_{l+1}$;

(ii) $e_i e_j = \delta_{ij} e_i$;

(iii) $\langle e_i, e_j \rangle = 0$ *if* $i \neq j$;

(iv) $c(e_i) = 1 - \dfrac{6}{(i+2)(i+3)}$, $1 \leq i \leq l$;

(v) $c(e_{l+1}) = \dfrac{2l}{l+3}$.

Remark. The significance of (iv), of course, is that the series of values $1 - \frac{6}{(i+2)(i+3)}$, $i = 1, 2, \ldots$ correspond to the central charges of the discrete series representations of the Virasoro algebra. These are the unitary highest weight representations of the Virasoro algebra with central charge c satisfying $0 < c < 1$. These values were first identified in [FQS], and the unitarity was proved in [GKO].

We owe to Jeff Harvey the observation that the central charge $\frac{2l}{l+3}$ also has some significance, namely it corresponds to the central charge of the parafermion algebras [ZF], [DL]. Both the discrete series and parafermionic representations of the Virasoro algebra arise from the GKO "coset construction," indeed our arguments leading to Theorem 2.7 amount to a coset construction at the level of root systems. We pursue these ideas in the following sections.

It goes without saying that (i)–(iii) and a modification of (iv) and (v) also hold if A_l is replaced by root systems of type D or E.

3. The vertex operator algebra $V^+_{\sqrt{2}R}$

Suppose that L is a positive-definite even lattice. The vertex operator algebra V_L associated with L is one of the most fundamental examples of a vertex operator algebra. We refer the reader to [FLM] for the construction of V_L. We wish to focus on the case in which R is the root lattice corresponding to a simple root system Φ as in Section 2, and where $L = \sqrt{2}R$ (later, we will relax our conditions to allow R to be semi-simple). Thus L is indeed a positive-definite even lattice which has, in addition, no vectors of squared length 2.

The lattice L has a canonical automorphism t of order 2, namely the reflection automorphism $t : x \mapsto -x$ for $x \in L$. Then t lifts to a canonical automorphism of V_L (loc. cit.), and we denote by V_L^+ the sub vertex operator algebra of t-fixed points of V_L. In the case that $L = \sqrt{2}R$, V_L^+ has no vectors of weight 1 owing to the absence of vectors in L of squared length 2. Thus

$$V^+_{\sqrt{2}R} = \mathbb{C}\mathbf{1} \oplus B^+ \oplus \cdots \tag{3.1}$$

where in (3.1) we have set $B^+ = (V^+_{\sqrt{2}R})_2$, i.e., B^+ is the space of vectors in $V^+_{\sqrt{2}R}$ of weight 2.

There is a canonical structure of commutative, non-associative algebra on B^+, namely

$$ab = a_1 b, \quad a, b \in B^+$$

where the vertex operator for a is given by

$$Y(a, z) = \sum_{n \in \mathbb{Z}} a_n z^{-n-1}.$$

Similarly, there is a non-degenerate, invariant inner product $\langle \cdot, \cdot \rangle$ on B^+ given by

$$\langle a, b \rangle \mathbf{1} = a_3 b.$$

Conveniently, the algebra structure of B^+ has been written down in Theorem 8.9.5 and Remark 8.9.7 of [FLM]. To describe it, let $H = \mathbb{Q} \otimes \sqrt{2}R$. Then

$$B^+ = S^2(H) \oplus \oplus_{\alpha \in \Phi^+} \mathbb{Q}x_\alpha$$

where

$$x_\alpha = e^{\sqrt{2}\alpha} + e^{-\sqrt{2}\alpha}$$

and $e^{\sqrt{2}\alpha}$ is the standard notation for the element in the group algebra $\mathbb{C}[L]$ corresponding to $\sqrt{2}\alpha$. (Note that in this case $\hat{L}$ is a direct product of L and $\langle \pm 1 \rangle$.) The relations are as follows:

$$h^2 \cdot k^2 = 4(h, k)hk, \quad h, k \in H$$

$$h^2 \cdot x_\alpha = 2(h, \alpha)^2 x_\alpha$$

$$x_\alpha \cdot x_\beta = \begin{cases} 0, & \beta \in \Delta_2(\alpha) \\ x_\gamma, & \alpha \sim \beta \sim \gamma \sim \alpha \\ 2\alpha^2, & \alpha = \beta. \end{cases}$$

Moreover

$$\langle h^2, k^2 \rangle = 2(h, k)^2$$

$$\langle h^2, x_\alpha \rangle = 0$$

$$\langle x_\alpha, x_\beta \rangle = \begin{cases} 0, & \alpha \neq \beta \\ 2, & \alpha = \beta. \end{cases}$$

Theorem 3.1. *Let Φ be a simple root system of type* A, D, E, *let $A = A(\Phi)$ be as in Section 2, and let B^+ be as above. Then there is an isometric surjection of algebras $A \to B^+$ given by*

$$\phi : t(\alpha) \mapsto \frac{1}{2}\alpha^2 - x_\alpha$$

$$u(\alpha) \mapsto \frac{1}{2}\alpha^2 + x_\alpha.$$

Corollary 3.2. *If Φ is of type A_1 then the map $\Phi : A \to B^+$ is an isometric isomorphism of algebras*

Note that $\dim B^+ = \frac{l(l+1)}{2} + N$. If Φ is of type A_l then $N = \frac{l(l+1)}{2}$, so that $\dim B^+ = 2N = \dim A$. So in this case, Theorem 3.1 implies the corollary.

Remark. In the other cases, i.e., Φ of type D or E, we similarly see that the surjection $A \to B^+$ is *not* an isomorphism. Since it is an isometry (by the theorem), the kernel is precisely the radical of the form $\langle , \rangle$ on A.

Proof of Theorem. This is quite straightforward. For example if $\alpha \sim \beta \sim \gamma \sim \alpha$ then

$$\phi(u(\alpha))\phi(t(\beta)) = (\frac{1}{2}\alpha^2 + x_\alpha)(\frac{1}{2}\beta^2 - x_\beta) = (\alpha, \beta)\alpha\beta - x_\gamma + \frac{1}{2}(\beta^2 x_\alpha - \alpha^2 x_\beta). \quad (3.2)$$

In this case we have $(\alpha, \beta) = -1$ and $\alpha\beta = \frac{1}{2}(\gamma^2 - \alpha^2 - \beta^2)$ (since $\alpha + \beta - \gamma = 0$). So (3.2) reads

$$\frac{1}{2}(\alpha^2 + \beta^2 - \gamma^2) - x_\gamma + x_\alpha - x_\beta = \phi(u(\alpha)) + \phi(t(\beta)) - \phi(u(\gamma))$$

as required. The other relations which show that ϕ is an algebra morphism are proved similarly. It is clear that ϕ is an epimorphism, so we only need check that it is also an isometry. This is also easy; for example

$$\langle \phi(u(\alpha)), \phi(t(\beta)) \rangle = \langle \frac{1}{2}\alpha^2 + x_\alpha, \frac{1}{2}\beta^2 - x_\beta \rangle = \frac{1}{2}(\alpha, \beta)^2 - 2\delta_{\alpha,\beta} = \langle u(\alpha), t(\beta) \rangle.$$

The other cases are similar.   □

Denote by $L(c, 0)$ the simple vertex operator algebra associated to the Virasoro algebra with central charge c. We have

Theorem 3.3. *Let Φ be the simple root system of type A_l, with R the corresponding root lattice. Then the vertex operator algebra $V^+_{\sqrt{2}R}$ contains a sub vertex operator algebra isomorphic to a tensor product*

$$\bigotimes_{i=1}^{l+1} L(c_i, 0)$$

where $c_i = 1 - \frac{6}{(i+2)(i+3)}$ for $1 \leq i \leq l$ and $c_{l+1} = \frac{2l}{l+3}$.

Proof. By a simple calculation, Miyamoto has shown (Theorem 4.1 of [M]) that if e is an idempotent in B^+ then $2e$ is such that the components of the vertex operator $Y(2e, z)$ generate a copy of the Virasoro algebra with central charge $c(e)$ in the sense of (2.4). Moreover, orthogonal idempotents e_i, e_j satisfying $e_i e_j = 0$ are such that the corresponding Virasoro algebras commute, that is they generate a tensor product of the two algebras (cf. Lemma 5.1 of (loc. cit.) and Lemma 2.4 of [DLM]). Now the theorem follows from Theorem 2.7.   □

Remark. The theorem has an obvious extension to semi-simple root systems, all of whose simple components are of type A_l (varying l).

4. Niemeier lattices and the moonshine module

Recall that a *Niemeier lattice** is a self-dual, even, positive-definite lattice L of rank 24. The set of vectors L_2 of squared-length 2 forms a root system which is either empty (in which case L is the Leech lattice) or else itself has rank 24. In this latter case the root system L_2 is such that its simple components have a common Coxeter number. We call

---

*   Somewhat unconventionally, according to this terminology the Leech lattice falls under the rubric of Niemeier lattice.

such a root system a Niemeier root system; then the map

$$L \rightarrow L_2$$

sets up a bijection between Niemeier lattices and Niemeier root systems (possibly empty). See [Nie] and [V] for background.

We now discuss the following result, which was hinted at in [Nor, P305] and which extends the results of [Cur].

Theorem 4.1. *Let L be any Niemeier lattice. Then there is at least one (and in general several) isometric embedding $\sqrt{2}L \rightarrow \Lambda$, where Λ is the Leech lattice.*

We want to know how many sublattices of the Leech lattice Λ there are which are versions of each Niemeier lattice, including the Leech lattice itself, rescaled by a factor of 2. For each type of lattice the mass of sublattices of that type is shown in Table 1; the number of sublattices can be obtained from this by multiplying by the order of the Conway group Co_1, which is $2^{21}3^9 5^4 7^2 11.13.23$.

We now explain how the results in Table 1 may be obtained. First, it is easily seen (see Lemma 4.3 below) that if $N = \sqrt{2}L \subset \Lambda$ with L a Niemeier lattice then N corresponds to a Lagrangian (i.e., maximal totally isotropic) subspace of $\Lambda/2\Lambda$, under the orthogonal form whose value on a vector is 0 or 1 according as the squared length of the corresponding Leech lattice vector is or is not divisible by 4. This form is of type $+$, so that a Lagrangian subspace has dimension 12.

Now the number of extensions of any isotropic subspace to a Lagrangian one is known. To be precise, if the space has codimension n (i.e. dimension $12 - n$) the number will be

$$\prod_{i=0}^{n-1}(2^i + 1) \tag{4.1}$$

this being interpreted as 1 if $n = 0$. The problem now is to determine how to use this information to work out the number of Lagrangian subspaces of each type.

We do this by induction on the codimension. Let us call an isotropic subspace *connected* if it is generated by its minimal (norm 4) vectors, and these cannot be split into components that generate disjoint subspaces. It is well known that connected spaces correspond to Dynkin diagrams of type A_n $(n \geq 1)$, D_n $(n \geq 4)$ or E_n $(n = 6, 7, 8)$. We classify all subspaces of each of these types and, for each of them, determine how many extensions it has to connected spaces of the next higher dimension (or, equivalently, how many subspaces of the next lower dimension belong to each type.

This gives us the information we need. For by induction we already know how many Lagrangian spaces contain any connected subspace of higher dimension, and we know by (4.1) the total number of extensions. The difference between these will, therefore, be the number of extensions which contain a subspace of the relevant type as a connected component.

To take some examples, any Lagrangian subspace in which A_8 appears as a connected component must correspond to a Niemeier lattice of type A_8^3. If D_6 appears as a component then it will be of type D_6^4 or $A_9^2 D_6$; we can distinguish between these by counting the number of extensions of an A_9.

Table 1

Λ	153715/123771648
A_1^{24}	141985575/58032128
A_2^{12}	469525/19008
A_3^8	3077275/86016
A_4^6	39821/2560
$A_5^4 D_4$	24653/5120
D_4^6	58025/524288
A_6^4	16813/20160
$A_7^2 D_5^2$	6087/20480
A_8^3	1765/64512
$A_9^2 D_6$	4037/276480
D_6^4	791/294912
$A_{11} D_7 E_6$	67/46080
E_6^4	13/184320
A_{12}^2	61/366080
D_8^3	575/14680064
$A_{15} D_9$	41/3870720
$A_{17} E_7$	1/645120
$D_{10} E_7^2$	11/5160960
D_{12}^2	19/259522560
A_{24}	1/244823040
$D_{16} E_8$	1/660602880
E_8^3	1/1981808640
D_{24}	1/501397585920

The computations were done in **MAPLE**.

Table 2 shows the complete list of connected isotropic subspaces. The first column shows the corresponding symbol, with Greek letters used to distinguish subspaces with the same Dynkin diagram. The second column shows the dimension. The third column gives the maximal intersection with the lattice of type A_{23} generated by vectors of type $(4, -4, 0^{22})$ (parametrized by the corresponding subset of the coordinate positions, on which the Conway group determines a Golay code), plus the number of "spare" vectors. The symbols s_{12}^+, s_n and u_n are taken from [C1]: the first denotes the union of two special octads which intersect in a tetrad; the second a set of cardinality n in an ascending chain containing a special octad and the complement of another such, and *not* containing an s_{12}^+; and the third a set of the right size in an ascending chain linking a non-special hexad (i.e., *not* an s_6), a special dodecad, and the complement of a non-special hexad.

The fourth column shows the structure of the stabilizer of the subspace in Co_1. The subgroup to the left of the aligned column of dots is the pointwise stabilizer. (Square brackets denote an unspecified group of the relevant order.) The last column gives a list of lattices one dimension lower, with the number of extensions and containments of the relevant type. For example, the entry "$30:2(A_8^\alpha)$" in the row beginning "A_9^α" means that an A_9^α contains 2 A_8^α's, while an A_8^α extends to 30 A_9^α's.

Table 2

0	0		$Co_1.1$	
A_1	1	s_2	$Co_2.1$	$98280:1(0)$
A_2	2	s_3	$U_6(2).S_3$	$2300:3(A_1)$
A_3	3	s_4	$2^9.M_{21}.S_4$	$891:4(A_2)$
D_4	4	s_4+1	$2^{4+8}.A_5.2^{1+4}.(S_3 \times 3)$	$21:12(A_3)$
A_4	4	s_5	$2^{1+8}.A_5.S_5$	$336:5(A_3)$
D_5	5	s_5+1	$2^7.2^4.3.2^4.S_5$	$60:5(D_4),\ 5:16(A_4)$
A_5^α	5	s_6	$2^{1+8}.S_3.S_6$	$10:6(A_4)$
A_5^β	5	u_6	$2^6.3.S_6$	$160:6(A_4)$
E_6	6	s_6+1	$2^{1+8}.3.U_4(2).2$	$4:27(D_5),\ 1:36(A_5^\alpha)$
D_6^α	6	s_6+1	$2^6.2^4.2^5.S_6$	$3:6(D_5),\ 3:32(A_5^\alpha)$
D_6^β	6	u_6+1	$2^6.3.2^5.S_6$	$16:6(D_5),\ 1:32(A_5^\beta)$
A_7^α	6	s_8	$2^{1+8}.S_8$	$3:28(A_5^\alpha)$
A_6^α	6	u_7	$2^5.S_6$	$96:1(A_5^\alpha),\ 36:6(A_5^\beta)$
A_6^β	6	u_6+1	$3.S_7$	$64:7(A_5^\beta)$
E_7	7	s_8+1	$2^{1+8}.S_6(2)$	$3:28(E_6),\ 2:63(D_6^\alpha),\ 1:36(A_7^\alpha)$
D_8^α	7	s_8+1	$2^{1+8}.2^6.A_8$	$1:28(D_6^\alpha),\ 2:64(A_7^\alpha)$
D_7	7	u_7+1	$2^5.2^6.S_6$	$16:1(D_6^\alpha),\ 18:6(D_6^\beta),\ 1:64(A_6^\alpha)$
A_8^α	7	s_9	$2^4.A_8$	$64:1(A_7^\alpha),\ 2:28(A_6^\alpha)$
A_7^β	7	u_8	$2^4.2^4.S_4$	$30:8(A_6^\alpha)$
A_7^γ	7	u_7+1	$1.S_6 \times 2$	$32:2(A_6^\alpha),\ 63:6(A_6^\beta)$
E_8	8	s_8+2	$2^{1+8}.O_8^+(2)$	$1:120(E_7),\ 1:135(D_8^\alpha)$
D_9^α	8	s_9+1	$2^4.2^7.A_8$	$16:1(D_8^\alpha),\ 1:28(D_7),\ 1:128(A_8^\alpha)$
D_8^β	8	u_8+1	$2^4.2^7.(2^4 S_4)$	$15:8(D_7),\ 1:128(A_7^\beta)$
A_9^α	8	s_{10}	$2^3.2^4.L_3(2)$	$30:2(A_8^\alpha),\ 8:28(A_7^\beta)$
A_9^γ	8	s_9+1	$1.S_8$	$16:2(A_8^\alpha),\ 1:28(A_7^\gamma)$

A_8^β	8	u_9	$2^3.3^2.2S_4$	$16:9(A_7^\beta)$
A_8^γ	8	u_8+1	$1.2^4.S_4$	$16:1(A_7^\beta)$, $30:8(A_7^\gamma)$
A_8^δ	8	u_7+2	$1.S_3\wr S_3$	$10:9(A_7^\gamma)$
D_{10}^α	9	$s_{10}+1$	$2^3.2^8.(2^4.L_3(2))$	$15:2(D_9^\alpha)$, $4:28(D_8^\beta)$, $1:256(A_9^\alpha)$
D_9^β	9	u_9+1	$2^3.2^8.(3^2.2S_4)$	$8:9(D_8^\beta)$, $1:256(A_8^\beta)$
A_{11}^α	9	s_{12}^+	$2^2.2^6.3^{1+2}.2^2$	$14:18(A_9^\alpha)$, $8:64(A_8^\beta)$
A_{10}^α	9	$s_{10}+1$	$1.2^4.L_3(2)$	$8:1(A_9^\alpha)$, $30:2(A_9^\gamma)$, $4:28(A_8^\gamma)$
A_9^β	9	u_{10}	$2^2.S_6.2$	$6:10(A_8^\beta)$
A_9^δ	9	u_9+1	$1.3^2.2S_4$	$8:1(A_8^\beta)$, $8:9(A_8^\gamma)$
A_9^ϵ	9	u_8+2	$1.2^5.S_5$	$1:10(A_8^\gamma)$
A_9^ζ	9	u_8+2	$1.[2^63]$	$12:6(A_8^\gamma)$, $27:4(A_8^\delta)$
D_{12}^α	10	s_{12}^++1	$2^2.2^9.(2^6.3^{1+2}.2^2)$	$7:18(D_{10}^\alpha)$, $4:64(D_9^\beta)$, $1:512(A_{11}^\alpha)$
D_{10}^β	10	$u_{10}+1$	$2^2.2^9.S_6.2$	$3:10(D_9^\beta)$, $1:512(A_9^\beta)$
A_{15}^α	10	s_{16}	$2.2^4.A_8$	$6:140(A_{11}^\alpha)$, $4:448(A_9^\beta)$
A_{12}^α	10	s_{12}^++1	$1.2^6.3^{1+2}.2^2$	$4:1(A_{11}^\alpha)$, $7:18(A_{10}^\alpha)$, $4:64(A_9^\delta)$
A_{11}^β	10	u_{12}	$2.M_{12}$	$2:66(A_9^\beta)$
A_{11}^γ	10	$s_{10}+2$	$1.[2^93]$	$7:4(A_{10}^\alpha)$, $10:4(A_9^\epsilon)$, $3:24(A_9^\zeta)$
A_{10}^β	10	$u_{10}+1$	$1.S_6.2$	$4:1(A_9^\beta)$, $3:10(A_9^\delta)$
A_{10}^γ	10	u_9+2	$1.[2^43^2]$	$6:2(A_9^\delta)$, $12:9(A_9^\zeta)$
D_{16}	11	$s_{16}+1$	$2.2^{10}.(2^4.A_8)$	$3:140(D_{12}^\alpha)$, $2:448(D_{10}^\beta)$, $1:1024(A_{15}^\alpha)$
D_{12}^β	11	$u_{12}+1$	$2.2^{10}.M_{12}$	$1:66(D_{10}^\beta)$, $1:1024(A_{11}^\beta)$
A_{23}	11	s_{24}	$1.M_{24}$	$2:759(A_{15}^\alpha)$, $2:2576(A_{11}^\beta)$
A_{16}	11	$s_{16}+1$	$1.2^4.A_8$	$2:1(A_{15}^\alpha)$, $3:140(A_{12}^\alpha)$, $2:448(A_{10}^\beta)$
A_{13}	11	s_{12}^++2	$1.[2^93^2]$	$3:2(A_{12}^\alpha)$, $6:18(A_{11}^\gamma)$, $2:64(A_{10}^\gamma)$
A_{12}^β	11	$u_{12}+1$	$1.M_{12}$	$2:1(A_{11}^\beta)$, $1:66(A_{10}^\beta)$
A_{11}^δ	11	$u_{10}+2$	$1.S_6\times 2$	$2:2(A_{10}^\beta)$, $1:10(A_{10}^\gamma)$
A_{11}^ϵ	11	u_9+3	$1.3^2.2S_4$	$4:12(A_{10}^\gamma)$
D_{24}	12	$s_{24}+1$	$1.2^{11}.M_{24}$	$1:759(D_{16})$, $1:2576(D_{12}^\beta)$
A_{24}	12	$s_{24}+1$	$1.M_{24}$	$1:1(A_{23})$, $1:759(A_{16})$
A_{17}	12	$s_{16}+2$	$1.2^4.A_8\times 2$	$1:2(A_{16})$, $1:140(A_{13})$, $1:448(A_{11}^\delta)$
A_{15}^β	12	s_{12}^++4	$1.[2^{12}3^3]$	$1:24(A_{13})$, $1:256(A_{11}^\epsilon)$
A_{12}^γ	12	u_9+4	$1.L_3(3)$	$1:13(A_{11}^\epsilon)$

Let us fix a Niemeier lattice L *not* equal to the Leech lattice, and let R denote the root lattice of the corresponding root system L_2. So there are isometric embeddings

$$\sqrt{2}R \to \sqrt{2}L \to \Lambda$$

and correspondingly there are vertex operator algebra embeddings

$$V^+_{\sqrt{2}R} \to V^+_{\sqrt{2}L} \to V^+_\Lambda \to V^\natural \tag{4.2}$$

where $V^\natural$ is the moonshine module [FLM].

Suppose that L_2 is one of the following type: A_1^{24}, A_2^{12}, or A_{24}. Then we may use (4.2) together with Theorem 3.3 to see that $V^\natural$ contains sub vertex operator algebras of the following kind:

$$L(\tfrac{1}{2}, 0)^{\otimes 48} \tag{4.3}$$

$$\bigotimes_{i=1}^{24} L(c_i, 0) \otimes L(\tfrac{16}{9}, 0) \tag{4.4}$$

as well as the type in equation (1.1).

Type (4.3) was already constructed in [DMZ]. Of course, we get many analogous tensor products by using other Niemeier lattices. Type (1.1) is of interest because, along with (4.3), it seems to be the only tensor product of Virasoro algebras that one can obtain in this way which is both of central charge 24 and has only discrete series as factors. The factor $L(\tfrac{116}{117}, 0)$ in (4.4) (corresponding to $i = 24$) corresponds to the discrete series with largest value of c which we know occurs via an idempotent of the Griess algebra. It would be interesting to know if it is indeed the maximal such value of c.

Of course, the embeddings (4.2) imply embeddings of the corresponding homogeneous spaces. In particular, if we let $B^+(L_2)$ be the weight 2 subspace of $V^+_{\sqrt{2}R}$ then we get $B^+(L_2) \subset B$. Then application of Theorem 3.3 (and the remark following it) yields embeddings of associative subalgebras into the Griess algebra B. For example if L_2 is of type A_1^{24}, corresponding to type (4.3), we get a maximal associative subalgebra of B of dimension 48. This was first constructed in [MN]. Similarly type (1.1) gives a maximal associative subalgebra of B of dimension 36. In the general case we have

Lemma 4.2. *Let L be a Niemeier lattice with root system L_2 which has k simple components. Then L determines (in several ways) an associative subalgebra of B of dimension $24 + k$.*

Proof. The procedure of Theorem 2.7 yields an $(l + 1)$-dimensional associative algebra for any simple root system of rank l, and it maps isomorphically into the corresponding B^+ of theorem 3.1 because ϕ is an isometry. Thus our Niemeier lattice L affords an associative algebra of dimension $\sum_{i=1}^k (1 + l_i)$ where $\{l_i\}$ are the ranks of the simple components of L_2. Since $\sum_i l_i = 24$, the lemma follows. ☐

There is another way to look at these matters which involves the Monster more directly and which was briefly touched on above and in [Nor]. Namely, let $\bar\Lambda = \Lambda/2\Lambda$ be the Leech lattice mod 2 equipped with the non-degenerate form which was described earlier. Then $\bar\Lambda$ has type $+$, that is the Lagrangian subspaces have rank 12. If $\bar U \subset \bar\Lambda$ is such a space then the full inverse image $U \subset \Lambda$ has index 2^{12}, so that $\tfrac{1}{\sqrt{2}}U = L$, say, is a unimodular lattice. Moreover since $\bar{U}$ is totally isotropic then L is integral and hence is a Niemeier lattice. Thus again we see how to embed $U = \sqrt{2}L \to \Lambda$. Indeed, for a Niemeier lattice L together with an isometric embedding $\sqrt{2}L \to \Lambda$ we have already observed that $|\Lambda : \sqrt{2}L| = 2^{12}$ and $\sqrt{2}L$ maps to a totally isotropic subspace of $\bar{\Lambda}$, hence is Lagrangian. We have shown

Lemma 4.3. *There is a natural bijection between embedded and re-scaled Niemeier lattices $\sqrt{2}L \to \Lambda$ and Lagrangian subspaces of $\Lambda/2\Lambda$.*

Next we lift $\Lambda/2\Lambda$ to an extra-special group $Q \simeq 2_+^{1+24}$. As is well-known [G], the subgroup C of the Monster which leaves invariant the "untwisted" part V_Λ^+ of $V^\natural$ (cf. (4.1)) contains Q as a normal subgroup with quotient $C/Q \simeq Co_1$.

The Lagrangian subspaces of $\bar{\Lambda}$ are precisely those which lift to (maximal) elementary abelian subgroups of Q isomorphic to $\mathbb{Z}_2^{13}$. Now in the Monster there are two classes of involutions, of types $2A$ $(2+)$ and $2B$ $(2-)$ respectively, and one may ask how they distribute themselves among the $\mathbb{Z}_2^{13}$ subgroups. There is a pretty answer, which runs as follows: one knows (cf. [C1], for example) that the $2A$ involutions of Q map onto elements of $\Lambda/2\Lambda$ which themselves lift to elements $\lambda \in \Lambda$ satisfying $(\lambda, \lambda) = 4$, whereas the $2B$ involutions correspond to λ such that $(\lambda, \lambda) = 8$. Thus for a fixed maximal elementary abelian subgroup $E \leq Q$ corresponding to the re-scaled Niemeier lattice $\sqrt{2}L \subset \Lambda$, the elements in E of type $2A$ correspond precisely to the elements of squared-length 2, i.e., to the elements of L_2. Of course, these form a Niemeier root system. So we have proved

Proposition 4.4. *Let $E \subset Q$ be a maximal elementary abelian subgroup which corresponds (via the bijection of Lemma 4.3) to the re-scaled Niemeier lattice $\sqrt{2}L$. Then there is a natural bijection*

$$\{2A \text{ involutions in } E\} \to \text{root system of type } L_2.$$

Let E be as in Proposition 4.4, corresponding to $\sqrt{2}L$. We come full circle with the next result, which identifies the sub vertex operator algebra of $V^\natural$ fixed pointwise by E.

Proposition 4.5. *There is a natural isomorphism of vertex operator algebras*

$$(V^\natural)^E \simeq V_{\sqrt{2}L}^+.$$

Proof. Consider V_Λ, which is linearly isomorphic to $S(H_-) \otimes \mathbb{C}[\Lambda]$ (cf. [FLM]) where $H = \mathbb{C} \otimes \Lambda$, $H_- = H_{-1} \oplus H_{-2} \oplus \cdots$ with each $H_i \simeq H$, and $\mathbb{C}[\Lambda]$ is the group algebra of Λ. Now the center $Z(Q)$ of Q acts on $V^\natural$ with fixed-point sub vertex operator algebra naturally isomorphic to V_Λ^+, so we can study the action of $E/Z(Q)$ on V_Λ^+ to prove the Proposition.

Now elements γ of $\mathbb{R} \otimes_\mathbb{Z} \Lambda$ act on V_Λ by fixing $S(H_-)$ identically and acting on basis vectors e^α, $\alpha \in \Lambda$, via

$$\gamma \cdot e^\alpha = e^{2\pi i(\gamma, \alpha)} e^\alpha.$$

This induces an action of the Leech torus $\mathbb{R}^{24}/\Lambda$ on V_Λ in which E corresponds to $(\sqrt{2}L)^*/\Lambda = \frac{1}{\sqrt{2}}L/\Lambda$. Thus if $\alpha \in \Lambda$, $\gamma \cdot e^\alpha = e^\alpha$ for all $\gamma \in \frac{1}{\sqrt{2}}L$ if, and only if, $\alpha \in \sqrt{2}L$.

This shows that $V_\Lambda^{\frac{1}{\sqrt{2}}L/\Lambda} \simeq V_{\sqrt{2}L}$, and therefore also

$$(V^\natural)^E \simeq (V_\Lambda^+)^{\frac{1}{\sqrt{2}}L/\Lambda} \simeq (V_\Lambda^{\frac{1}{\sqrt{2}}L/\Lambda})^+ \simeq V_{\sqrt{2}L}^+. \quad\square$$

Finally, let us continue to let E and L be as above. Restricting Proposition 4.5 to the Griess algebra yields an isomorphism of algebras

$$B^E \simeq (V_{\sqrt{2}L}^+)_2$$

and in particular $B^E \supset B^+(L_2) = (V_{\sqrt{2}R}^+)_2$ where R is the root lattice associated to L_2. This refines the containment of $B^+(L_2)$ in B found earlier.

We have seen in Theorem 3.1 (and its extension to semi-simple root systems) that there is a surjection of $A(L_2)$ onto $B^+(L_2)$ and in particular $B^+(L_2)$ is generated by the images of $t(\alpha), u(\alpha)$ for $\alpha \in L_2^+$.

Now as $t(\alpha)^2 = 8t(\alpha)$ and $\langle t(\alpha), t(\alpha) \rangle = 4$ then each $t(\alpha)/8$ corresponds to an idempotent with central charge $1/2$ (cf. (2.4)). Of course α itself corresponds to an involution of type $2A$. It turns out that the bijection $\alpha \mapsto t(\alpha)/8$ is precisely the correspondence between *transpositions* (i.e., $2A$ involutions of the Monster) and *transposition axes* in the Griess algebra. See [C2] and [M] for more information on this point. Using this perspective, one can read off the relations satisfied by the images of the $t(\alpha)/8$, from Table 1 of [Nor]. This was the original motivation for introducing the algebra $A(\Phi)$.

Bibliography

[B]   N. Bourbaki, Groupes et algèbres de Lie, Chaps. 4, 5, 6, Hermann, Paris, 1968.

[C1]   J. H. Conway, Three lectures on exceptional groups, Chap. 10 of J. H. Conway and N. J. A. Sloane, Sphere packing, lattices and groups, 2nd ed., Springer-Verlag, New York, 1993, 267–298.

[C2]   J. H. Conway, A simple construction for the Fischer–Griess monster group, Invent. Math. **79** (1985), 513–540.

[Cur]   R. T. Curtis, On subgroups of $\cdot 0$. II. Local structure, J. Algebra **63** (1980), 413–434.

[D]   C. Dong, Representations of the moonshine module vertex operator algebra, in: Mathematical aspects of conformal and topological field theories and quantum groups, P. J. Sally et al. (Ed.), Contemp. Math. **175** (1994), 27–36.

[DL]   C. Dong, J. Lepowsky, Generalized vertex algebras and relative vertex operators, Progr. Math. 112, Birkhäuser, Boston 1993.

[DLM]   C. Dong, H. Li, G. Mason, Some twisted modules for the moonshine vertex operator algebras, in: Moonshine, the monster, and related topics, C. Dong, G. Mason (Ed.), Contemp. Math. **193**, Amer. Math. Soc., Providence, RI, 1996, 25–43.

[DMZ] C. Dong, G. Mason, Y. Zhu, Discrete series of the Virasoro algebra and the moonshine module, in: Algebraic groups and their generalizations, Part 2, Proc. Sympos. Pure. Math. **56** II (1994), 295–316.

[FLM] I. B. Frenkel, J. Lepowsky, A. Meurman, Vertex operator algebras and the monster, Pure and Appl. Math. **134**, Academic Press, Boston 1988.

[FQS] D. Friedan, Z. Qiu , S. Shenker, Conformal invariance, unitarity and two-dimensional critical exponents, in: Vertex operators in mathematics and physics, J. Lepowsky, S. Mandelstam, I. M. Singer (Ed.), Math. Sci. Res. Inst. Publ. 3, Springer-Verlag, 419–449 (1985).

[GKO] P. Goddard, A. Kent , D. Olive, Unitary representations of the Virasoro algebra and super-Virasoro algebras, Comm. Math. Phys. 103, 105–119 (1986).

[G]   R. Griess, Jr., The friendly giant, Invent. Math. **69** (1982), 1–102.

[H]   Y. Huang, A non-meromorphic extension of the moonshine module vertex operator algebra, in: Moonshine, the monster, and related topics, C. Dong, G. Mason (Ed.), Contemp. Math. **193**, Amer. Math. Soc., Providence, RI, 1996, 123–148.

[MN]   W. Meyer, W. Neutsch, Associative subalgebras of the Griess algebra, J. Algebra **158** (1993),1–17.

[M]   M. Miyamoto, Griess algebras and conformal vectors in vertex operator algebras, J. Algebra **179** (1996), 523–548.

[Nie] H.-V. Niemeier, Definite quadratische Formen der Dimension 24 und Diskriminante 1, J. Number Theory **5** (1973), 142–178.

[Nor] S. Norton, The monster algebra: some new formulae, in: Moonshine, the monster, and related topics, C. Dong, G. Mason (Ed.), Contemp. Math. **193**, Amer. Math. Soc., Providence, RI, 1996, 297–306.

[V]   B. B. Venkov, The classification of integral even unimodular 24-dimensional quadratic forms, Proc. Steklov Inst. Math. **4** (1980), 63–74; Even unimodular 24-dimensional lattices, Chap. 18 of J. H. Conway and N. J. A. Sloane, Sphere Packing, Lattices and Groups, 2nd ed., Springer-Verlag, New York, 1993, 427–438.

[ZF]   A. B. Zamolodchikov and V. A. Fateev, Nonlocal (parafermion) currents in two-dimensional conformal quantum field theory and self-dual critical points in Z_N-symmetric statistical systems, Soviet Phys. JETP **62** (1985), 215–225.

C. Dong, H. Li, G. Mason
Mathematics Department
University of California
Santa Cruz, CA 95064, U.S.A.
E-mail: dong@cats.ucsc.edu
hli@crab.rutgers.edu
gem@cats.ucsc.edu

S. P. Norton
DPMMS
16 Mill Lane
Cambridge CB2 1SB, U.K.
E-mail: simon@dpmms.cam.ac.uk

A vertex operator algebra related to E_8 with automorphism group $O^+(10, 2)$

Robert L. Griess, Jr.

Abstract. We study a particular VOA which is a subVOA of the E_8-lattice VOA and determine its automorphism group. Some of this group may be seen within the group $E_8(\mathbb{C})$, but not all of it. The automorphism group turns out to be the 3-transposition group $O^+(10, 2)$ of order $2^{21}3^5 5^2 7.17.31$ and it contains the simple group $\Omega^+(10, 2)$ with index 2. We use a recent theory of Miyamoto to get involutory automorphisms associated to conformal vectors. This VOA also embeds in the moonshine module and has stabilizer in $\mathbb{M}$, the monster, of the form $2^{10+16}.\Omega^+(10, 2)$.

Hypotheses

We review some definitions, based on the usual definitions for the elements, products and inner products for lattice VOAs; see [FLM].

Notation 1.2. Φ is a root system whose components have types ADE, $\mathfrak{g}$ is a Lie algebra with root system Φ, $Q := Q_\Phi$, the root lattice and $V := V_Q := \mathbb{S}(\hat{H}_-) \otimes \mathbb{C}[Q]$ is the lattice VOA in the usual notation.

Remark 1.3. We display a few graded pieces of V ($\otimes$ is omitted, and here Q can be any even lattice). We write H_m for $H \otimes t^{-m}$ in the usual notation for lattice VOAs (2.1) and $Q_m := \{x \in Q \mid (x, x) = 2m\}$, the set of lattice vectors of type m.

$$V_0 = \mathbb{C}, \quad V_1 = H_1,$$
$$V_2 = [S^2 H_1 + H_2] + H_1 \mathbb{C} Q_1 + \mathbb{C} Q_2,$$
$$V_3 = [S^3 H_1 + H_1 H_2 + H_3] + [S^2 H_1 + H_2] \mathbb{C} Q_1 + H_1 \mathbb{C} Q_2 + \mathbb{C} Q_3,$$
$$V_4 = [S^4 H_1 + S^2 H_1 H_2 + H_1 H_3 + S^2 H_2 + H_4]$$
$$+ [S^3 H_1 + H_1 H_2 + H_3] \mathbb{C} Q_1 + [S^2 H_1 + H_2] \mathbb{C} Q_2 + H_1 \mathbb{C} Q_3 + \mathbb{C} Q_4.$$

Remark 1.4. Let F be a subgroup of $\text{Aut}(\mathfrak{g})$, where $\mathfrak{g}$ is the Lie algebra $V_1 = H_1 + \mathbb{C} Q_1$ with 0th binary composition. The fixed points V^F of F on V form a subVOA. We have an action of $N(F)/F$ as automorphisms of this sub VOA.

Notation 1.5. *For the rest of this article, we take Q to be the E_8-lattice.* Take F to be a $2B$-pure elementary abelian 2-group of rank 5 in $\text{Aut}(\mathfrak{g}) \cong E_8(\mathbb{C})$; it is fixed point free. Let $E := F \cap T$ where T is the standard torus and where F is chosen to make rank$(E) = 4$. Let $\theta \in F \setminus E$; we arrange for θ to interchange the standard Chevalley generators x_α and $x_{-\alpha}$. See [Gr91]. The Chevalley generator x_α corresponds to the standard generator e^α of the lattice VOA V_Q.

Notation 1.7. $L := Q^{[E]} \cong \sqrt{2}Q$ denotes the common kernel of the lattice characters associated to the elements of E; in the [Carter] notation, these characters are $h^{-1}(E)$; in the root lattice modulo 2, they correspond to the sixteen vectors in a maximal totally singular subspace. Then

$$V_1^F = 0 \tag{1.7.1}$$

and

$$V_2^F = S^2 H_1 + 0 + \mathbb{C}L_2^\theta, \tag{1.7.2}$$

where the latter summand stands for the span of all $e^\lambda + e^{-\lambda}$, where λ runs over all the $15 \cdot 16 = 240$ norm 4 lattice vectors in L. Thus, V_2^F has dimension $\binom{9}{2} + \frac{240}{2} = 36 + 120 = 156$ and has a commutative algebra structure invariant under $N(F) \cong 2^{5+10} \cdot \mathrm{GL}(5, 2)$. We note that $N(F)/F \cong 2^{10}: \mathrm{GL}(5, 2)$ [Gr76], [CoGr], [Gr91].

We will show (6.10) that $\mathrm{Aut}(V^F) \cong O^+(10, 2)$.

2. Inner product

Definition 2.1. The inner product on $S^n H_m$ is $\langle x^n, x^n \rangle = n! m^n \langle x, x \rangle^n$. This is based on the adjointness requirement for $h \otimes t^k$ and $h \otimes t^{-k}$ (see (1.8.15), FLM, p. 29). When $k > 0$, $h \otimes t^{-k}$ acts like multiplication by $h \otimes t^{-k}$ and, when h is a root, $h \otimes t^k$ acts like k times differentiation with respect to h.

When $n = 2$, this means $\langle x^2, x^2 \rangle = 2m^2 \langle x, x \rangle$. In V_2^F, $m = 1$.

Definition 2.2. *The Symmetric Bilinear Form.* Source: [FLM], p. 217. This form is associative with respect to the product (Section 3). We write H for H_1. The set of all g^2 and x_α^+ spans V_2.

$$\langle g^2, h^2 \rangle = 2\langle g, h \rangle^2, \tag{2.2.1}$$

whence

$$\langle pq, rs \rangle = \langle p, r \rangle \langle q, s \rangle + \langle p, s \rangle \langle q, r \rangle, \text{ for } p, q, r, s \in H. \tag{2.2.2}$$

$$\langle x_\alpha^+, x_\beta^+ \rangle = \begin{cases} 2 & \alpha = \pm\beta \\ 0 & \text{else} \end{cases} \tag{2.2.3}$$

$$\langle g^2, x_\beta^+ \rangle = 0. \tag{2.2.4}$$

Notation 2.3. In addition, we have the distinguished Virasoro element ω and identity $\mathbb{I} := \frac{1}{2}\omega$ on V_2 (see Section 3). If h_i is a basis for H and h_i^* the dual basis, then $\omega = \frac{1}{2}\sum_i h_i h_i^*$.

Remark 2.4.

$$\langle g^2, \omega \rangle = \langle g, g \rangle \tag{2.4.1}$$

$$\langle g^2, \mathbb{I} \rangle = \frac{1}{2} \langle g, g \rangle \tag{2.4.2}$$

$$\langle \mathbb{I}, \mathbb{I} \rangle = \dim(H)/8 \tag{2.4.3}$$

$$\langle \omega, \omega \rangle = \dim(H)/2. \tag{2.4.4}$$

If $\{x_i \mid i = 1, \ldots \ell\}$ is an ON basis,

$$\mathbb{I} = \frac{1}{4} \sum_{i=0}^{\ell} x_i^2 \tag{2.4.5}$$

$$\omega = \frac{1}{2} \sum_{i=0}^{\ell} x_i^2. \tag{2.4.6}$$

3. The product on V_2^F

Definition 3.1. *The product on* V_2^F comes from the vertex operations. We give it on standard basis vectors, namely $xy \in S^2 H_1$, for $x, y \in H_1$ and $v_\lambda := e^\lambda + e^{-\lambda}$, for $\lambda \in L_2$. Note that (3.1.1) give the Jordan algebra structure on $S^2 H_1$, identified with the space of symmetric 8×8 matrices, and with $\langle x, y \rangle = \frac{1}{8} \mathrm{tr}(xy)$. The function ε below is a standard part of notation for lattice VOAs.

$$x^2 \times y^2 = 4\langle x, y \rangle xy, \quad pq \times y^2 = 2\langle p, y \rangle qy + 2\langle q, y \rangle py,$$
$$pq \times rs = \langle p, r \rangle qs + \langle p, s \rangle qr + \langle q, r \rangle ps + \langle q, s \rangle pr; \tag{3.1.1}$$

$$x^2 \times v_\lambda = \langle x, \lambda \rangle^2 v_\lambda, \quad xy \times v_\lambda = \langle x, \lambda \rangle \langle y, \lambda \rangle v_\lambda \tag{3.1.2}$$

$$v_\lambda \times v_\mu = \begin{cases} 0 & \langle \lambda, \mu \rangle \in \{0, \pm 1, \pm 3\}; \\ \varepsilon \langle \lambda, \mu \rangle v_{\lambda+\mu} & \langle \lambda, \mu \rangle = -2; \\ \lambda^2 & \lambda = \mu. \end{cases} \tag{3.1.3}$$

Convention 3.2. Recall that $L = Q^{[E]}$. Since $(L, L) \leq 2\mathbb{Z}$, we may and do assume that ε is trivial on $L \times L$.

4. Some calculations with linear combinations of the v_λ

Notation 4.1. For a subset M of H, there is a unique element ω_M of $S^2 H$ which satisfies (1) $\omega_M \in S^2(\mathrm{span}(M))$; (2) for all $x, y \in \mathrm{span}(M)$, $\langle x, y \rangle = \langle \omega_M, xy \rangle$. We define $\mathbb{I}_M := \frac{1}{2} \omega_M$. If M and N are orthogonal sets, we have $\omega_{M \cup N} = \omega_M + \omega_N$. Define $\omega'_M := \omega - \omega_M$ and $\mathbb{I}'_M := \mathbb{I} - \mathbb{I}_M$. This element can be written as $\omega_M = \frac{1}{2} \sum_i x_i^2$, where the x_i form an orthonormal basis of $\mathrm{span}(M)$. We have $\langle \omega_M, \omega_M \rangle =$

$\frac{1}{2}$ dim span(M) and $\langle \mathbb{I}_M, \mathbb{I}_M \rangle = \frac{1}{8}$ dim span(M). Also, $\langle \omega_M, xy \rangle = \langle \omega_M, x'y' \rangle = \langle \omega, x'y' \rangle$, where priming denotes orthogonal projection to span(M).

Notation 4.2. $e_\lambda^\pm := f_\lambda^\mp := \frac{1}{32}[\lambda^2 \pm 4v_\lambda]$, $e_\lambda = e_\lambda^+$, $f_\lambda := e_\lambda^-$. If $a \in \mathbb{Z}$ or $\mathbb{Z}_2$, define $e_{\lambda,a}$ to be e_λ^+ or e_λ^-, as $a \equiv 0, 1 \pmod 2$, respectively; see (4.7). Also, let $e'_{\lambda,a} = e_{\lambda,a+1}$. We define $e_{\lambda,\mu}$ to be $e_{\lambda,a}$, where a is $\frac{1}{2}\langle \lambda, \mu \rangle$ in case μ is a vector in L, and a is $[\hat\lambda, \mu]$, where $[.,.]$ is the nonsingular bilinear form on $\mathrm{Hom}(L, \{\pm 1\})$ gotten from $2\langle ., . \rangle$ by thinking of $\mathrm{Hom}(L, \{\pm 1\})$ as $\frac{1}{2}L/L$ and where $\hat\lambda$ is the character gotten by reducing the inner product with $\frac{1}{2}\lambda$ modulo 2. Finally. let q be the quadratic form on $\mathrm{Hom}(L, \{\pm 1\})$ gotten by reducing $x \mapsto \langle x, x \rangle$ modulo 2, for $x \in \frac{1}{2}L$.

Lemma 4.3.

(i)   The $e_\lambda^\pm$ are idempotents.

(ii) $\langle e_\lambda^\pm, e_\mu^\pm \rangle = \begin{cases} \frac{1}{16} & \lambda = \mu; \\ \frac{1}{128} & \langle \lambda, \mu \rangle = -2; \\ 0 & \langle \lambda, \mu \rangle = 0. \end{cases}$

(iii) $\langle e_\lambda^\pm, e_\mu^\mp \rangle = \begin{cases} 0 & \lambda = \mu; \\ \frac{1}{128} & \langle \lambda, \mu \rangle = -2; \\ 0 & \langle \lambda, \mu \rangle = 0. \end{cases}$

Proof. (i) $(e_\lambda^\pm)^2 = \frac{1}{1024}[4 \cdot 4\lambda^2 + 16\lambda^2 \pm 8 \cdot 4^2 v_\lambda] = e_\lambda^\pm$. (ii) and (iii) follow trivially from (2.2).

Notation 4.4. For finite $X \subseteq L$, define $s(X) := \sum_{x \in \pm X/\{\pm 1\}} x^2$.

Lemma 4.5. *If* $X \subseteq L_2$, $\langle \omega, s(X) \rangle = 4|(\pm X)/\{\pm 1\}|$ *and so* $s(L_2) = \frac{|L_2|}{2}\omega = 120\omega$.

Proof. (2.2.5)

Corollary 4.6.

(i)   For $\alpha \in L_2$, $\langle \omega_\alpha, s(\alpha) \rangle = \langle \omega, s(\alpha) \rangle = 4$ and $\langle \omega_\alpha, \omega_\alpha \rangle = \frac{1}{2}$, whence $s(\alpha) = 8\omega_\alpha = 16\mathbb{I}_\alpha$ and $\mathbb{I}_\alpha = \frac{1}{16}\alpha^2$.

(ii) $\langle \omega_{E_7}, s(\Phi_{E_7}) \rangle = \langle \omega, s(\Phi_{E_7}) \rangle = 63$, whence $s(\Phi_{E_7}) = 18\omega_{\Phi_{E_7}}$;

(iii) $\langle \omega, s(\Phi_{D_8}) \rangle = 56$, whence $s(\Phi_{D_8}) = 56\omega_{\Phi_{E_7}}$.

Notation 4.7. For $\varphi \in \mathrm{Hom}(L, \{\pm 1\})$, define $f(\varphi) := \sum_{\lambda \in L_2/\{\pm 1\}} \varphi(\lambda)v_\lambda$, $u(\varphi) := \sum_{\lambda \in L_2/\{\pm 1\}} \varphi(\lambda)\lambda^2$ and $e(\varphi) := \frac{1}{16}\mathbb{I} + \frac{1}{64}f(\varphi)$. These arguments may come from other domains, as in (4.2), and we allow mixing as in $e(\varphi\lambda)$, for a character φ and lattice vector λ. We prove later that $e(\varphi)$ is an idempotent.

Lemma 4.8. *Let* $r, s \in L$, $a, b \in \mathbb{Z}$ *and let*

$$n(r, s, a, b) := \frac{1}{2} |\{t \in \Phi \mid \langle r, t \rangle \equiv 2a \pmod 2, \langle s, t \rangle \equiv 2b \pmod 2\}|.$$

(i) *Suppose that the images of r and s in $L/2L$ are nonzero and distinct. The values of $n(r, s, a, b)$ depend only on the isometry type of the images of the ordered pair $\langle r, s \rangle$ in $L/2L$ and are listed below.*

$\frac{1}{4}\langle r, r \rangle$	$\frac{1}{4}\langle s, s \rangle$	$\frac{1}{2}\langle r, s \rangle$	$2n(rs00)$	$2n(rs01)$	$2n(rs10)$	$2n(rs11)$
0	0	0	48	64	64	64
0	0	1	56	56	56	72
0	1	0	64	48	64	64
0	1	1	56	56	72	56
1	0	0	64	64	48	64
1	0	1	56	72	56	56
1	1	0	64	64	64	48
1	1	1	72	56	56	56

(ii) *If $s = 0$ and $\langle r, r \rangle = 4$, then $2n(r, s, 0, 0) = 128$ and $2n(r, s, 1, 0) = 112$. If $s = 0$ and $\langle r, r \rangle = 8$, then $2n(r, s, 0, 0) = 112$ and $2n(r, s, 1, 0) = 128$.*

Lemma 4.9. *The $f(\varphi)$, as φ ranges over all nonsingular characters of L of order 2, form a basis for $\mathbb{C}L_2^\theta$.*

Proof. Use the action of the subgroup of the Weyl group stabilizing the maximal totally singular subspace $L/2Q$ of $Q/2Q$ (its shape is $2^{-5}2^8.2_+^{1+6}$. GL(4, 2)); it also stabilizes the maximal totally singular subspace $2Q/2L$ of $L/2L$ (halve the quadratic form on L, then reduce modulo 2). Since W_{E_8} induces the group $O^+(8, 2)$ on $Q/2Q$, Witt's theorem implies that the stabilizer of a maximal isotropic subspace is transitive on the nonsingular vectors outside it. The action of this group on $L/2L$ has the analogous property.

Notation 4.10. $u(\varphi) := \sum_{\lambda \in L_2/\{\pm 1\}} \varphi(\lambda)\lambda^2$. This also makes sense for $\varphi \in L$ by the identification in (4.2).

Proposition 4.11. *Let $\alpha \in \mathrm{Hom}(L, \{\pm 1\})$.*
(i)

$$\langle u(\alpha), \omega \rangle = \sum \alpha(\lambda)\lambda^2 = \begin{cases} 480 & \alpha = 1; \\ -32 & \alpha \text{ singular;} \\ 32 & \alpha \text{ nonsingular.} \end{cases}$$

(ii)

$$u(\alpha) = \begin{cases} 240\mathbb{I} & \alpha = 1; \\ -16\mathbb{I} & \alpha \text{ is singular;} \\ -208\mathbb{I}_\alpha + 48\mathbb{I}'_\alpha = -256\mathbb{I}_\alpha + 48\mathbb{I} = -16\alpha^2 + 48\mathbb{I} & \alpha \text{ is nonsingular;} \end{cases}$$

(in the third case, α is taken to be a norm 4 lattice vector in L^θ; it is well defined up to its negative, and this suffices). Their respective norms are 57600, 256 and $\frac{1}{8}208^2 + \frac{7}{8}48^2 = 7424 = 2^8 29$.

Proof. We deal with cases, making use of inner product results (2.2) and (2.3); at once, we get (i). If $u(\alpha)$ were known to be a multiple of $\mathbb{I}$, this inner product information would be enough to determine $u(\alpha)$. This is so for $u(1)$ since the linear group (isomorphic to the Weyl group of E_8) stabilizing L_2 is irreducible and so fixes a subspace of dimension just 1 in the symmetric square of H. It follows that $u(1) = 240\mathbb{I}$.

Notice that in all cases $u(\alpha) = 2u'(\alpha) - u(1)$, where $u'(\alpha) := \sum_{\lambda \in \Phi'/\{\pm1\}} \lambda^2$ and $\Phi' := \{\lambda \in \Phi \mid \alpha(\lambda) = 1\}$.

Now to evaluate $u' := u'(\alpha)$ for $\alpha \neq 1$. If Φ' has type D_8, we have an irreducible group as above and conclude that $u'(\alpha) = b\mathbb{I}$, where $b = \langle u', \mathbb{I} \rangle = \frac{1}{2}\langle u', \omega \rangle = \frac{1}{2}56 \cdot 4 = 112$. If Φ' has type $A_1 E_7$, we have a reducible group with two constituents and conclude that $u' = c\mathbb{I}_\alpha + d\mathbb{I}_{\alpha^\perp}$, where we interpret α as an element of L_2 and moreover as a root in the A_1-component of Φ'. Since $\langle \mathbb{I}_\alpha, \mathbb{I}_\alpha \rangle = \frac{1}{8}$ and $\langle \alpha^2, \alpha^2 \rangle = 32$, $c = 16$. Since $\mathbb{I} = \mathbb{I}_\alpha + \mathbb{I}_{\alpha^\perp}$, $\frac{1}{8}c + \frac{7}{8}d = \langle u', \mathbb{I} \rangle = 128$, whence $d = 144$. Thus, $2u' - 240\mathbb{I} = 32\mathbb{I}_\alpha + 288\mathbb{I}'_\alpha - 240\mathbb{I} = -208\mathbb{I}_\alpha + 48\mathbb{I}'_\alpha = -256\mathbb{I}_\alpha + 48\mathbb{I}$.

Lemma 4.12. $f(\varphi) \times f(\psi) =$

$$
\begin{cases}
(-1)^{1+q(\varphi\psi)}4(f(\varphi) + f(\psi)) + (-1)^{1+\langle\varphi,\psi\rangle}64v_{\varphi\psi} + u(\varphi\psi), & \text{if } \varphi \neq \psi; \\
\text{furthermore, this equals} \\
\quad \begin{cases}
-4(f(\varphi) + f(\psi)) - 16\mathbb{I} & \text{if } \varphi\psi \text{ singular,} \\
4(f(\varphi) + f(\psi)) + 48\mathbb{I} - 512e_{\alpha,\langle\varphi,\psi\rangle}, & \text{if } \varphi\psi \text{ nonsingular;}
\end{cases} \\
56f(\varphi) + u(1), & \text{if } \varphi = \psi.
\end{cases}
$$

Proof. The left side is

$$\sum_\lambda \sum_{\mu:\langle\mu,\lambda\rangle=-2} \varphi(\lambda)\psi(\mu)v_{\lambda+\mu} + u(\varphi\psi)$$

$$= \sum_\nu \psi(\nu) \sum_{\lambda:\langle\nu,\lambda\rangle=2} (\varphi\psi)(\lambda)v_\nu \quad (\text{for } \nu = \lambda + \mu)$$

$$= \sum_\nu \psi(\nu)(n(\nu, \varphi\psi, 1, 0) - n(\nu, \varphi\psi, 1, 1))v_\nu + u(\varphi\psi).$$

We use (4.8) and (4.9). The coefficent of v_ν is 0 if $\varphi\psi(\nu) \equiv 1$ (mod 2). If $\varphi\psi(\nu) \equiv 0$ (mod 2), then $\varphi(\nu) = \psi(\nu)$; the coefficient is $56\psi(\nu)$ if $\varphi\psi = 1$, $(-1)^{1+q(\varphi\psi)}8$ if $\varphi\psi \neq 1$ or ν and, if $\varphi\psi = \nu$, it is $-56(-1)^{\langle\varphi,\psi\rangle}$.

Corollary 4.13. $e(\varphi)^2 = e(\varphi)$.

The 256 $f(\varphi)$ live in $\mathbb{C}L^\theta$, a space of dimension 120, so they are linearly dependent. There is a natural subset which forms a basis.

Proposition 4.14. If $\varphi\psi$ is singular, $f(\varphi) \times f(\psi) = -4(f(\varphi) + f(\psi)) - 16\mathbb{I}$ and $e(\varphi) \times e(\psi) = 0$.

Proof. It suffices to show that $(4\mathbb{I} + f(\varphi)) \times (4\mathbb{I} + f(\psi)) = 0$, or $16\mathbb{I} + 4(f(\varphi) + f(\psi)) + 4(-1)^{1+q(\varphi\psi)}(f(\varphi) + f(\psi)) + u(\varphi\psi) = 0$. This follows from $q(\varphi\psi) = 0$ and $u(\varphi\psi) = -16\mathbb{I}$; see (4.12.i).

Lemma 4.15.

(i) $\quad \langle f(\varphi), f(\psi) \rangle = \begin{cases} 240 & \text{if } \varphi = \psi; \\ -16 & \text{if } \varphi\psi \text{ singular}; \\ 16 & \text{if } \varphi\psi \text{ nonsingular}. \end{cases}$

(ii) $\quad \langle e(\varphi), e(\psi) \rangle = \begin{cases} \frac{1}{16} & \text{if } \varphi = \psi; \\ 0 & \text{if } \varphi\psi \text{ singular}; \\ \frac{1}{128} & \text{if } \varphi\psi \text{ nonsingular}. \end{cases}$

Proof. (i) This inner product is $2 \sum_\lambda \varphi\psi(\lambda)$, so consider the cases that $\varphi\psi$ is 1, singular or nonsingular. One can also use (4.12) and associativity of the form. We leave (ii) as an exercise, with (i) and (2.2).

Theorem 4.16. *The $2e(\varphi)$ are conformal vectors of conformal weight ($=$ central charge) $\frac{1}{2}$.*

Proof. By (4.10) and [Miy], Theorem 4.1, these are conformal vectors. Fix φ. Choose a maximal, totally singular subspace, J, of L modulo $2L$. Let $\mathfrak{J}$ be the set of distinct linear characters of L which contain J in their kernel. The $e(\psi)$, for $\psi \in \varphi\mathfrak{J}$, are pairwise orthogonal idempotents (4.12) which sum to $\mathbb{I}$ (to prove this, use the orthogonality relations for this set of 16 distinct characters). We use the fact that conformal weight of $2e(\varphi)$ is at least $\frac{1}{2}$ (see Proposition 6.1 of [Miy]). Since their conformal weights add to 8, the conformal weight of ω, we are done.

Notation 4.17. In an integral lattice, an element of norm 2 is called a *root* and an element of norm 4 is called a *quoot* (suggested by the term "quartic" for degree 4).

Notation 4.18. The idempotents $e_\lambda^\pm$ are called *idempotents of quoot type* or *quooty idempotents* and the $e(\varphi)$ are called idempotents of *tout type* or *tooty idempotents* (suggested by "tout" or "tutti"). The set of all such is denoted $\mathcal{Q}\mathfrak{J}$, $\mathcal{T}\mathfrak{J}$, respectively. Set $\mathcal{Q}\mathcal{T}\mathfrak{J} := \mathcal{Q}\mathfrak{J} \cup \mathcal{T}\mathfrak{J}$.

5. Eigenspaces

Notation 5.1. For an element x of a ring, ad_x, $\text{ad}(x)$ denotes the endomorphism: right multiplication by x. If the ring is a finite dimensional algebra over a field, the *spectrum of x* means the spectrum of the endomorphism $\text{ad}(x)$.

The main result of this section is the following.

Theorem 5.2. *If e is one of the idempotents e_λ, f_λ or $e(\varphi)$ of Section 4, its spectrum is $(1^1, \frac{1}{4}^{35}, 0^{120})$.*

We prove (5.2) in steps, treating the quooty and tooty cases separately.

Table 5.3. The action of $\mathrm{ad}(e_\lambda)$ on a spanning set. Recall that $e_\lambda = \frac{1}{32}[\lambda^2 + 4v_\lambda]$.

vector	image under $\mathrm{ad}(e_\lambda)$	dimension
μ^2	$\frac{1}{32}[4\langle\lambda,\mu\rangle\lambda\mu + 4\langle\lambda,\mu\rangle^2 v_\lambda]$ $= \frac{1}{8}[\langle\lambda,\mu\rangle\lambda\mu + \langle\lambda,\mu\rangle^2 v_\lambda]$	36
μv	$\frac{1}{32}[2\langle\lambda,\mu\rangle\lambda v + 2\langle\lambda,v\rangle\lambda\mu + 4\langle\lambda,\mu\rangle\langle\lambda,v\rangle v_\lambda]$ $= \frac{1}{16}[\langle\lambda,\mu\rangle\lambda v + \langle\lambda,v\rangle\lambda\mu + 2\langle\lambda,\mu\rangle\langle\lambda,v\rangle v_\lambda]$	36
λ^2	$\frac{1}{32}[16\lambda^2 + 64v_\lambda] = 16e_\lambda$	1
$\lambda h,\ \langle\lambda,h\rangle = 0$	$\frac{1}{32}8\lambda h = \frac{1}{4}\lambda h$	7
$gh,\ \langle g,\lambda\rangle = \langle h,\lambda\rangle = 0$	0	28
v_λ	$4\frac{1}{32}[4\lambda^2 + 16v_\lambda] = 4e_\lambda$	1
$v_\mu,\ \langle\lambda,\mu\rangle = 0$	0	63
$v_\mu,\ \langle\lambda,\mu\rangle = -2$	$\frac{1}{32}[4v_{\lambda+\mu} + 4v_\mu] = \frac{1}{8}[v_{\lambda+\mu} + v_\mu]$	56

Table 5.4. The eigenspaces of $\mathrm{ad}(e_\lambda)$.

eigenvalue	basis element(s)	dimension
1	e_λ	1
$\frac{1}{4}$	$\lambda h,\ \langle\lambda,h\rangle = 0$	7
$\frac{1}{4}$	$v_{\lambda+\mu} + v_\mu,\ \langle\lambda,\mu\rangle = -2$	28
0	$v_{\lambda+\mu} - v_\mu,\ \langle\lambda,\mu\rangle = -2$	28
0	$gh,\ \langle g,\lambda\rangle = \langle h,\lambda\rangle = 0$	28
0	$v_\mu,\ \langle\lambda,\mu\rangle = 0$	63
0	f_λ	1

Table 5.5. The action of $\mathrm{ad}(f_\lambda)$ on a spanning set. Recall that $f_\lambda = \frac{1}{32}[\lambda^2 - 4v_\lambda] = -e_\lambda + \frac{1}{16}\lambda^2$, so the table below may be deduced from Table (5.3) and (3.1.1).

vector	image under $\mathrm{ad}(f_\lambda)$	dimension
μ^2	$\frac{1}{4}\langle\lambda,\mu\rangle\lambda\mu - \frac{1}{8}[\langle\lambda,\mu\rangle\lambda\mu + \langle\lambda,\mu\rangle^2 v_\lambda]$ $= \frac{1}{8}[\langle\lambda,\mu\rangle\lambda\mu - \langle\lambda,\mu\rangle^2 v_\lambda]$	36
$\mu\nu$	$\frac{1}{8}[\langle\lambda,\mu\rangle\lambda\nu + \langle\lambda,\nu\rangle\lambda\mu]$ $-\frac{1}{16}[\langle\lambda,\mu\rangle\lambda\nu + \langle\lambda,\nu\rangle\lambda\mu + 2\langle\lambda,\mu\rangle\langle\lambda,\nu\rangle v_\lambda]$ $= \frac{1}{16}[\langle\lambda,\mu\rangle\lambda\nu + \langle\lambda,\nu\rangle\lambda\mu - 2\langle\lambda,\mu\rangle\langle\lambda,\nu\rangle v_\lambda]$	36
λ^2	$\lambda^2 - 16e_\lambda = 16f_\lambda$	1
$\lambda h,\ \langle\lambda, h\rangle = 0$	$\frac{1}{32}8\lambda h = \frac{1}{4}\lambda h$	7
$gh,\ \langle g,\lambda\rangle = \langle h,\lambda\rangle = 0$	0	28
v_λ	$v_\lambda - 4e_\lambda = -4f_\lambda$	1
$v_\mu,\ \langle\lambda,\mu\rangle = 0, \pm 1$	0	63
$v_\mu,\ \langle\lambda,\mu\rangle = -2$	$\frac{1}{4}v_\mu - \frac{1}{8}[v_{\lambda+\mu} + v_\mu] = \frac{1}{8}[-v_{\lambda+\mu} + v_\mu]$	56

Table 5.6. The eigenspaces of $\mathrm{ad}(f_\lambda)$.

eigenvalue	basis element(s)	dimension
1	f_λ	1
$\frac{1}{4}$	$\lambda h,\ \langle\lambda, h\rangle = 0$	7
0	$v_{\lambda+\mu} + v_\mu,\ \langle\lambda,\mu\rangle = -2$	28
$\frac{1}{4}$	$v_{\lambda+\mu} - v_\mu,\ \langle\lambda,\mu\rangle = -2$	28
0	$gh,\ \langle g,\lambda\rangle = \langle h,\lambda\rangle = 0$	28
0	$v_\mu,\ \langle\lambda,\mu\rangle = 0$	63
0	e_λ	1

Table 5.7. The action of $\mathrm{ad}(f(\varphi))$ on a spanning set.

vector	image under $\mathrm{ad}(f(\varphi))$	dimension
$f(\varphi)$	$56f(\varphi) + u(1)$	1
$f(\psi),\ \psi\varphi$ singular	$-4(f(\varphi) + f(\psi)) - 16\mathbb{I}$	120
$f(\psi),\ \psi\varphi$ nonsingular	$4(f(\varphi) + f(\psi)) + 48\mathbb{I} - 512 e_{\alpha,\langle\varphi,\psi\rangle}$	120
$u(\alpha),\ \alpha$ nonsingular	$\sum_\mu \varphi(\mu) \sum_\lambda \alpha(\lambda)\langle\lambda,\mu\rangle^2 v_\mu$ $= \sum_\mu \varphi(\mu)[48 - 16\langle\mu,\alpha\rangle^2]v_\mu$	36
$\mathbb{I}$	$f(\varphi)$	1

(5.7.1). Proofs of the above are straightforward. We give a proof only of the formula for $\xi := f(\varphi) \times u(\alpha)$. Clearly, ξ is a linear combination of the v_λ, so we just get its coefficent at v_λ as $\frac{1}{2}\langle\xi, v_\lambda\rangle$. By associativity of the form, this is $\frac{1}{2}\langle u(\alpha), f(\varphi) \times v_\lambda\rangle = \frac{1}{2}\langle u(\alpha), \varphi(\lambda)\lambda^2\rangle$. By (4.12.ii), we have an expression for $u(\alpha)$. Since $\langle\mathbb{I}, \lambda^2\rangle = 2$ and $\langle\mathbb{I}_\alpha, \lambda^2\rangle = \frac{1}{2}\langle\lambda, \alpha\rangle^2\langle\alpha, \alpha\rangle^{-1} = \frac{1}{8}\langle\lambda, \alpha\rangle^2$, the respective cases of (4.12.ii) lead to $\frac{1}{2}\langle u(\alpha), \varphi(\lambda)\lambda^2\rangle = \varphi(\lambda)240,\ -\varphi(\lambda)16$ and $\varphi(\lambda)[48 - 16\langle\lambda, \alpha\rangle^2]$. Only the latter case is recorded in the table since $u(\alpha)$ is otherwise a multiple of $\mathbb{I}$.

Table 5.8. The action of $\mathrm{ad}(e(\varphi))$ on a spanning set. Recall that $e(\varphi) = \frac{1}{16}\mathbb{I} + \frac{1}{64}f(\varphi)$. We use the notation $\alpha := \varphi\psi$, when $\varphi\psi$ is nonsingular. Note that the set of such α^2 span $S^2(H)$.

vector	image under $\mathrm{ad}(e(\varphi))$	dimension
$f(\varphi)$	$\frac{15}{16}f(\varphi) + \frac{15}{4}\mathbb{I}$	1
$f(\psi),\ \varphi\psi$ singular	$-\frac{1}{16}f(\psi) - \frac{1}{4}\mathbb{I}$	120
$f(\psi),\ \varphi\psi$ nonsingular	$4e(\varphi) + 8e(\psi) - 8e_{\alpha,\langle\varphi,\psi\rangle}$	120
α^2 if $\alpha := \varphi\psi$ nonsingular	$2[e(\varphi) - e(\psi)] + 2e_{\alpha,\varphi(\alpha)}$	36
$u(\alpha),\ \alpha$ nonsingular	$\frac{1}{16}u(\alpha) + \sum_\mu \varphi(\mu)[\frac{3}{4} - \frac{1}{4}\langle\mu,\alpha\rangle^2]v_\mu$	36

$\mathbb{I}$	$e(\varphi)$	1
$e(\varphi)$	$e(\varphi)$	1
$e(\psi)$ if $\varphi\psi$ singular	0	120
$e(\psi)$ if $\alpha := \varphi\psi$ nonsingular	$\frac{1}{8}[e(\varphi)+e(\psi) - e_{\alpha,\langle\varphi,\psi\rangle}]$	120
$v_\alpha = 4(e_\alpha^+ - e_\alpha^-)$	$\varphi(\alpha)\frac{1}{2}[e_{\alpha,\langle\varphi,\psi\rangle} + e(\varphi) - e(\psi)]$	120
$e_{\alpha,\varphi}$	$\frac{1}{8}[e_{\alpha,\langle\varphi,\psi\rangle} + e(\varphi) - e(\psi)]$	35
$e'_{\alpha,\varphi}$	0	120

Table 5.9. Eigenspaces of $\mathrm{ad}(e(\varphi))$. In the table, we use the convention that $\alpha := \varphi\psi$ is nonsingular. Recall that $e_\lambda^\pm = \frac{1}{32}(\lambda^2 \pm 4e_\lambda)$. Recall that $v_\alpha = 4(e_\alpha^+ + e_\alpha^-)$.

eigenvalue	basis elements	dimension
1	$e(\varphi)$	1
0	$e'_{\alpha,\varphi}$	120
$\frac{1}{4}$	$-e_{\alpha,\varphi} + e(\psi)$	35

Table 5.10. Action of idempotents on idempotents. Recall the definitions $e_\lambda^\pm = \frac{1}{32}[\lambda^2 \pm 4v_\lambda]$, $e_{\lambda,\varphi} = e_{\lambda,\langle\varphi,\lambda\rangle}$, $e(\varphi) = \frac{1}{16}\mathbb{I} + \frac{1}{64}f(\varphi)$. In expressions below, a and b are integers modulo 2.

$$e_{\lambda,a} \times e_{\mu,b} = \begin{cases} 0 & \text{if } \langle\lambda,\mu\rangle = 0 \\[6pt] \begin{aligned} & 2^{-10}[-8\lambda\mu + 16((-1)^a v_\lambda \\ & + (-1)^b v_\mu + 16(-1)^{a+b} v_{\lambda+\mu}] \\ &= 2^{-10}[-4(\lambda+\mu)^2 - (-1)^{a+b}4v_{\lambda+\mu} \\ & + 4((\lambda^2 + 4(-1)^a v_\lambda) \\ & + 4(\mu^2 + 4(-1)^b v_\mu)] \\ &= 2^{-3}[e_{\lambda+\mu,a+b+1} + e_{\lambda,a} + e_{\mu,b}] \end{aligned} & \text{if } \langle\lambda,\mu\rangle = -2 \\[6pt] e_{\lambda,a} & \begin{aligned}\text{if } (\langle\lambda,\mu\rangle, (-1)^{a+b}) \\ = (4,0), (-4,1)\end{aligned} \\[6pt] 0 & \begin{aligned}\text{if } (\langle\lambda,\mu\rangle, (-1)^{ab}) \\ = (4,1), (-4,0)\end{aligned} \end{cases}$$

$$e(\varphi) \times e(\psi) = \begin{cases} e(\varphi) & \text{if } \varphi = \psi \\ 0 & \text{if } \varphi\psi \text{ singular} \\ 2^{-3}[e(\varphi) + e(\psi) - e_{\varphi\psi,\varphi}] & \text{if } \varphi\psi \text{ nonsingular} \end{cases}$$

$$e_{\lambda,a} \times e(\psi) = \begin{cases} 0 & [\lambda, \psi] = a + 1 \\ 2^{-3}[e(\psi\lambda) - e(\psi) - e_{\lambda,\psi}] & [\lambda, \psi] = a. \end{cases}$$

Table 5.11. Inner products of idempotents

See the basic inner products in Section 2. We also need $(f(\varphi), f(\psi))$ from (4.15).

$$(e_{\lambda,a}, e_{\mu,b}) = 2^{-9}\langle\lambda, \mu\rangle^2 + 2^{-5}(-1)^{a+b}\delta_{\lambda,\mu} = \begin{cases} 2^{-4} & \lambda = \mu \\ 0 & \lambda\mu \text{ singular} \\ 2^{-7} & \lambda\mu \text{ nonsingular} \end{cases}$$

$$(e_{\lambda,a}, e(\varphi)) =$$
$$2^{-8} + 2^{-8}(-1)^a\varphi(\alpha) = \begin{cases} 2^{-7} & \text{if } (-1)^a\varphi(\alpha) = 1, \text{ i.e., } a + [\varphi, \alpha] = 0 \\ 0 & \text{if } (-1)^a\varphi(\alpha) = -1, \text{ i.e., } a + [\varphi, \alpha] = 1 \end{cases}$$

$$(e(\varphi), e(\psi)) = 2^{-8} + 2^{-12} \begin{cases} 240 & \varphi = \psi \\ -16 & \varphi\psi \text{ singular} \\ 16 & \varphi\psi \text{ nonsingular} \end{cases} = \begin{cases} 2^{-4} \\ 0 \\ 2^{-7} \end{cases}$$

6. Idempotents and involutions

Notation 6.1. The polynomial $p(t) := \frac{32}{3}t^2 - \frac{32}{3} + 1$ takes values $p(0) = p(1) = 1$ and $p(\frac{1}{4}) = -1$. For an idempotent e such that $\mathrm{ad}(e)$ is semisimple with eigenvalues $0, \frac{1}{4}$ and 1, we define $t(e) := p(\mathrm{ad}(e))$, an involution which is 1 on the $0-$ and 1-eigenspaces and is -1 on the $\frac{1}{4}$-eigenspace. Let $E_\pm = E_\pm(e) = E_\pm(t(e))$ denote the ± 1 eigenspace of this involution.

The main results of this section are the following.

Theorem 6.2. *For a quooty or tooty idempotent, e, $t(e)$ is an automorphism of V^F.*

This follows from the theory in [Miy] and (5.2). In this section, we shall verify this directly on the algebra V_2^F only, for the $e_\lambda^\pm$ and $e(\varphi)$ and prove that these elements are all the idempotents whose doubles are conformal vectors of conformal weight $\frac{1}{2}$. See (6.5) and (6.6).

Theorem 6.3. *The subgroup of* $\mathrm{Aut}(V^F)$ *generated by all* $t(e)$ *as in* (6.2) *is isomorphic to* $O^+(10, 2)$.

The Miyamoto theory proves that the $t(e)$ are in $\mathrm{Aut}(V^F)$. It turns out that the group they generate restricts faithfully to V_2^F, and there we can identify it.

Theorem 6.4. *The group generated by the* $t(e_\lambda^{\pm})$ *is isomorphic to the maximal 2-local subgroup of* $O^+(10, 2)$ *of shape* $2^8 : O^+(8, 2)$. *The normal subgroup of order* 2^8 *is generated by all* $t(e_\lambda^+)t(e_\lambda^-)$ *and acts regularly on the set of tooty idempotents. A complement to this normal subgroup is the stabilizer of any* $e(\varphi)$, *for example, the stabilizer of* $e(1)$ *(1 means the trivial character) is generated by all* $t(e_{\lambda,1})$. *Such a complement is isomorphic to the Weyl group of type* E_8, *modulo its center.*

To verify that the involution $t(e)$ is an automorphism of V_2^F, it suffices to check that $E_+E_+ + E_-E_- \le E_+$ and $E_+E_- \le E_-$.

Proposition 6.5. *If* e *is quooty,* $t(e)$ *is an automorphism of* V_2^F.

Proof. This is straightforward with (6.4) and (5.4).

Proposition 6.6. *If* e *is tooty,* $t(e)$ *is an automorphism of* V_2^F.

Proof. This is harder. We use (5.1), (6.4), (5.8) and (5.11). It is easy to verify that $E^+ \times E^+ \le E^+$. To prove $E^- \times E^- \le E^+$, we verify that $(E^- \times E^-, E^-) = 0$ (this suffices since the eigenspaces are nonsingular and pairwise orthogonal); the verification is a straightforward checking of cases. To prove that $E^- \times E^+ \le E^-$, we use the previous result, commutativity of the product and associativity of the form.

Table 6.7. (i) The action of $t(e_{\lambda,a})$ on $\mathfrak{QTI}$:

fixed are
$$e_{\mu,b} \text{ if } \langle \mu, \lambda \rangle = 0 \text{ or } \pm 4; \quad e(\varphi) \text{ if } [\lambda, \varphi] = 0;$$

interchanged are
$$e_{\mu,b} \text{ and } e_{\lambda+\mu,a+b+1} \text{ if } \langle \lambda, \mu \rangle = -2; \quad e(\varphi) \text{ and } e(\varphi\lambda) \text{ if } [\lambda, \varphi] = 1.$$

(ii) The action of $t(e(\varphi))$ on $\mathfrak{QTI}$:

fixed are
$$\text{all } e'_{\lambda,\varphi} \text{ and all } e(\psi) \text{ with } \varphi = \psi \text{ or } \varphi\psi \text{ singular;}$$

interchanged are
$$\text{all } e(\varphi\lambda) \text{ and } e_{\lambda,\varphi} \text{ with } \lambda \text{ a quoot.}$$

Proof. For an involution t to interchange vectors x and y in characteristic not 2, it is necessary and sufficient that t fix $x + y$ and negate $x - y$. The following is a useful observation: since the $+1$ eigenspace for $t = t(e)$ is the sum of the 0-eigenspace and the $\frac{1}{4}$-eigenspace for $\mathrm{ad}(e)$, a vector u is fixed by $t(e)$ iff $e \times u$ is in the $\frac{1}{4}$-eigenspace. Another useful observation is that if $x - y$ is negated, then $(x - y)^2$ is fixed. The proof of (i) and (ii) is an exercise in checking cases.

The identification of G, the group generated by all such $t(e)$, $e \in \mathfrak{Q}\mathfrak{J}\mathfrak{J}$, is based on a suitable identification of this set of involutions with nonsingular points in $\mathbb{F}_2^{10}$ with a maximal index nonsingular quadratic form.

Notation 6.8. Let $T := \mathbb{F}_2^{10}$ have a quadratic form q of maximal Witt index. Decompose $T = U \perp W$, with $\dim(U) = 2$, $\dim(W) = 8$, both of plus type. Let $U = \{0, r, s, f\}$, where $q(f) = 1$, $q(r) = q(s) = 0$. Identify W with $\mathrm{Hom}(L, \{\pm 1\})$. For $x \in V$ nonsingular, write $x = p + y$, for $p \in U$, $y \in W$. If $p = 0$, correspond x to $e_{y,1}$. If $p = r$, correspond x to $e_{y,0}$. If $p \in \{s, f\}$, correspond $e(y)$ to x. This correspondence is G-equivariant; use (6.7).

So, we have a map of G onto $O^+(10, 2)$ by restriction to V_2^F. Its kernel fixes all of our idempotents, which span V_2^F. By Corollary 6.2 of [DGH], this kernel is trivial. So, $G \cong O^+(10, 2)$ and (6.3) is proven.

Proposition 6.9. G acts irreducibly on $\mathbb{I}^\perp$ (dimension 155).

Proof. This follows from the character table of $\Omega^+(10, 2)$, but we can give an elementary proof.

(1) The subgroup H of (6.2) has an irreducible constituent P of dimension 120 with monomial basis v_α, $\alpha \in L_2$;

(2) the squares of the v_α generate the 36-dimensional orthogonal complement, $P^\perp$. The action fixes $\mathbb{I}$ and the action on the 35-dimensional space $P^\perp \cap \mathbb{I}^\perp$ is nontrivial, hence irreducible (the subgroup $O_2(H) \cong 2^8$ acts trivially and the quotient $H/O_2(H) \cong O^+(8, 2)$ acts transitively on the spanning set of 120 elements $v_\alpha^2 = \alpha^2$, so acts faithfully. Now, the subgroup $2^6 : O^+(6, 2) \cong 2^6 : Sym_8$ has smallest faithful irreducible degrees 28 and 35; if H is reducible on $P^\perp \cap \mathbb{I}^\perp$, then 28 occurs and H has an irreducible R of dimension d, $28 \le d \le 34$ and so $P^\perp \cap R^\perp$ is a trivial module of dimension $36 - d \ge 2$. This is impossible since $P^\perp$ is an H-constituent of a transitive permutation module of degree 120, contradiction).

(3) We now have $V_2^F = 1 + 35 + 120$ as a decomposition into H-irreducibles. But each $t(e(\varphi))$ fixes $\mathbb{I}$ and does not fix the 120-dimensional constitutent, whence irreduciblity of G on $\mathbb{I}^\perp$.

Theorem 6.10. $\mathrm{Aut}(V^F) = G \cong O^+(10, 2)$.

Proof. Set $A := \mathrm{Aut}(V^F)$. We quote Theorem (6.13) of [Miy], which says that if $\mathfrak{X}$ is the set of conformal vectors of central charge $\frac{1}{2}$, then $|t(x)t(y)| \in \{1, 2, 3\}$. So, if $\mathfrak{X}$ is a conjugacy class, it is a set of 3-transpositions. If it is not a conjugacy class, we have a nontrivial central product decomposition of $\langle \mathfrak{X} \rangle$, which is clearly impossible since A acts faithfully and G acts irreducibly on V_2^F. Now, the classification of groups generated by a class of 3-transpositions [Fi69], [Fi71] may be invoked to identify A. It is a fairly straightforward exercise to eliminate any 3-transposition group which properly contains $O^+(10, 2)$.

7. A related subVOA of $V^\natural$

The VOA defined in [FLM], denoted $V^\natural$, has the monster as its automorphism group. One of the parabolics, $P \cong 2^{10+16}\Omega^+(10, 2)$, acts on the subVOA V' of fixed points of $O_2(P)$; the degree 2 part V'_2 contains V_2^F. In fact, V'_2 is isomorphic to the direct sum of algebras V_2^F (with $\times$) and $\mathbb{C}$. The proper subVOA V'' of V' generated by the V_2^F-part is isomorphic to V^F (this is so because we can see our $L = Q^{[E]}$ embedded in the Leech lattice, as the fixed point sublattice of an involution). This subVOA V'' contains idempotents given by formulas like ours for quooty and tooty ones, but these idempotents have $\frac{1}{16}$ in their spectrum on $V^\natural$, so the involutions associated to them by the Miyamoto theory act trivially on V'. The involutory automorphisms of V' given by our formulas in Section 6 do not extend to automorphisms of $V^\natural$ since otherwise the stabilizer of this subVOA in M, the monster, would induce $O^+(10, 2)$ on it, contrary to the above structure of the maximal 2-local P; we mention that the maximal 2-locals have been classifed [Mei].

Acknowledgements. We thank Chongying Dong, Gerald Höhn, Geoffrey Mason, Masahiko Miyamoto and Steve Smith for discussions on this topic. This article was written with financial support from NSF grant DMS-9623038 and University of Michigan Faculty Recognition Grant (1993–96).

References

[Carter] R. Carter, Simple groups of Lie type, John Wiley, London 1989.

[CoGr] A. Cohen, R. L. Griess, Jr., On finite simple subgroups of the complex Lie group of type E_8, in: The Arcata conference on representations of finite groups, Part 2, Proc. Sympos. Pure Math. **47**, Amer. Math. Soc, Providence 1987, 367–405.

[DGH] C. Dong, G. Höhn, R. L. Griess, Jr., Framed Vertex operator algebras and the moonshine module, to appear in Comm. Math. Phys.

[Fi69] B. Fischer, Finite groups generated by 3-transpositions. Univ. of Warwick, Preprint 1969.

[Fi71] B. Fischer, Finite groups generated by 3-transpositions. Invent. Math. **13** (1971), 232–246.

[FLM] I. Frenkel, J. Lepowsky, A. Meurman, Vertex operator algebras and the monster, Pure Appl. Math. **134**, Academic Press, Boston 1988.

[Gr76] R.L. Griess, Jr., A subgroup of order $2^{15}|\, GL(5, 2)|$ in $E_8(\mathbb{C})$, the Dempwolff group and Aut($D_8 \circ D_8 \circ D_8$), J. Algebra **40** (1976), 271–279.

[Gr91] R. L. Griess, Jr., Elementary abelian p-subgroups of algebraic groups, Geom. Dedicata **39** (1991), 253–305.

[Mei] U. Meierfrankenfeld, The maximal 2-locals of the monster, preprint.

58   R. L. Griess, Jr.

[Miy]   M. Miyamoto, Griess algebras and conformal vectors in vertex operator algebras, to appear in J. Algebra.

Department of Mathematics
University of Michigan
Ann Arbor, Michigan 48109-1109, U.S.A.
E-mail: rlg@math.lsa.umich.edu

Modular forms associated with the Monster module

Koichiro Harada and Mong Lung Lang

Introduction

In Harada–Lang [HL], we associated to each irreducible character χ of the monster simple group $\mathbb{M}$ a modular function $t_\chi(z)$, called in [HL], the McKay–Thompson series for χ. $t_\chi(z)$ is a weighted average of all McKay–Thompson series $t_g(z)$ for the element g of $\mathbb{M}$ as g ranges over $\mathbb{M}$:

$$t_\chi(z) = \frac{1}{|\mathbb{M}|} \sum_{g \in \mathbb{M}} \chi(g) t_g(z).$$

If Γ_χ is the invariance subgroup of $t_\chi(z)$, then we showed

$$\Gamma_\chi = \Gamma_0(N_\chi) = \bigcap_{g \in \mathbb{M}} \Gamma_g$$

where g ranges over all the elements of $\mathbb{M}$ such that $\chi(g) \neq 0$ and

$$N_\chi = \mathrm{lcm}\{n_g h_g : \text{for all } g \in \mathbb{M} \text{ with } \chi(g) \neq 0\}.$$

As shown in Conway–Norton [CN], the invariance group Γ_g of $t_g(z)$ is a certain subgroup of index h of the conjugate by

$$\begin{pmatrix} h & 0 \\ 0 & 1 \end{pmatrix}$$

of

$$\Gamma_0(\frac{n}{h}) + e, f, \dots$$

where e, f, etc. denote the Atkin–Lehner involutions. In [CN], such a conjugate is denoted by

$$n|h + e, f, \dots.$$

The numbers n, h depend on g, hence our notation n_g, h_g. Obviously every $t_\chi(z)$ is invariant by

$$\bigcap_{g \in \mathbb{M}} \Gamma_g = \Gamma_0(N_0)$$

where $N_0 = 2^6 3^3 5^2 7 \cdot 11 \cdot 13 \cdot 17 \cdot 19 \cdot 23 \cdot 29 \cdot 31 \cdot 41 \cdot 47 \cdot 59 \cdot 71 \sim 10^{21}$. The level N_χ can be very large or relatively small. For example,

$$N_{\chi_1} = N_0, \quad N_{\chi_{166}} = 2^6 3^3 7 = 4032$$

where $\chi_1 = 1$ is the trivial character and the character numbering such as χ_{166} is taken from the Atlas. In this paper, we will investigate the relation between $t_\chi(z)$ and the generating functions of the highest weight vectors (also called singular vectors, primary fields or lowest weight vectors.)

The Monster module as a *Vir* module

The monster module $\mathbb{V}$ is constructed in Frenkel–Lepowsky–Meurman [FLM] as a vertex operator algebra and is denoted by $\mathbb{V}^\natural$ there. Let V be a vertex operator algebra. Then V possesses two distinguished elements 1 and ω, called the vacuum and the conformal vector (or the Virasoro element) of V, respectively.

If $Y(\omega, z) = \sum \omega_n z^{-n-1}$ is the vertex operator corresponding to the conformal vector ω and if we set $L(n) = \omega_{n+1}$ for $n \in \mathbb{Z}$, then $L(n)$ satisfies the commutation relation

$$[L(n), L(m)] = (n - m)L_{n+m} + \frac{1}{12}(n^3 - n)c\delta_{n+m,0}$$

where c is a constant called the central charge of V. For the monster module $\mathbb{V}$, $c = 24$. c is also called the rank of the vertex operator algebra V.

Let $\mathcal{L}$ be the Lie algebra generated by $L(n)$, $n \in \mathbb{Z}$. $\mathcal{L}$ is denoted by *Vir* elsewhere. The subalgebras $\mathcal{L}^+$ and $\mathcal{L}^-$ are generated by $L(n)$, $n \in \mathbb{Z}^+$ and $L(n)$, $n \in \mathbb{Z}^-$, respectively. It is known that $\mathbb{V}$ possesses a positive-definite invariant bilinear form and so $\mathbb{V}$ is completely reducible as an $\mathcal{L}$-module and is a sum of highest weight modules.

Let $M(h, c)$ be the Verma module of the Virasoro algebra of central charge c generated by the highest weight vector v of height h, i.e.

$$M(h, c) = \mathcal{L}v, \quad \mathcal{L}^+ v = 0, \quad L(0)v = hv.$$

The module structure of $M(h, c)$ has been determined by Feigin–Fuchs [FF]. We will use their results to determine the module structure of $\mathbb{V}$ as an $\mathcal{L}$ module. Feigin–Fuchs showed that every submodule of $M(h, c)$ is a sum of submodules that are also Verma modules. Therefore, the knowledge of all embeddings among Verma modules gives all submodules of a given Verma module. The main theorem of Feigin–Fuchs states that there are six types of embeddings of the Verma modules into other Verma modules. Let

$$\begin{cases} p\alpha - q\beta = m \\ c = 24 = \frac{(3p-2q)(3q-2p)}{pq} \\ h = \frac{m^2 - (p-q)^2}{4pq} \end{cases} \tag{1}$$

where p, q and m are complex numbers. The submodule structure of $M(h, c)$ is determined by the solutions of α and β of the equations above. See [FF] for the details. We will write below only the necessary calculation for our case. Let us now solve for integers α and β. Let

$$\epsilon = \frac{-11 \pm i\sqrt{23}}{12}, \quad \bar{\epsilon} = \frac{-11 \mp i\sqrt{23}}{12}.$$

We compute

$$\epsilon\bar{\epsilon} = 1, \quad \epsilon + \bar{\epsilon} = -\frac{11}{6}, \quad \epsilon^2 + \bar{\epsilon}^2 = \frac{49}{36}.$$

Using the second equality of (1), we obtain

$$(p\alpha - q\beta)^2 = m^2 = 4pq + (q - p)^2,$$

which may be rewritten as

$$(\alpha - \epsilon\beta)^2 = 4\epsilon h + (\epsilon - 1)^2.$$

We therefore obtain two equations:

$$\alpha^2 - 2\epsilon\alpha\beta + \epsilon^2\beta^2 = 4\epsilon h + (\epsilon - 1)^2,$$

and

$$\alpha^2 - 2\bar{\epsilon}\alpha\beta + \bar{\epsilon}^2\beta^2 = 4\bar{\epsilon}h + (\bar{\epsilon} - 1)^2.$$

Taking the sum of them, we get

$$72\alpha^2 + 132\alpha\beta + 49\beta^2 = -264h + 253.$$

By subtracting one from the other, we get

$$-12\alpha\beta - 11\beta^2 = 24h - 23.$$

Therefore

$$\alpha^2 - \beta^2 = 0,$$

or $\alpha = \pm\beta$. Setting $\alpha = \delta\beta$ with $\delta = \pm 1$, we have

$$\beta^2 = \frac{24h - 1}{11 - 12\delta}.$$

If $h = 0$, then we must have $\delta = 1$ and so $\alpha = \beta = \pm 1$. In particular, $\alpha\beta = 1 > 0$. On the other hand, if $h \in \mathbb{Z}^+$, then $\delta = -1$ and so $\alpha = -\beta = \pm 1$, and hence $\alpha\beta = -1 < 0$. Using the results of Feigin–Fuchs [FF], we conclude (which must be well known to experts):

Theorem. $M(0, 24)$ *has a unique submodule, which is isomorphic to* $M(1, 24)$. *For all positive integers* h, $M(h, 24)$ *is irreducible.*

Let $L(h, c)$ be the unique irreducible highest weight $\mathcal{L}$-module of central charge c and height h. Then

Corollary. *We have*
(i) $L(0, 24) = M(0, 24)/M(1, 24)$, *and,*
(ii) $L(h, 24) = M(h, 24)$ *if* $h \in \mathbb{Z}^+$.

Let us now express the monster module $\mathbb{V}$ as a sum of $L(h, 24)$'s as follows:

$$\mathbb{V} = \sum_{h=0}^{\infty} s_h L(h, 24).$$

Then s_h is the number of linearly independent singular vectors v_h of height h (a non-zero vector v of $\mathbb{V}$ is called singular if $L(n)v = 0$ for all $n \in \mathbb{Z}^-$), hence $v_h \in \mathbb{V}_h$. Since the Virasoro algebra $\mathcal{L}$ commutes with the action of the monster $\mathbb{M}$, we can actually split s_h into the sum of s_h^k where the index k corresponds to the irreducible character χ_k. More precisely, let

$$\mathbb{V}_h^k = c_{hk}\chi_k$$

where c_{hk} is the multiplicity of χ_k in $\mathbb{V}_h$ and

$$\mathbb{V}^k = \coprod_{h=0}^{\infty} \mathbb{V}_h^k.$$

Thus $\mathbb{V}^k$ is the $\mathbb{M}$-submodule of $\mathbb{V}$ consisting entirely of irreducible submodules isomorphic to χ_k and $\mathbb{V}_h^k$ is the $\mathbb{M}$-submodule of $\mathbb{V}^k$ of height h. We also define

$$W_h^k = \mathcal{L}\left(\coprod_{0 \le i < h} \mathbb{V}_i^k \right) \cap \mathbb{V}_h^k,$$

which is the $\mathbb{M}$-submodule of $\mathbb{V}_h^k$ that is generated by elements of lesser height. Let

$$s_h^k = \dim \mathbb{V}_h^k / W_h^k.$$

Then s_h^k is the number of linearly independent singular vectors in $\mathbb{V}_h^k$. Obviously

$$s_h = \sum_{k=1}^{194} s_h^k.$$

For a graded module $X = \sum_{h \in \mathbb{Z}} X_h$, the character of X (or the partition function of X) is defined to be a formal sum

$$\operatorname{char}(X) = \sum_{h \in \mathbb{Z}} \dim X_h x^h.$$

Using this notation, we have, as is well known,

$$\operatorname{char} M(h, c) = x^h \sum_{n \ge 0} p(n)x^n$$

where $p(n)$ is the partition function of n. For convenience, set $p(0) = 1$, and $p(n) = 0$ if $n \in \mathbb{Z}^-$. Let us consider the $\mathcal{L}$ submodule generated by the vacuum 1. We have $V_1 = 0$ but the height 1 component of $M(0, 24)$ is nonzero, we conclude that

$$\mathcal{L} \cdot 1 \simeq M(0, 24)/M(1, 24).$$

Hence

$$\operatorname{char}(\mathcal{L} \cdot 1) = \sum_{n \geq 0} p(n)x^n - x \sum_{n \geq 0} p(n)x^n.$$

Writing

$$\operatorname{char}(\mathcal{L} \cdot 1) = \sum_{h \geq 0} a_{h1}x^h,$$

we get a partial list:

h	0	2	3	4	5	6	7	8	9	10	11
a_{h1}	1	1	1	2	2	4	4	7	8	12	14

In [HL], we had a partial list of c_{h1} where c_{h1} is the multiplicity of the trivial character χ_1 occuring in $\mathbb{V}_h$.

h	0	2	3	4	5	6	7	8	9	10	11
c_{h1}	1	1	1	2	2	4	4	7	8	12	14

The coincidence $c_{h1} = a_{h1}$ stops there and we have

h	12
a_{h1}	21
c_{h1}	22

This means $s_{12}^1 = 1$, namely, $\mathbb{V}_{12}^1$ contains a singular vector, while $\mathbb{V}_h^1$, $0 < h \leq 11$, do not. The number s_h^1 of linearly independent singular vectors occuring in $\mathbb{V}_h^1$ ($0 \leq h \leq 30$) is as follows:

h	12	16	18	20	22	24	26	27	28	29	30
s_h^1	1	1	1	1	1	3	2	1	4	2	6

We are now lead to consider its generating function for each k, $1 \leq k \leq 194$. Define

$$G^k(x) = \sum_{h \geq 0} s_h^k x^h.$$

The character of $\mathbb{V}^k$ is

$$\operatorname{char}(\mathbb{V}^k) = \sum_{h \geq 0} c_h^k (\deg \chi_k) x^h = x \deg \chi_k t_\chi(x)$$

where $t_\chi(z)$ is the McKay–Thompson series for the irreducible character χ. On the other hand, using the expression

$$\mathbb{V}^k = \sum_{h \geq 0} s_h^k L(h, 24),$$

we obtain

$$\operatorname{char}(\mathbb{V}^k) = \sum_{h \geq 0} s_h^k \operatorname{char} L(h, 24).$$

Suppose $k > 0$. Then $s_0^k = 0$ and so

$$\text{char}(\mathbb{V}^k) = \sum_{h \geq 1} s_h^k x^h \sum_{n \geq 0} p(n) x^n.$$

On the other hand if $k = 1$, then $L(0, 24)$ occurs only once as a constitient of $\mathbb{V}^1$. Therefore

$$\text{char}(\mathbb{V}^1) = (1 - x + \sum_{h \geq 2} s_h^1 x^h) \sum_{n \geq 0} p(n) x^n.$$

Using the Dedekind eta-function and replacing x by $q = e^{2\pi i z}$, we obtain, by setting $s_1^1 = -1$ for convenience,

$$\deg \chi_k t_{\chi_k}(q) = \frac{q^{-1}(\sum_{h \geq 0} s_h^k q^h) q^{\frac{1}{24}}}{\eta(q)}.$$

Hence

$$\deg \chi_k t_{\chi_k}(q) \eta(q) = q^{-\frac{23}{24}} \sum_{h \geq 0} s_h^k q^h,$$

which implies

$$q^{-\frac{23}{24}} G^k(q) = \deg \chi_k t_{\chi_k}(q) \eta(q)$$

where as defined before $G^k(q)$ is the generating function of the singular vectors in $\mathbb{V}^k$. Writing $G^k = G^\chi$ in general, we obtain:

Theorem. $q^{-\frac{23}{24}} G^\chi(q)$ is a meromorphic modular form of weight $\frac{1}{2}$ and level N_χ.

Corollary. $q^{-\frac{23}{24}} G^\chi(q) \eta(q)^{23}$ is a holomorphic modular form of weight 12 and level N_χ.

The first 52 coefficients t_h of $\frac{G^{\chi_i}(q)}{\deg \chi_i}$, $1 \leq i \leq 194$, can be found in the following pages.

The following observation has been made by the referee of this paper.

h	12	14	16	18	20	22	24	26	27	28	29	30
s_h^1	1	0	1	1	1	1	3	2	1	4	2	6
#	1	0	1	1	1	1	2	1	0	2	0	2

where # is the dimension of the space of cusp forms on $SL(2, \mathbb{Z})$ of weight h.

References

[CN]   J. H. Conway, S. P. Norton, Monstrous moonshine, Bull. London Math. Soc. **11** (1979), 308–338.

[FF]   B. L. Feigin, D. B. Fuchs, Verma modules over the Virasoro algebras, in: Topology, general and algebraic topology and applications, L. D. Faddeev, A. A. Mal'cev (Ed.), Lecture Notes in Math. **1060**, 230–245. Springer-Verlag, Berlin–Heidleberg–New York 1984,

[FLM] I. Frenkel, J. Lepowsky, A. Meurman, Vertex operator algebras and the monster, Pure and Appl. Math. **134**, Academic Press, Boston 1988.

[HL]   K. Harada, M. L. Lang, The McKay–Thompson Series associated with the Irreducible Characters of the Monster, in: Moonshine, the monster, and related topics, C. Dong, G. Mason (Ed.), Contemp. Math. **193**, Amer. Math. Soc., Providence, RI, 1996, 93–111.

Department of Mathematics
The Ohio State University
Columbus, Ohio 43210, USA
E-mail: haradako@math.ohio-state.edu

Department of Marhematics
National University of Singapore
Kent Ridge, Singapore
Republic of Singapore
E-mail: matlml@leonis.nus.sg

t_h	χ_1	χ_2	χ_3	χ_4	χ_5	χ_6	χ_7	χ_8	χ_9	χ_{10}	χ_{11}	χ_{12}
t_0	1	0	0	0	0	0	0	0	0	0	0	0
t_1	−1	0	0	0	0	0	0	0	0	0	0	0
t_2	0	1	0	0	0	0	0	0	0	0	0	0
t_3	0	0	1	0	0	0	0	0	0	0	0	0
t_4	0	0	0	1	0	0	0	0	0	0	0	0
t_5	0	0	0	0	0	1	0	0	0	0	0	0
t_6	0	0	0	0	1	0	1	0	0	0	0	0
t_7	0	0	0	0	0	1	0	1	0	0	0	0
t_8	0	0	0	1	1	0	0	1	1	0	0	0
t_9	0	0	1	0	0	1	1	1	0	1	0	1
t_{10}	0	1	0	1	1	1	1	1	1	1	1	0
t_{11}	0	0	1	1	0	2	1	2	1	1	0	1
t_{12}	1	1	1	2	2	1	2	2	2	2	1	1
t_{13}	0	1	2	1	1	3	2	4	2	2	0	2
t_{14}	0	2	1	3	3	3	3	4	4	4	2	2
t_{15}	0	1	3	2	2	5	4	7	4	5	1	5
t_{16}	1	3	2	5	5	5	5	9	8	7	3	4
t_{17}	0	2	4	4	4	9	7	13	8	10	3	8
t_{18}	1	4	4	7	9	9	11	15	14	15	7	10
t_{19}	0	3	6	8	7	15	12	24	16	20	6	16
t_{20}	1	6	6	13	14	16	17	30	26	28	12	19
t_{21}	0	5	11	12	14	25	23	42	31	38	13	31
t_{22}	1	9	10	20	23	29	31	53	48	53	23	36
t_{23}	0	8	16	22	24	42	39	75	59	71	25	57
t_{24}	3	13	18	33	39	48	54	95	87	99	42	72
t_{25}	0	14	25	37	41	70	68	132	108	133	49	105
t_{26}	2	21	28	53	64	83	92	169	156	182	77	133
t_{27}	1	21	41	60	72	115	119	230	197	244	93	195
t_{28}	4	31	45	87	105	139	156	297	279	331	140	247
t_{29}	2	34	63	100	121	192	202	400	356	443	174	351
t_{30}	6	48	74	137	175	233	269	514	494	600	255	456
t_{31}	3	53	99	165	204	315	342	692	636	797	320	633
t_{32}	8	73	116	224	286	388	450	893	872	1072	457	825
t_{33}	5	82	158	266	343	517	583	1185	1126	1427	586	1140
t_{34}	10	112	184	358	469	643	761	1534	1528	1910	821	1484
t_{35}	8	128	245	435	569	849	978	2029	1982	2532	1058	2027
t_{36}	17	170	296	574	771	1057	1278	2623	2667	3379	1464	2660
t_{37}	11	198	383	702	942	1388	1641	3456	3467	4472	1899	3595
t_{38}	21	261	463	919	1259	1739	2134	4470	4637	5947	2599	4719
t_{39}	19	304	604	1130	1556	2262	2747	5861	6034	7862	3389	6358
t_{40}	31	396	729	1473	2054	2845	3555	7588	8029	10421	4597	8343
t_{41}	27	470	940	1817	2553	3687	4574	9918	10460	13755	6013	11174
t_{42}	45	604	1151	2344	3352	4642	5922	12826	13857	18201	8109	14696
t_{43}	40	720	1468	2914	4181	5991	7599	16737	18065	23984	10623	19592
t_{44}	63	920	1800	3746	5451	7558	9811	21644	23855	31671	14248	25765
t_{45}	62	1102	2296	4653	6835	9719	12611	28162	31102	41701	18707	34268
t_{46}	89	1402	2817	5962	8859	12283	16249	36418	40962	54973	24988	45046
t_{47}	91	1688	3573	7436	11137	15752	20859	47318	53418	72303	32832	59753
t_{48}	133	2129	4412	9483	14390	19914	26864	61155	70199	95204	43727	78593
t_{49}	132	2579	5566	11851	18120	25493	34466	79358	91549	125119	57497	104015
t_{50}	185	3244	6878	15076	23330	32262	44333	102544	120111	164575	76390	136788
t_{51}	200	3932	8677	18858	29450	41208	56884	132903	156644	216162	100508	180819

χ13	χ14	χ15	χ18	χ19	χ20	χ21	χ22	χ23	χ24	χ25
0	0	0	0	0	0	0	0	0	0	0
0	0	0	0	0	0	0	0	0	0	0
0	0	0	0	0	0	0	0	0	0	0
0	0	0	0	0	0	0	0	0	0	0
0	0	0	0	0	0	0	0	0	0	0
0	0	0	0	0	0	0	0	0	0	0
0	0	0	0	0	0	0	0	0	0	0
0	0	0	0	0	0	0	0	0	0	0
0	0	0	0	0	0	0	0	0	0	0
0	0	0	0	0	0	0	0	0	0	0
0	0	1	0	0	0	0	0	0	0	0
1	0	1	1	0	0	0	0	0	0	0
1	1	1	1	1	1	0	0	0	0	0
1	0	3	2	1	0	0	1	1	0	0
1	1	4	2	2	1	1	2	1	1	0
3	1	5	4	3	2	0	2	3	1	1
3	2	8	6	5	4	2	4	3	2	1
6	2	12	10	6	4	1	6	7	4	2
7	5	16	13	11	8	4	10	9	7	3
11	4	24	21	15	10	4	15	16	11	5
14	9	34	28	24	18	10	23	21	18	8
23	10	47	43	33	23	10	34	37	27	14
28	17	68	59	50	37	21	52	49	43	20
44	20	94	88	69	49	25	75	79	63	33
57	35	129	119	104	78	45	110	109	98	49
83	40	182	174	142	103	56	160	167	144	75
108	65	251	237	207	156	94	233	231	218	109
158	81	343	339	287	213	122	330	347	315	168
206	122	474	464	411	312	195	475	479	467	241
293	154	648	654	565	424	258	674	704	674	360
388	231	882	891	801	614	394	957	978	985	519
541	294	1205	1247	1101	837	532	1348	1409	1408	761
718	427	1636	1697	1544	1193	794	1900	1953	2036	1090
996	556	2214	2350	2118	1630	1076	2659	2789	2897	1584
1318	786	3003	3195	2945	2292	1574	3729	3858	4150	2254
1812	1031	4054	4398	4031	3133	2151	5192	5451	5874	3244
2410	1448	5462	5964	5575	4378	3095	7232	7536	8353	4607
3278	1900	7364	8166	7609	5970	4242	10041	10565	11775	6573
4359	2635	9901	11055	10465	8279	6037	13929	14577	16650	9292
5911	3487	13286	15066	14264	11296	8285	19252	20324	23380	13198
7854	4784	17833	20367	19535	15577	11693	26613	27982	32897	18589
10595	6341	23889	27660	26564	21216	16060	36689	38826	46057	26267
14088	8659	31965	37321	36266	29147	22502	50550	53386	64561	36912
18924	11484	42754	50545	49237	39652	30925	69516	73781	90141	51956
25151	15592	57117	68110	67031	54283	43117	95546	101287	125972	72836
33705	20729	76233	92012	90879	73787	59230	131117	139598	175522	102248
44762	28007	101719	123852	123444	100730	82260	179873	191387	244703	143060
59844	37257	135595	167013	167163	136801	113005	246441	263152	340398	200386
79476	50210	180651	224568	226702	186408	156467	337531	360459	473727	280048
106029	66802	240598	302400	306694	252958	214924	461928	494762	658238	391639
140765	89792	320270	406322	415386	344151	296968	631998	677230	914987	546869
187582	119568	426126	546554	561634	466865	407915	864160	928510	1270500	764184

X_{28}	X_{29}	X_{30}	X_{31}	X_{32}	X_{33}	X_{34}	X_{35}	X_{36}	X_{37}	X_{38}
0	0	0	0	0	0	0	0	0	0	0
0	0	0	0	0	0	0	0	0	0	0
0	0	0	0	0	0	0	0	0	0	0
0	0	0	0	0	0	0	0	0	0	0
0	0	0	0	0	0	0	0	0	0	0
0	0	0	0	0	0	0	0	0	0	0
0	0	0	0	0	0	0	0	0	0	0
0	0	0	0	0	0	0	0	0	0	0
0	0	0	0	0	0	0	0	0	0	0
0	0	0	0	0	0	0	0	0	0	0
0	0	0	0	0	0	0	0	0	0	0
0	0	0	0	0	0	0	0	0	0	0
0	0	0	0	0	0	0	0	0	0	0
0	0	0	0	0	0	0	0	0	0	0
0	0	0	0	0	0	0	0	0	0	0
0	0	0	0	0	0	0	0	0	0	0
1	0	1	0	0	0	0	0	0	0	0
2	1	1	1	0	1	0	0	0	0	0
3	1	3	2	0	1	0	0	0	1	1
5	3	4	3	1	3	0	0	2	1	2
8	4	9	6	0	4	0	0	1	4	4
14	8	12	10	2	9	1	0	4	6	7
21	11	23	16	2	12	0	0	6	12	14
34	21	32	27	5	24	2	0	12	18	23
52	30	55	44	6	34	2	0	16	34	39
80	50	79	67	13	59	6	0	33	51	62
119	73	128	107	17	86	6	0	45	87	104
180	117	183	163	32	141	15	0	80	131	163
265	169	287	249	44	205	18	0	117	215	258
394	264	412	376	74	326	38	0	191	323	396
573	382	627	567	105	473	50	0	279	509	616
840	578	899	838	172	728	91	1	449	758	935
1213	836	1341	1251	244	1058	126	0	650	1171	1425
1760	1244	1918	1836	384	1595	217	3	1009	1735	2136
2522	1789	2819	2699	554	2308	304	2	1478	2625	3214
3627	2631	4019	3936	845	3431	499	7	2239	3866	4774
5169	3769	5839	5741	1219	4949	717	9	3262	5779	7099
7381	5479	8300	8302	1833	7265	1134	23	4890	8464	10464
10464	7827	11949	12028	2642	10448	1639	30	7096	12514	15433
14855	11286	16932	17308	3921	15202	2537	65	10505	18242	22607
20974	16062	24211	24908	5649	21785	3683	95	15232	26759	33101
29636	23021	34211	35694	8290	31486	5609	179	22335	38850	48233
41696	32670	48642	51136	11932	44994	8162	272	32286	56628	70255
58686	46576	68569	72979	17398	64663	12269	487	47083	81924	101949
82334	65946	97075	104174	24987	92196	17882	751	67875	118856	147854
115531	93657	136559	148262	36228	131956	26660	1291	98473	171494	213856
161707	132322	192690	210974	51995	187758	38858	2020	141781	247896	309191
226369	187394	270618	299632	75076	267927	57578	3367	204934	356965	446165
316299	264367	380910	425496	107640	380704	84001	5308	294657	514746	643632
442002	373589	534341	603326	155043	542085	123965	8719	425017	740288	927366
616839	526554	750825	855621	222201	769649	180992	13777	610691	1065918	1336177
860993	743134	1052524	1212253	319590	1094568	266561	22420	879754	1532269	1924034
1200744	1046943	1477500	1718020	458190	1553736	389774	35613	1264463	2205293	2771428

χ_{39}	χ_{41}	χ_{43}	χ_{44}	χ_{46}	χ_{47}	χ_{49}	χ_{50}	χ_{51}	χ_{52}	χ_{53}	χ_{55}	χ_{57}
0	0	0	0	0	0	0	0	0	0	0	0	0
0	0	0	0	0	0	0	0	0	0	0	0	0
0	0	0	0	0	0	0	0	0	0	0	0	0
0	0	0	0	0	0	0	0	0	0	0	0	0
0	0	0	0	0	0	0	0	0	0	0	0	0
0	0	0	0	0	0	0	0	0	0	0	0	0
0	0	0	0	0	0	0	0	0	0	0	0	0
0	0	0	0	0	0	0	0	0	0	0	0	0
0	0	0	0	0	0	0	0	0	0	0	0	0
0	0	0	0	0	0	0	0	0	0	0	0	0
0	0	0	0	0	0	0	0	0	0	0	0	0
0	0	0	0	0	0	0	0	0	0	0	0	0
0	0	0	0	0	0	0	0	0	0	0	0	0
0	0	0	0	0	0	0	0	0	0	0	0	0
0	0	0	0	0	0	0	0	0	0	0	0	0
0	0	0	0	0	0	0	0	0	0	0	0	0
0	0	0	0	0	0	0	0	0	0	0	0	0
0	0	0	0	0	0	0	0	0	0	0	0	0
0	0	0	0	0	0	0	0	0	0	0	0	0
0	0	0	0	0	0	0	0	0	0	0	0	0
0	0	0	0	0	0	0	0	0	0	0	0	0
0	0	1	0	1	0	0	0	0	0	0	0	0
0	0	2	0	1	0	0	0	0	2	0	0	0
0	0	3	0	3	0	1	0	0	4	0	0	1
0	0	7	0	7	0	1	1	0	7	0	0	2
0	0	11	0	12	0	2	0	0	13	0	0	5
0	0	21	0	21	0	3	2	0	25	0	0	8
0	0	35	0	38	0	8	2	0	43	0	0	17
0	0	59	0	60	0	10	5	0	75	0	0	28
0	0	94	0	102	0	21	7	0	123	0	0	52
0	0	155	0	165	0	33	17	0	205	0	0	85
0	0	242	0	265	0	57	21	0	329	0	0	147
0	0	384	0	416	0	87	43	0	530	0	0	236
0	0	594	0	660	0	149	63	0	831	0	0	393
0	0	921	0	1015	0	224	111	0	1312	0	0	620
0	0	1404	0	1574	0	366	165	0	2030	0	0	1000
0	0	2149	0	2405	0	559	283	0	3139	0	0	1562
0	0	3237	0	3665	0	881	419	0	4800	0	0	2468
0	0	4888	0	5535	0	1337	688	0	7334	0	0	3811
0	0	7311	0	8364	0	2083	1040	2	11096	0	0	5928
0	0	10929	0	12512	0	3134	1651	2	16780	0	0	9077
0	0	16224	0	18726	0	4810	2494	6	25191	0	0	13954
0	0	24084	0	27862	0	7231	3909	16	37778	0	0	21215
0	0	35541	0	41391	1	10967	5887	36	56370	0	0	32327
0	0	52434	0	61240	5	16415	9104	68	84006	0	0	48873
0	0	77054	0	90541	9	24746	13735	147	124713	0	0	73990
0	0	113157	0	133381	32	36907	21026	283	185003	0	0	111386
0	0	165707	0	196395	71	55336	31694	563	273668	1	0	167844
2	0	242576	0	288534	166	82412	48303	1074	404562	9	0	251982
11	2	354369	6	423656	360	123111	72780	2054	597013	41	10	378615
33	10	517625	24	621219	810	183173	110592	3863	880792	140	47	567613
98	47	755213	107	910885	1679	273314	166965	7301	1298177	429	198	851926
269	147	1102014	330	1334634	3593	406796	253636	13626	1913903	1248	678	1277559
685	477	1607414	1025	1956465	7392	607295	384221	25591	2821306	3371	2160	1919182
1676	1302	2346206	2812	2868528	15386	906291	585754	48083	4162826	8771	6233	2884882

X_{58}	X_{59}	X_{61}	X_{62}	X_{63}	X_{64}	X_{65}	X_{66}	X_{67}	X_{68}	X_{69}
0	0	0	0	0	0	0	0	0	0	0
0	0	0	0	0	0	0	0	0	0	0
0	0	0	0	0	0	0	0	0	0	0
0	0	0	0	0	0	0	0	0	0	0
0	0	0	0	0	0	0	0	0	0	0
0	0	0	0	0	0	0	0	0	0	0
0	0	0	0	0	0	0	0	0	0	0
0	0	0	0	0	0	0	0	0	0	0
0	0	0	0	0	0	0	0	0	0	0
0	0	0	0	0	0	0	0	0	0	0
0	0	0	0	0	0	0	0	0	0	0
0	0	0	0	0	0	0	0	0	0	0
0	0	0	0	0	0	0	0	0	0	0
0	0	0	0	0	0	0	0	0	0	0
0	0	0	0	0	0	0	0	0	0	0
0	0	0	0	0	0	0	0	0	0	0
0	0	0	0	0	0	0	0	0	0	0
0	0	0	0	0	0	0	0	0	0	0
0	0	0	0	0	0	0	0	0	0	0
0	0	0	0	0	0	0	0	0	0	0
0	0	0	0	0	0	0	0	0	0	0
1	0	0	0	0	0	0	0	0	0	0
2	0	0	1	1	0	0	0	0	0	0
4	0	1	2	1	1	1	0	2	0	0
9	0	2	6	4	2	1	1	4	1	0
16	0	4	10	8	4	5	1	10	2	0
29	0	9	21	16	9	8	5	17	5	0
51	0	16	36	29	16	19	6	37	10	0
87	0	30	68	54	30	30	16	64	22	0
143	0	54	112	90	54	63	25	119	38	0
239	0	93	197	160	92	100	53	201	74	0
384	0	159	320	263	158	185	85	355	127	0
617	0	267	537	440	266	299	160	582	228	0
977	0	440	857	713	437	519	257	984	381	0
1539	0	718	1395	1164	714	829	455	1588	657	0
2391	0	1162	2196	1845	1157	1391	730	2603	1077	0
3715	0	1858	3499	2952	1851	2198	1238	4146	1800	0
5705	0	2952	5445	4628	2943	3585	1976	6656	2910	0
8742	0	4661	8530	7272	4652	5633	3262	10469	4753	1
13306	0	7306	13158	11289	7297	9000	5168	16558	7599	0
20201	0	11400	20367	17537	11397	14033	8366	25812	12222	3
30494	0	17710	31172	26987	17724	22134	13191	40337	19354	6
45970	0	27390	47816	41557	27437	34300	21062	62437	30771	18
68989	0	42228	72766	63560	42351	53546	33047	96771	48415	35
103391	0	64923	110862	97197	65190	82638	52299	148967	76353	101
154497	1	99552	167984	148001	100086	128114	81810	229545	119590	208
230610	5	152385	254744	225331	153415	197128	128755	352123	187678	535
343506	23	232927	384927	342055	234871	304405	201224	540664	293308	1246
511443	85	355691	582147	519468	359272	467848	316045	828112	459417	3066
760563	291	542958	878626	787727	549546	721435	494757	1270153	718418	7293
1131101	903	829017	1327792	1195664	841133	1110256	778647	1946953	1127044	17771
1681860	2689	1266777	2005663	1815025	1289231	1715028	1225773	2991932	1770153	42347
2502839	7533	1938934	3036098	2760934	1981167	2651226	1943312	4605230	2794895	101467
3727955	20410	2975374	4602110	4207860	3056268	4120694	3095438	7120281	4434008	240033

χ70	χ71	χ73	χ74	χ76	χ77	χ78	χ79	χ80	χ81	χ83
0	0	0	0	0	0	0	0	0	0	0
0	0	0	0	0	0	0	0	0	0	0
0	0	0	0	0	0	0	0	0	0	0
0	0	0	0	0	0	0	0	0	0	0
0	0	0	0	0	0	0	0	0	0	0
0	0	0	0	0	0	0	0	0	0	0
0	0	0	0	0	0	0	0	0	0	0
0	0	0	0	0	0	0	0	0	0	0
0	0	0	0	0	0	0	0	0	0	0
0	0	0	0	0	0	0	0	0	0	0
0	0	0	0	0	0	0	0	0	0	0
0	0	0	0	0	0	0	0	0	0	0
0	0	0	0	0	0	0	0	0	0	0
0	0	0	0	0	0	0	0	0	0	0
0	0	0	0	0	0	0	0	0	0	0
0	0	0	0	0	0	0	0	0	0	0
0	0	0	0	0	0	0	0	0	0	0
0	0	0	0	0	0	0	0	0	0	0
0	0	0	0	0	0	0	0	0	0	0
0	0	0	0	0	0	0	0	0	0	0
0	0	0	0	0	0	0	0	0	0	0
0	0	0	0	0	0	0	0	0	0	0
0	0	0	0	0	0	0	0	0	0	0
0	0	0	0	0	0	0	0	0	0	0
0	0	0	0	0	0	0	0	0	0	0
0	0	0	0	1	1	0	0	0	0	0
0	0	0	0	3	2	1	0	2	0	0
0	0	0	0	6	6	2	2	5	0	0
0	0	0	0	15	13	6	3	11	0	0
0	0	0	0	29	28	11	9	23	0	0
0	0	0	0	57	51	26	17	48	0	0
0	0	0	0	106	99	46	38	91	0	0
0	0	0	0	194	176	92	66	170	0	0
0	0	3	0	339	319	161	133	305	0	0
0	0	2	0	595	548	296	230	543	0	0
0	0	10	0	1012	951	505	425	941	0	0
0	0	19	0	1712	1594	889	724	1617	0	0
0	0	46	0	2849	2685	1488	1275	2727	0	0
0	0	81	0	4715	4419	2534	2138	4574	0	0
4	0	179	0	7700	7282	4184	3659	7576	0	0
6	0	317	0	12539	11826	6964	6052	12469	0	0
24	0	636	0	20223	19194	11363	10132	20361	0	0
55	4	1139	0	32510	30831	18632	16613	33104	1	0
144	14	2153	1	51950	49498	30138	27413	53498	8	1
306	50	3841	13	82808	78920	48895	44625	86200	46	7
722	140	7090	46	131436	125770	78653	72993	138383	150	46
1508	400	12575	182	208390	199592	126807	118392	221791	521	191
3299	1018	22835	552	329621	316823	203504	192814	354895	1500	717
6844	2578	40668	1672	521393	501973	327559	312805	567930	4283	2333
14472	6203	73558	4602	824624	796597	526721	509922	909450	11410	7190
29909	14857	132129	12476	1306649	1265056	851211	831670	1460102	29945	20665
62712	34638	241081	32209	2075513	2016658	1380465	1366442	2353200	75632	57174
130290	80399	440946	82015	3312772	3228408	2258698	2257782	3816367	188680	151862
273986	184107	820710	203055	5321185	5207004	3731422	3774641	6242657	460125	392584
576333	420235	1543715	496080	8627638	8476067	6257769	6388609	10333106	1109339	988361

χ_{85}	χ_{87}	χ_{88}	χ_{89}	χ_{91}	χ_{92}	χ_{93}	χ_{94}	χ_{95}	χ_{96}
0	0	0	0	0	0	0	0	0	0
0	0	0	0	0	0	0	0	0	0
0	0	0	0	0	0	0	0	0	0
0	0	0	0	0	0	0	0	0	0
0	0	0	0	0	0	0	0	0	0
0	0	0	0	0	0	0	0	0	0
0	0	0	0	0	0	0	0	0	0
0	0	0	0	0	0	0	0	0	0
0	0	0	0	0	0	0	0	0	0
0	0	0	0	0	0	0	0	0	0
0	0	0	0	0	0	0	0	0	0
0	0	0	0	0	0	0	0	0	0
0	0	0	0	0	0	0	0	0	0
0	0	0	0	0	0	0	0	0	0
0	0	0	0	0	0	0	0	0	0
0	0	0	0	0	0	0	0	0	0
0	0	0	0	0	0	0	0	0	0
0	0	0	0	0	0	0	0	0	0
0	0	0	0	0	0	0	0	0	0
0	0	0	0	0	0	0	0	0	0
0	0	0	0	0	0	0	0	0	0
0	0	0	0	0	0	0	0	0	0
0	0	0	0	0	0	0	0	0	0
0	0	0	0	0	0	0	0	0	0
0	0	0	0	0	0	0	0	0	0
0	0	0	0	0	0	1	0	0	0
0	0	0	0	0	1	3	0	1	0
0	0	0	0	0	3	8	1	2	0
0	0	0	0	0	7	17	3	5	0
0	0	0	0	0	15	38	7	12	0
0	0	0	0	0	31	72	17	24	1
0	0	0	0	0	61	141	35	48	1
0	0	0	0	1	115	259	70	94	3
0	0	0	0	1	214	473	140	174	5
0	0	0	0	4	384	832	264	320	13
0	0	0	0	9	683	1466	486	571	20
0	0	0	0	19	1190	2507	887	1009	44
0	0	0	0	40	2055	4285	1583	1753	76
0	3	0	0	85	3495	7199	2794	3028	147
0	4	0	3	161	5919	12054	4880	5149	253
0	21	2	11	320	9909	19952	8447	8728	472
0	50	13	43	614	16531	32957	14494	14664	810
2	151	51	121	1170	27380	53990	24777	24561	1482
18	358	171	353	2223	45257	88340	42098	40912	2579
75	951	506	919	4250	74484	143861	71381	68123	4715
313	2222	1416	2353	8124	122607	234222	120793	113168	8510
1039	5470	3744	5784	15771	201608	380675	204521	188444	16147
3298	12725	9597	14077	31057	332510	619906	346712	314483	30994
9652	30150	23903	33499	62227	550246	1011030	590446	528165	62711
27194	69636	58545	79245	127273	917414	1657465	1011336	894243	130101
73200	162141	141005	185407	265407	1544048	2734112	1748606	1533379	281192
191843	372591	335833	431815	563819	2635887	4555888	3059265	2670540	620304
488400	858667	791626	999294	1216596	4578224	7687351	5435769	4746410	1399545
1218333	1964610	1851089	2303725	2659383	8128645	13192303	9835083	8633692	3185945

χ_{97}	χ_{98}	χ_{99}	χ_{101}	χ_{102}	χ_{104}	χ_{105}	χ_{107}	χ_{109}	χ_{110}
0	0	0	0	0	0	0	0	0	0
0	0	0	0	0	0	0	0	0	0
0	0	0	0	0	0	0	0	0	0
0	0	0	0	0	0	0	0	0	0
0	0	0	0	0	0	0	0	0	0
0	0	0	0	0	0	0	0	0	0
0	0	0	0	0	0	0	0	0	0
0	0	0	0	0	0	0	0	0	0
0	0	0	0	0	0	0	0	0	0
0	0	0	0	0	0	0	0	0	0
0	0	0	0	0	0	0	0	0	0
0	0	0	0	0	0	0	0	0	0
0	0	0	0	0	0	0	0	0	0
0	0	0	0	0	0	0	0	0	0
0	0	0	0	0	0	0	0	0	0
0	0	0	0	0	0	0	0	0	0
0	0	0	0	0	0	0	0	0	0
0	0	0	0	0	0	0	0	0	0
0	0	0	0	0	0	0	0	0	0
0	0	0	0	0	0	0	0	0	0
0	0	0	0	0	0	0	0	0	0
0	0	0	0	0	0	0	0	0	0
0	0	0	0	0	0	0	0	0	0
0	0	0	0	0	0	0	0	0	0
0	0	0	0	0	0	0	0	0	0
0	0	0	0	0	0	0	0	0	0
0	0	0	0	0	0	0	0	0	0
2	0	0	1	0	0	0	0	0	0
4	0	0	1	0	1	0	0	0	1
11	0	0	6	0	4	0	0	0	5
24	0	0	11	0	6	0	0	0	9
50	0	0	29	0	17	0	0	0	26
96	0	0	51	0	31	0	0	0	51
189	4	0	110	0	66	0	0	5	112
342	3	0	196	0	122	0	0	6	214
626	16	0	382	0	238	0	0	25	425
1106	28	0	667	0	421	0	0	48	778
1943	74	0	1225	0	777	0	0	121	1460
3340	133	0	2111	0	1355	0	0	233	2606
5731	298	0	3734	0	2400	0	0	515	4701
9668	543	0	6348	0	4127	0	0	983	8241
16272	1113	1	10953	3	7151	0	0	2010	14516
27121	2039	9	18446	21	12138	3	3	3806	25099
45075	3967	46	31343	85	20740	26	26	7460	43569
74477	7281	175	52484	296	35021	136	136	14066	74836
122944	13859	604	88478	936	59434	514	514	27098	129038
202346	25554	1886	148095	2723	100499	1761	1761	51225	221635
333401	48409	5591	249678	7563	171215	5427	5427	98594	383114
549786	90717	15633	420811	20260	292759	15920	15920	189119	664032
910390	173977	42402	716699	52637	507251	44178	44178	369296	1163714
1515705	334919	111122	1229757	133710	890248	118700	118700	726398	2060522
2546756	659976	284770	2144617	333338	1594972	308656	308656	1456587	3712178
4329221	1316915	713525	3801040	817725	2921759	784698	784698	2959329	6810236
7477233	2686648	1758319	6893845	1978892	5499070	1951765	1951765	6124162	12785353
13163453	5563486	4263252	12801184	4731646	10634085	4773670	4773669	12848040	24565376

χ_{111}	χ_{112}	χ_{113}	χ_{114}	χ_{115}	χ_{116}	χ_{117}	χ_{118}	χ_{119}
0	0	0	0	0	0	0	0	0
0	0	0	0	0	0	0	0	0
0	0	0	0	0	0	0	0	0
0	0	0	0	0	0	0	0	0
0	0	0	0	0	0	0	0	0
0	0	0	0	0	0	0	0	0
0	0	0	0	0	0	0	0	0
0	0	0	0	0	0	0	0	0
0	0	0	0	0	0	0	0	0
0	0	0	0	0	0	0	0	0
0	0	0	0	0	0	0	0	0
0	0	0	0	0	0	0	0	0
0	0	0	0	0	0	0	0	0
0	0	0	0	0	0	0	0	0
0	0	0	0	0	0	0	0	0
0	0	0	0	0	0	0	0	0
0	0	0	0	0	0	0	0	0
0	0	0	0	0	0	0	0	0
0	0	0	0	0	0	0	0	0
0	0	0	0	0	0	0	0	0
0	0	0	0	0	0	0	0	0
0	0	0	0	0	0	0	0	0
0	0	0	0	0	0	0	0	0
0	0	0	0	0	0	0	0	0
0	0	0	0	0	0	0	0	0
0	0	0	0	0	0	0	0	0
0	0	0	0	0	0	0	0	0
0	0	0	0	0	0	0	0	0
0	0	0	0	0	0	0	0	0
0	0	0	0	0	0	0	0	0
0	0	0	0	0	0	0	0	0
0	0	0	0	0	0	0	0	0
0	0	0	0	0	0	0	0	0
2	2	0	0	0	0	0	0	0
5	4	0	0	0	1	1	1	0
15	13	0	0	0	3	3	3	0
31	28	0	0	0	8	9	10	0
74	64	0	0	0	20	22	25	0
144	126	0	0	0	45	50	57	0
295	258	0	0	1	95	105	124	0
551	484	0	3	3	194	216	259	0
1054	921	0	6	12	383	425	515	0
1904	1676	0	27	28	736	826	1012	1
3481	3054	0	60	82	1390	1554	1931	2
6144	5411	2	164	183	2584	2902	3637	16
10905	9599	6	362	441	4735	5322	6740	37
18939	16719	30	870	959	8601	9703	12395	134
32977	29108	104	1860	2143	15516	17502	22561	373
56661	50176	356	4169	4563	27854	31529	40944	1109
97610	86502	1091	8834	9871	49981	56604	73998	3005
166986	148461	3256	19156	20906	89853	102026	133925	8269
286817	255591	9155	40589	44794	162587	184695	242966	21672
492247	440530	25174	87466	95535	297209	338083	444013	56604
851162	764954	66899	187210	206033	551661	627083	819010	144111
1480599	1339198	174239	406450	445855	1043446	1185268	1532122	363358
2609337	2377102	443887	884221	975163	2019450	2287578	2913431	901132
4662599	4287592	1112493	1945417	2145685	4004938	4522134	5650041	2212555
8497213	7896691	2743423	4299814	4760353	8143878	9152846	11186020	5364868
15814298	14877219	6675835	9576576	10613730	16946351	18953912	22622211	12885469
30158915	28736994	16039355	21400281	23777156	35982090	40031881	46665631	30632762

χ_{120}	χ_{121}	χ_{122}	χ_{123}	χ_{124}	χ_{126}	χ_{127}	χ_{128}	χ_{130}
0	0	0	0	0	0	0	0	0
0	0	0	0	0	0	0	0	0
0	0	0	0	0	0	0	0	0
0	0	0	0	0	0	0	0	0
0	0	0	0	0	0	0	0	0
0	0	0	0	0	0	0	0	0
0	0	0	0	0	0	0	0	0
0	0	0	0	0	0	0	0	0
0	0	0	0	0	0	0	0	0
0	0	0	0	0	0	0	0	0
0	0	0	0	0	0	0	0	0
0	0	0	0	0	0	0	0	0
0	0	0	0	0	0	0	0	0
0	0	0	0	0	0	0	0	0
0	0	0	0	0	0	0	0	0
0	0	0	0	0	0	0	0	0
0	0	0	0	0	0	0	0	0
0	0	0	0	0	0	0	0	0
0	0	0	0	0	0	0	0	0
0	0	0	0	0	0	0	0	0
0	0	0	0	0	0	0	0	0
0	0	0	0	0	0	0	0	0
0	0	0	0	0	0	0	0	0
0	0	0	0	0	0	0	0	0
0	0	0	0	0	0	0	0	0
0	0	0	0	0	0	0	0	0
0	0	0	0	0	0	0	0	0
0	0	0	0	0	0	0	0	0
0	0	0	0	0	0	0	0	0
0	0	0	0	0	0	0	0	0
1	1	0	0	0	0	0	0	0
4	5	0	0	0	0	0	0	0
12	13	0	0	0	0	0	0	1
28	34	0	0	0	0	0	0	2
66	75	0	3	0	0	4	0	6
139	163	0	5	0	0	8	0	15
291	332	0	19	0	0	28	0	38
572	668	0	42	0	0	66	0	81
1120	1286	0	104	0	0	161	0	182
2116	2461	1	221	0	4	351	0	381
3978	4581	12	496	0	16	793	0	807
7312	8481	34	1011	4	58	1644	8	1638
13400	15453	135	2134	27	189	3492	51	3379
24254	28080	414	4334	129	589	7175	229	6832
43879	50585	1278	8951	521	1700	14852	878	14009
78971	91232	3608	18305	1848	4802	30401	2951	28579
142681	164306	10145	38126	5995	13119	62979	9280	59282
258390	297772	27114	79788	18216	35123	130561	27396	123669
472453	543111	71712	170261	52617	92160	274645	77623	262640
872856	1003584	184544	367939	145975	237937	583082	211748	563990
1639253	1881618	468863	808612	391924	604598	1256303	561897	1230956
3133988	3598204	1170223	1798946	1024135	1515562	2737697	1453707	2716606
6122592	7025278	2887279	4048562	2615616	3749363	6039512	3686131	6066205
12223180	14035510	7028980	9179756	6550659	9165270	13441189	9176737	13646829
24947456	28658535	16935961	20926386	16129203	22151465	30141654	22495571	30892256
51933475	59730244	40366524	47817157	39123510	52974037	67901554	54369815	70137454

χ_{131}	χ_{132}	χ_{133}	χ_{134}	χ_{135}	χ_{137}	χ_{138}	χ_{139}	χ_{140}
0	0	0	0	0	0	0	0	0
0	0	0	0	0	0	0	0	0
0	0	0	0	0	0	0	0	0
0	0	0	0	0	0	0	0	0
0	0	0	0	0	0	0	0	0
0	0	0	0	0	0	0	0	0
0	0	0	0	0	0	0	0	0
0	0	0	0	0	0	0	0	0
0	0	0	0	0	0	0	0	0
0	0	0	0	0	0	0	0	0
0	0	0	0	0	0	0	0	0
0	0	0	0	0	0	0	0	0
0	0	0	0	0	0	0	0	0
0	0	0	0	0	0	0	0	0
0	0	0	0	0	0	0	0	0
0	0	0	0	0	0	0	0	0
0	0	0	0	0	0	0	0	0
0	0	0	0	0	0	0	0	0
0	0	0	0	0	0	0	0	0
0	0	0	0	0	0	0	0	0
0	0	0	0	0	0	0	0	0
0	0	0	0	0	0	0	0	0
0	0	0	0	0	0	0	0	0
0	0	0	0	0	0	0	0	0
0	0	0	0	0	0	0	0	0
0	0	0	0	0	0	0	0	0
0	0	0	0	0	0	0	0	0
0	0	0	0	0	0	0	0	0
1	0	1	2	0	0	0	0	1
2	0	5	5	0	0	0	0	4
8	0	13	15	0	0	0	4	14
21	0	34	37	0	2	0	9	36
48	0	80	86	0	6	0	27	87
107	0	179	185	0	18	0	62	194
233	0	376	397	0	52	0	155	423
477	0	778	799	0	127	2	332	876
965	2	1550	1599	0	303	6	738	1780
1908	21	3050	3111	2	697	36	1521	3528
3726	70	5883	6005	23	1551	102	3197	6916
7191	273	11264	11410	108	3386	339	6471	13368
13850	813	21359	21651	446	7313	965	13243	25735
26553	2459	40459	40795	1538	15618	2779	26657	49341
51219	6796	76539	77189	4883	33272	7542	54346	94932
99393	18662	145651	146504	14552	70932	20429	110570	183653
195289	49016	279386	281010	41321	151753	53562	228303	359341
389344	127573	543062	545515	113121	326848	139108	475324	713071
790821	323948	1072348	1077696	300750	709895	354530	1006211	1440613
1636517	814191	2158354	2168327	780400	1555789	893243	2156657	2966514
3453698	2014235	4431005	4454003	1984152	3440224	2219303	4693584	6230549
7417316	4934162	9279882	9329596	4958092	7669212	5454816	10331723	13327578
16176919	11944104	19789644	19905755	12201433	17209242	13253442	22989424	28970516
35708318	28648476	42864534	43126773	29625584	38807117	31881329	51525740	63789385
79529997	68038799	93989745	94597292	71062047	87777806	75927421	116110182	141808908

χ_{141}	χ_{142}	χ_{143}	χ_{144}	χ_{145}	χ_{146}	χ_{147}	χ_{148}
0	0	0	0	0	0	0	0
0	0	0	0	0	0	0	0
0	0	0	0	0	0	0	0
0	0	0	0	0	0	0	0
0	0	0	0	0	0	0	0
0	0	0	0	0	0	0	0
0	0	0	0	0	0	0	0
0	0	0	0	0	0	0	0
0	0	0	0	0	0	0	0
0	0	0	0	0	0	0	0
0	0	0	0	0	0	0	0
0	0	0	0	0	0	0	0
0	0	0	0	0	0	0	0
0	0	0	0	0	0	0	0
0	0	0	0	0	0	0	0
0	0	0	0	0	0	0	0
0	0	0	0	0	0	0	0
0	0	0	0	0	0	0	0
0	0	0	0	0	0	0	0
0	0	0	0	0	0	0	0
0	0	0	0	0	0	0	0
0	0	0	0	0	0	0	0
0	0	0	0	0	0	0	0
0	0	0	0	0	0	0	0
0	0	0	0	0	0	0	0
0	0	0	0	0	0	0	0
0	0	0	0	0	0	0	0
0	0	0	0	0	0	0	0
0	0	0	0	0	0	0	0
0	0	0	0	0	0	0	0
0	1	3	0	2	0	0	0
0	5	9	0	6	1	0	0
6	14	26	0	19	2	0	0
12	44	67	0	48	11	0	0
43	105	161	5	119	28	0	0
101	249	354	16	264	81	5	0
253	551	775	64	585	191	20	7
553	1204	1611	170	1227	473	74	27
1255	2523	3316	475	2560	1066	241	127
2632	5265	6673	1187	5207	2449	723	401
5640	10755	13373	2992	10607	5414	2031	1319
11695	21949	26516	7174	21411	12123	5521	3824
24545	44544	52818	17342	43630	26827	14510	11000
51038	90866	105290	41096	89336	60297	37365	29947
107735	186158	212263	97872	185845	135727	94625	80583
228226	385951	432579	231636	392101	310066	236593	210386
491599	809244	896407	550107	843066	712752	585363	542760
1069864	1722492	1887395	1303484	1842378	1655617	1435838	1374466
2363122	3717460	4045892	3092214	4092747	3863461	3494465	3441750
5273174	8136218	8807841	7319517	9206188	9059551	8443650	8503628
11888548	18013600	19448096	17300534	20920146	21262382	20262072	20797570
26972662	40272034	43405831	40751346	47845726	49902704	48292644	50315286
61491807	90647085	97667985	95644345	109849023	116838212	114324172	120597234
140450542	204960846	220857196	223412908	252481498	272667843	268807682	286334504

χ_{149}	χ_{150}	χ_{151}	χ_{152}	χ_{153}	χ_{154}	χ_{155}	χ_{156}
0	0	0	0	0	0	0	0
0	0	0	0	0	0	0	0
0	0	0	0	0	0	0	0
0	0	0	0	0	0	0	0
0	0	0	0	0	0	0	0
0	0	0	0	0	0	0	0
0	0	0	0	0	0	0	0
0	0	0	0	0	0	0	0
0	0	0	0	0	0	0	0
0	0	0	0	0	0	0	0
0	0	0	0	0	0	0	0
0	0	0	0	0	0	0	0
0	0	0	0	0	0	0	0
0	0	0	0	0	0	0	0
0	0	0	0	0	0	0	0
0	0	0	0	0	0	0	0
0	0	0	0	0	0	0	0
0	0	0	0	0	0	0	0
0	0	0	0	0	0	0	0
0	0	0	0	0	0	0	0
0	0	0	0	0	0	0	0
0	0	0	0	0	0	0	0
0	0	0	0	0	0	0	0
0	0	0	0	0	0	0	0
0	0	0	0	0	0	0	0
0	0	0	0	0	0	0	0
0	0	0	0	0	0	0	0
0	0	0	0	0	0	1	0
0	3	5	3	0	0	3	0
5	9	15	9	0	0	11	0
15	32	50	32	3	4	31	0
53	79	131	79	12	16	79	0
142	211	338	211	50	63	195	1
392	491	793	492	150	184	464	14
953	1165	1853	1176	458	544	1078	89
2340	2623	4136	2648	1231	1443	2479	328
5439	5971	9207	6060	3317	3798	5698	1209
12740	13345	20132	13562	8448	9539	13081	3835
29181	30234	44146	30805	21476	23814	30325	11741
67365	68435	96531	69720	53192	58271	70686	33748
154631	156988	212859	160032	131452	142097	166397	94500
357971	362293	471897	368845	320830	343417	393966	255254
829424	844661	1057139	859204	781149	828280	938588	675735
1934183	1979272	2386816	2010165	1888473	1988057	2241331	1748152
4518169	4662394	5436209	4729027	4550849	4760382	5360189	4450169
10585617	10999332	12455417	11140427	10905330	11351856	12802182	11143396
24795878	25960621	28677871	26263447	26019761	26974239	30504611	27537319
58052983	61153694	66185043	61800036	61732274	63796113	72393427	67176278
135569926	143630088	152880767	145025538	145678353	150165108	170989044	162050158
315598916	335885737	352781947	338901911	341691746	351529414	401591136	386727360

χ157	χ158	χ159	χ160	χ161	χ162	χ163	χ164
0	0	0	0	0	0	0	0
0	0	0	0	0	0	0	0
0	0	0	0	0	0	0	0
0	0	0	0	0	0	0	0
0	0	0	0	0	0	0	0
0	0	0	0	0	0	0	0
0	0	0	0	0	0	0	0
0	0	0	0	0	0	0	0
0	0	0	0	0	0	0	0
0	0	0	0	0	0	0	0
0	0	0	0	0	0	0	0
0	0	0	0	0	0	0	0
0	0	0	0	0	0	0	0
0	0	0	0	0	0	0	0
0	0	0	0	0	0	0	0
0	0	0	0	0	0	0	0
0	0	0	0	0	0	0	0
0	0	0	0	0	0	0	0
0	0	0	0	0	0	0	0
0	0	0	0	0	0	0	0
0	0	0	0	0	0	0	0
0	0	0	0	0	0	0	0
0	0	0	0	0	0	0	0
0	0	0	0	0	0	0	0
0	0	0	0	0	0	0	0
0	0	0	0	0	0	0	0
0	0	0	0	0	0	0	0
0	0	0	0	0	0	0	0
0	0	0	0	0	0	0	0
0	0	0	0	0	0	0	0
0	0	0	0	0	0	0	0
0	2	0	0	0	3	0	0
0	7	0	0	1	9	0	0
0	23	0	0	5	31	0	0
0	58	2	5	14	80	5	0
0	155	10	19	42	208	18	0
6	358	37	73	105	487	74	3
30	838	123	216	273	1138	218	21
138	1868	366	640	686	2536	649	125
489	4177	1058	1705	1705	5656	1742	466
1662	9164	2925	4529	4253	12379	4666	1690
5056	20318	7971	11546	10600	27214	11950	5424
14934	44962	21181	29282	26491	59741	30569	16605
41861	100920	55706	72989	66175	132372	76604	48092
114489	228232	144187	181392	165478	295310	191513	135005
303958	523043	369562	446810	412317	666257	473945	366357
792316	1208863	935777	1097029	1024459	1516495	1168755	971914
2025074	2819592	2347016	2677063	2531097	3483520	2861542	2521474
5102031	6608725	5824768	6503975	6217973	8053923	6972894	6428447
12670719	15546602	14322053	15708277	15168005	18713380	16878542	16123316
31096071	36593778	34878266	37728051	36737953	43587681	40615178	39882981
75441216	86060772	84179601	90036459	88296456	101600012	97065571	97384678
181165403	201836899	201354131	213498808	210584837	236551985	230434859	235064871
430787280	471580150	477517847	502815970	498290619	549406303	543182129	561279879

χ_{165}	χ_{166}	χ_{167}	χ_{168}	χ_{169}	χ_{170}	χ_{171}	χ_{172}
0	0	0	0	0	0	0	0
0	0	0	0	0	0	0	0
0	0	0	0	0	0	0	0
0	0	0	0	0	0	0	0
0	0	0	0	0	0	0	0
0	0	0	0	0	0	0	0
0	0	0	0	0	0	0	0
0	0	0	0	0	0	0	0
0	0	0	0	0	0	0	0
0	0	0	0	0	0	0	0
0	0	0	0	0	0	0	0
0	0	0	0	0	0	0	0
0	0	0	0	0	0	0	0
0	0	0	0	0	0	0	0
0	0	0	0	0	0	0	0
0	0	0	0	0	0	0	0
0	0	0	0	0	0	0	0
0	0	0	0	0	0	0	0
0	0	0	0	0	0	0	0
0	0	0	0	0	0	0	0
0	0	0	0	0	0	0	0
0	0	0	0	0	0	0	0
0	0	0	0	0	0	0	0
0	0	0	0	0	0	0	0
0	0	0	0	0	0	0	0
0	0	0	0	0	0	0	0
0	0	0	0	0	0	0	0
0	0	0	0	0	0	0	0
0	0	0	0	0	0	0	0
0	0	0	0	0	0	0	0
0	0	0	0	0	0	0	0
0	0	0	0	0	0	0	0
0	0	0	0	0	0	0	0
0	0	0	0	0	0	0	0
0	0	0	0	0	0	0	0
0	0	0	0	0	0	0	0
0	0	2	0	0	0	0	0
4	0	10	0	0	0	0	0
13	0	32	8	1	0	6	6
46	1	100	31	9	0	25	26
133	14	263	120	36	9	99	101
390	70	685	348	140	44	305	310
1018	287	1666	1013	466	202	908	922
2678	980	4033	2676	1446	754	2520	2575
6703	3106	9491	7020	4236	2616	6849	7011
16767	9158	22409	17694	12013	8291	18046	18545
41097	25951	52478	44400	32897	25183	47012	48364
100641	70832	123664	109543	88305	72832	120622	124295
244241	188690	291599	269501	232205	204167	306819	316210
592482	491548	691470	657778	601225	555019	772857	796607
1430621	1259317	1642958	1601335	1534161	1474563	1931852	1990172
3448795	3177555	3914651	3878363	3866982	3833778	4788977	4930987
8282268	7917311	9327510	9360643	9632138	9793714	11780898	12121302
19826847	19497266	22210533	22480153	23739220	24613127	28750101	29561051
47247906	47516228	52753326	53735252	57910652	60989101	69614362	71529220
112074237	114661812	124873756	127728332	139920261	149158176	167225495	171724175
264396972	274156181	294251789	301884793	334945071	360516419	398552940	409056107
620211912	649734731	689814450	709077121	794726241	861825194	942429906	966834467

χ173	χ174	χ175	χ176	χ177	χ178	χ179	χ180
0	0	0	0	0	0	0	0
0	0	0	0	0	0	0	0
0	0	0	0	0	0	0	0
0	0	0	0	0	0	0	0
0	0	0	0	0	0	0	0
0	0	0	0	0	0	0	0
0	0	0	0	0	0	0	0
0	0	0	0	0	0	0	0
0	0	0	0	0	0	0	0
0	0	0	0	0	0	0	0
0	0	0	0	0	0	0	0
0	0	0	0	0	0	0	0
0	0	0	0	0	0	0	0
0	0	0	0	0	0	0	0
0	0	0	0	0	0	0	0
0	0	0	0	0	0	0	0
0	0	0	0	0	0	0	0
0	0	0	0	0	0	0	0
0	0	0	0	0	0	0	0
0	0	0	0	0	0	0	0
0	0	0	0	0	0	0	0
0	0	0	0	0	0	0	0
0	0	0	0	0	0	0	0
0	0	0	0	0	0	0	0
0	0	0	0	0	0	0	0
0	0	0	0	0	0	0	0
0	0	0	0	0	0	0	0
0	0	0	0	0	0	0	0
0	0	0	0	0	0	0	0
0	0	0	0	0	0	0	0
0	0	0	0	0	0	0	0
0	0	0	0	0	0	0	0
0	0	0	0	0	0	0	0
0	0	0	0	0	0	0	0
0	0	0	0	0	0	0	0
0	0	1	3	0	0	1	0
1	0	7	15	2	0	5	0
3	0	23	49	6	3	18	2
11	1	72	144	27	17	59	11
32	14	197	385	78	68	168	49
112	70	541	992	272	240	470	183
365	304	1404	2442	791	771	1250	608
1162	1069	3606	5914	2347	2291	3285	1874
3567	3548	9111	14070	6493	6529	8481	5509
10646	10843	22923	33346	17931	17963	21770	15606
30688	31896	57253	78688	47680	48176	55381	42919
86171	89760	142684	186069	125775	126677	140227	115338
235806	246062	354204	440584	324863	327908	352926	304076
631351	656306	876565	1046694	831242	837256	883425	788556
1656792	1716778	2160195	2491235	2097241	2113233	2197735	2015749
4272407	4406492	5299244	5938548	5243072	5277214	5432386	5086474
10842106	11139610	12929446	14154038	12964854	13049718	13335295	12684035
27122746	27757439	31360601	33695471	31777394	31969641	32500368	31284238
66960895	68308961	75576940	80008554	77154241	77623152	78620465	76367166
163330415	166133438	180905035	189304787	185741250	186839179	188744380	184599922
393942565	399763101	429980931	445889589	443292151	445939361	449644858	442073292
940279810	952244844	1014653642	1044847974	1049341911	1055588645	1062941447	1049201139

χ_{181}	χ_{182}	χ_{183}	χ_{184}	χ_{185}	χ_{186}	χ_{187}
0	0	0	0	0	0	0
0	0	0	0	0	0	0
0	0	0	0	0	0	0
0	0	0	0	0	0	0
0	0	0	0	0	0	0
0	0	0	0	0	0	0
0	0	0	0	0	0	0
0	0	0	0	0	0	0
0	0	0	0	0	0	0
0	0	0	0	0	0	0
0	0	0	0	0	0	0
0	0	0	0	0	0	0
0	0	0	0	0	0	0
0	0	0	0	0	0	0
0	0	0	0	0	0	0
0	0	0	0	0	0	0
0	0	0	0	0	0	0
0	0	0	0	0	0	0
0	0	0	0	0	0	0
0	0	0	0	0	0	0
0	0	0	0	0	0	0
0	0	0	0	0	0	0
0	0	0	0	0	0	0
0	0	0	0	0	0	0
0	0	0	0	0	0	0
0	0	0	0	0	0	0
0	0	0	0	0	0	0
0	0	0	0	0	0	0
0	0	0	0	0	0	0
0	0	0	0	0	0	0
0	0	0	0	0	0	0
0	0	0	0	0	0	0
0	0	2	0	0	2	1
0	1	15	2	2	11	7
4	16	58	19	14	51	41
36	80	202	91	78	192	163
188	360	640	380	314	661	585
776	1291	1972	1364	1193	2093	1918
2814	4343	5764	4530	4012	6309	5930
9312	13436	16519	14049	12792	18216	17416
28827	39807	45934	41594	38426	51010	49521
84895	112856	125368	118314	111259	139094	136412
240293	310796	335001	326371	310484	371303	367153
658666	832622	880790	876808	843863	972748	967633
1758256	2184407	2277570	2304698	2237080	2507744	2506111
4589563	5621644	5807361	5943424	5813597	6371523	6388302
11751829	14237909	14606053	15075624	14833778	15977886	16060720
29590320	35534728	36288251	37678585	37263310	39585753	39862635
73406600	87549483	89105335	92930306	92272428	96982448	97793255
179689401	213134661	216441863	226443556	225589197	235110291	237305268
434552068	513227288	520344745	545661140	545027959	564332517	570010172
1039254265	1223234064	1238862123	1301306496	1302589157	1341813900	1356005205

χ_{188}	χ_{189}	χ_{190}	χ_{191}	χ_{192}	χ_{193}	χ_{194}
0	0	0	0	0	0	0
0	0	0	0	0	0	0
0	0	0	0	0	0	0
0	0	0	0	0	0	0
0	0	0	0	0	0	0
0	0	0	0	0	0	0
0	0	0	0	0	0	0
0	0	0	0	0	0	0
0	0	0	0	0	0	0
0	0	0	0	0	0	0
0	0	0	0	0	0	0
0	0	0	0	0	0	0
0	0	0	0	0	0	0
0	0	0	0	0	0	0
0	0	0	0	0	0	0
0	0	0	0	0	0	0
0	0	0	0	0	0	0
0	0	0	0	0	0	0
0	0	0	0	0	0	0
0	0	0	0	0	0	0
0	0	0	0	0	0	0
0	0	0	0	0	0	0
0	0	0	0	0	0	0
0	0	0	0	0	0	0
0	0	0	0	0	0	0
0	0	0	0	0	0	0
0	0	0	0	0	0	0
0	0	0	0	0	0	0
0	0	0	0	0	0	0
0	0	0	0	0	0	0
3	0	0	0	0	0	0
11	7	0	2	2	2	5
35	33	3	16	16	16	28
113	133	25	67	70	76	114
345	424	133	250	267	292	397
1019	1291	529	824	888	998	1269
2925	3692	1895	2589	2802	3152	3824
8249	10247	6121	7672	8294	9403	11010
22740	27560	18670	22065	23739	26912	30755
61777	72974	54128	61347	65654	74608	83669
164803	189839	151627	166847	177503	201621	223233
433636	488181	411911	443832	469442	533637	585122
1124321	1240569	1093586	1160647	1221107	1387798	1511464
2878418	3122268	2844093	2985899	3126587	3554506	3851830
7276020	7781435	7272757	7575984	7900677	8981663	9699171
18182985	19221579	18315446	18971736	19716621	22418803	24147298
44932970	47056778	45515074	46954193	48658375	55330373	59489463
109885218	114218503	111728679	114922408	118812940	135125590	145090896
266030583	274879700	271230580	278387519	287263332	326733727	350504982
637935025	656056553	651597386	667754743	687981911	782608145	838975286
1515707382	1553001759	1550207654	1586853958	1632907259	1857663612	1990504962

Quilts, the 3-string braid group, and braid actions on finite groups: an introduction

Tim Hsu

Abstract. Quilts (developed by Norton, Parker, Conway, and the author) are 2-complexes which serve as specialized coset diagrams for subgroups of B_3 (the 3-string braid group). We describe some of the basic definitions and results in the theory of quilts, and show how quilts may be used to study a monodromy action of B_3 on pairs of elements (resp. triples of involutions) of a finite group which may be relevant to Norton's Generalized Moonshine conjectures. We also work out some small examples, both in terms of using quilts as coset diagrams for arbitrary subgroups of B_3, and in terms of using quilts to study orbits of the braid action on pairs.

The annotated 2-complexes known as *quilts* were initially developed by Norton [**Nor87**], Parker, Conway, and others, and were first formally defined in print by Conway and the author [**CH95**]. This article is an informal but condensed exposition of some of the basic material in the author's thesis [**Hsu94**]; complete definitions and proofs may be found there.

In Section 1, we discuss quilts purely as coset diagrams for subgroups of B_3 (the 3-string braid group). We build from a reduced version of quilts for subgroups of the modular group (1.1) to the full definition (1.2), discuss some specific features of the way in which braid actions are represented by the geometry of quilts (1.3), and then describe some results about subgroups of B_3 which can be obtained from the geometry of quilts (1.4). In Section 2, we discuss how to use quilts to study certain braid actions on pairs and triples of elements of finite groups. We define and motivate these actions (2.1), show how quilts may be used to picture these actions (2.2), describe how to draw a quilt arising from the action on pairs (2.3), and work out an example coming from the alternating group A_6 (2.4).

For further motivation and history, see Norton [**Nor**]. For further development of this basic material, and more examples, see [**Hsu94**], [**Hsu96**]. For recent and current work on this subject, see Norton [**Nor**], [**Nor96a**], [**Nor96b**], [**Nor96c**]. Note that we have repeated some basic material published elsewhere (e.g., in [**CH95**], [**Hsu96**], and [**Nor**]) to make this article more self-contained. We apologize for any needless duplication.

Notation. Let L and R be the braids shown in Figure 1, let $V = R^{-1}L$, and let $E = LR^{-1}L$. Then B_3 is given by the presentation

$$\langle V, E \mid V^3 = E^2 \rangle.$$

Furthermore, if

$$Z = E^2 = (LR^{-1}L)^2 = V^3 = (R^{-1}L)^3,$$

we have $B_3/\langle Z \rangle \cong \mathrm{PSL}_2(\mathbb{Z})$. By convention, we choose this projection so that L and R are sent to $\begin{pmatrix} 1 & 1 \\ 0 & 1 \end{pmatrix}$ and $\begin{pmatrix} 1 & 0 \\ 1 & 1 \end{pmatrix}$, respectively.

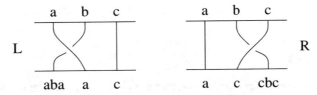

Figure 1. The braids L and R

Note that to convert our notation to the notation used in [**Nor**], we take $x = L^{-1}$, $y = R$, $s = V^{-1}$, $t = E^{-1}$, and $z = Z^{-1}$.

Section 1. Quilts as coset diagrams

(1.1) Modular quilts and $\mathrm{PSL}_2(\mathbb{Z})$. We begin with a principle from permutation groups.

Principle. *Let G be a group. The (finite) transitive permutation representations of G are equivalent to the conjugacy classes of subgroups (of finite index) of G, by the map which sends a transitive permutation representation of G to the conjugacy class of its point stabilizers.*

We can represent the subgroups, or transitive permuation representations, of $\mathrm{PSL}_2(\mathbb{Z})$ as follows. Consider the triangular "puzzle piece" shown in Figure 2. Such a puzzle piece is called a *seam*. The vertical line at the top of the seam is called a *hash mark* (in the sense of American football), and the object at the bottom of the seam is called a *dot*. More formally, we define a seam to be a 2-complex made of four 0-cells (the hash mark, the dot, and the two 0-cells at the tips of the seam), five 1-cells (the 4 dotted lines and the solid line), and two 2-cells (the two regions of the diagram enclosed by its 1-skeleton), glued together as shown.

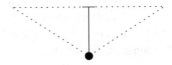

Figure 2. A seam

Let Q be a connected diagram (2-complex) made of seams attached to each other (identified topologically) in the following way: At each hash mark of Q, either 2 or 1 seams meet in what is known as an *edge* (Figure 3(a)) or a *collapsed edge* (Figure 3(b)), respectively; and at each dot of Q, either 3 or 1 seams meet in a *vertex* (Figure 3(c)) or a *collapsed vertex* (Figure 3(d)), respectively. Q is then an orientable connected surface without boundary, and we call Q a *modular quilt*. Note that a modular quilt is compact if and only if it consists of a finite number of seams.

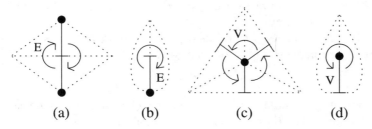

(a)   (b)   (c)   (d)

Figure 3. Edges and vertices

Now, the rules shown in Figure 3 determine a homomorphism from $PSL_2(\mathbb{Z})$ to the group of permutations on the seams of Q, sending E (resp. V) to a permutation of order dividing 2 (resp. 3). It follows that modular quilts are equivalent to transitive permutation representations (conjugacy classes of subgroups) of $PSL_2(\mathbb{Z})$. (Transitivity follows from connectedness.) For example, the modular quilt in Figure 4 (which, by convention, is assumed to be drawn on a sphere, with the dotted lines omitted) represents $\Gamma(4)$.

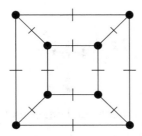

Figure 4. Modular quilt of $\Gamma(4)$

Figure 4 also illustrates the reason behind our choice of terminology, as the quilt in Figure 4 consists of *patches* (the regions bounded by the solid lines) held together by seams. It is easy to see that this will be true in general, though somewhat harder to draw for higher genus.

Remark. Modular quilts and similar objects have long been studied under many names. Most recently, the theory of *dessins d'enfants*, initiated by Grothendieck, has been concerned with the study of the number theory and algebraic geometry of such constructions. For an excellent introduction to *dessins* and related work, see Schneps [**Sch94**].

(1.2) Quilts.   Let B_3 act transitively on a domain Ω, which, for simplicity, we take to be finite. Since $\langle Z \rangle$ is central in B_3, each $\langle Z \rangle$-orbit of Ω has the same size, say M. We call M the *modulus* of Ω.

The action of B_3 on Ω determines a transitive action of $\mathrm{PSL}_2(\mathbb{Z})$ on the $\langle Z \rangle$-orbits of Ω. From our discussion of modular quilts, it follows that the $\mathrm{PSL}_2(\mathbb{Z})$ action is represented by some modular quilt Q, with each $\langle Z \rangle$-orbit of Ω corresponding to a seam of Q. In fact, we can recover the full B_3 action by decorating Q with integers mod M in the following way.

Let A, B, C, $\ldots$ be the $\langle Z \rangle$-orbits of Ω. Pick representatives A_0, B_0, C_0, $\ldots$ for these orbits, and name the other elements of Ω by the rule

$$A_0 Z^i = A_i,$$

where i is any integer mod M. We now decorate Q as follows. Suppose $AE = B$ in terms of $\langle Z \rangle$-orbits. Then for a unique integer i mod M,

$$A_0 E = B_i.$$

In the diagram, we then label the clockwise dotted line going from A to B with i (Figure 5). Note that by convention, we take E to act clockwise, as in Figure 3, and we orient arrows going away from the hash mark.

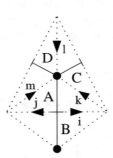

Figure 5. Decorated edges and vertices

Applying E again, we also get $B_0 E = A_j$ for a unique integer j mod M. However, since $Z = E^2$,

$$i + j \equiv 1 \pmod{M}. \tag{E}$$

Proceeding similarly with the vertex in Figure 5 consisting of the seams A, C, and D, we obtain

$$k + l + m \equiv 1 \pmod{M}. \tag{V}$$

Again by convention, we take V to act counterclockwise, as in Figure 3, and we orient arrows going into the dot.

The rules **(E)** and **(V)**, along with their analogues for collapsed edges and vertices, are called the *edge and vertex rules*. Note that, if we think of the arrows as flows on the diagram, we may restate these rules as:

(E) The flow out of an edge is 1.
(V) The flow into a vertex is 1.

To summarize, from any transitive permutation representation Ω of B_3, we get a well-defined decorated modular quilt whose decorations satisfy the edge and vertex rules. Conversely, define a *quilt diagram* with modulus M to be a modular quilt decorated with integers mod M satisfying the edge and vertex rules. Running the above discussion backwards, we see that any quilt diagram induces a well-defined transitive permutation representation of B_3.

Now, in general, there is more than one quilt diagram for a given transitive permutation representation of B_3, as the decorations on a quilt diagram depend on the choice of $\langle Z \rangle$-orbit representatives. However, suppose we think of a (finite) quilt diagram Q as being triangulated by its seams, and we orient the dotted lines as indicated by the arrows. Then we observe that the integers mod M on dotted lines of Q define a simplicial 1-chain with coefficients in $\mathbb{Z}/M$, satisfying the edge and vertex conditions.

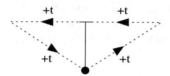

Figure 6. Effect of changing a representative

Examining our conventions carefully, especially with regards to clockwise and counterclockwise, we see that if we replace the representative A_0 of a given $\langle Z \rangle$-orbit with the representative A_t, then we change the corresponding quilt diagram precisely by adding the boundary 1-chain (in the sense of algebraic topology) shown in Figure 6. It follows that:

Theorem 1. *Homology classes of quilt diagrams with coefficients in $\mathbb{Z}/M$ are equivalent to transitive permutation representations of B_3 with modulus M.*   □

A *quilt* is therefore defined to be a homology class of quilt diagrams, and Theorem 1 just says that quilts are equivalent to transitive permutation representations of B_3.

(1.3) Left and right turns, and patches. Consider again the situation in Figure 5, retaining our previous notation. Now, in terms of the modular action on $\langle Z \rangle$-orbits, we have that $BL = BEV^{-1} = D$ and $BR = BE^{-1}V = C$. In other words, L acts as a left turn in the diagram, and R acts as a right turn. More precisely, using the arrows to recover the full B_3 action, we get:

$$B_0 L = B_0 E V^{-1} = A_j V^{-1} = D_{j-m} \tag{L}$$
$$B_0 R = B_0 E^{-1} V = A_{-i} V = C_{-i+k} \tag{R}$$

Applying **(L)** (resp. **(R)**) repeatedly, we also find that the $\langle L \rangle$-orbits (resp. $\langle R \rangle$-orbits) of a transitive permutation representation of B_3 correspond bijectively with the patches of its quilt.

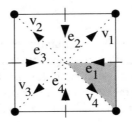

Figure 7. Patch, size 4, with inflow

The *size* of a patch is defined to be the number of edges it touches, counting multiplicities. A typical patch of size 4 is shown in Figure 7. Now, applying **(L)** repeatedly, it follows that if a patch has size n, then n divides the size of the corresponding $\langle L \rangle$-orbit, and is equal to the size of the corresponding $\langle L \rangle$-orbit of $\langle Z \rangle$-orbits. More precisely, consider the patch in Figure 7, and let i be the *inflow* of the patch, that is, $\sum e_j - \sum v_j$. Then if A is the seam whose left half is shaded in Figure 7, repeatedly applying **(L)** shows that $A_0 L^4 = A_i$, which means that the size of the $\langle L \rangle$-orbit of A_0 is 4 times the additive order of $i \pmod{M}$. We therefore say that a patch is *collapsed* if it has a nonzero inflow.

(1.4) Quilts and homology.   A key feature of quilt theory is that elementary facts from the homology of surfaces can be used to obtain interesting results about subgroups of B_3. For instance, we have the following theorem, in which we say that a quilt $\tilde{Q}$ has modular structure Q if omitting the arrow flows on $\tilde{Q}$ yields the modular quilt Q.

Theorem 2. *Let Q be a modular quilt with s seams, and with no collapsed edges or vertices. There exists a quilt $\tilde{Q}$ with the modular structure of Q, with modulus M, and with no collapsed patches, if and only if M divides $s/6$. Furthermore, if Q has genus 0, for a given M dividing $s/6$, $\tilde{Q}$ is uniquely determined.*   □

For example, consider the modular quilt Q shown in Figure 4. The first statement of Theorem 2 implies that if $\tilde{Q}$ is a quilt with the modular structure of Q, and $\tilde{Q}$ has no collapsed patches, then the modulus of $\tilde{Q}$ must divide 4. Since Q has genus 0, the second statement of Theorem 2 implies that for any M dividing 4, there is a unique quilt $\tilde{Q}$ with the modular structure of Q and no collapsed patches.

Remark.   The most important consequence of Theorem 2 is that in certain special cases, especially with regards to our main application of quilts (Section 2), we may ignore the subtleties of lifting $\mathrm{PSL}_2(\mathbb{Z})$ to B_3 and consider ourselves to be working with the modular group and its subgroups. See (2.4) for a related example.

Section 2. Quilts and monodromy actions

(2.1) Monodromy actions on triples and pairs. Let G be a group, and for simplicity, assume G is finite. We recall (Birman **[Bir74]**) that B_3 has a natural monodromy action on the set of triples (a, b, c) of involutions of G, as described by the following rules.

$$(a, b, c)L = (aba, a, c)$$
$$(a, b, c)R = (a, c, cbc)$$

Figure 1 gives a mnemonic for remembering this action: the involution on the overcrossing string is preserved, and the involution on the undercrossing string is conjugated.

Note that the monodromy action preserves the product abc of three involutions in a triple, which means that this product is invariant within each orbit. We also note that

$$(a, b, c)Z = (a^{cba}, b^{cba}, c^{cba}).$$

In other words, Z acts on a given orbit by conjugation by the inverse of the product preserved within that orbit.

If we think of the map sending (a, b, c) to $(\alpha, \beta) = (ab, bc)$ as the *reduction* of a triple of involutions to a pair of arbitrary elements of G, then reduction of the monodromy action induces an action on pairs of elements of G, as described by the following rules.

$$(\alpha, \beta)L = (\alpha, \alpha\beta)$$
$$(\alpha, \beta)R = (\alpha\beta, \beta)$$
$$(\alpha, \beta)Z = (\alpha^{cba}, \beta^{cba})$$

We call this induced action the *T-action*, and the orbits of the T-action are called *T-systems*. An elementary but lengthy computation **[Hsu94**, Ch. 8] shows that reduction of an orbit of the monodromy action gives a T-system of the same size. Therefore, to study the shapes of the orbits of the monodromy action, or equivalently, the stabilizers of the orbits of the monodromy action, it suffices to look at T-systems.

Remark. We are interested in the orbit shapes of T-systems, or equivalently, the stabilizers in B_3 of pairs (α, β), because of the following potential non-commutative extension of Generalized Moonshine **[Nor87]**. Let $\mathbb{H}$ denote the complex upper half-plane. We say that a function $F((\alpha, \beta), \tau) : (G \times G) \times \mathbb{H} \mapsto \mathbb{C}$ is *automorphic* if it satisfies appropriate analytic conditions and for each $\pi \in B_3$,

$$F((\alpha, \beta)\pi, \tau) = F((\alpha, \beta), \pi(\tau)),$$

where the action in the first factor is the T-action, and the action in the second factor is obtained by projecting π down to $PSL_2(\mathbb{Z})$. Now, for any automorphic function F, fix (α, β), and let Γ be the projection of the stabilizer of (α, β) down to $PSL_2(\mathbb{Z})$. As a function on $\mathbb{H}$, $F((\alpha, \beta), \tau)$ will then be automorphic over Γ in the usual sense. Furthermore, when G is a 6-transposition group (e.g., the Monster), in many cases, we can almost be assured that Γ will have genus 0 (see **[Nor87]**, **[Nor]** or (2.3)). However, since Γ will usually be a noncongruence subgroup unless α and β commute, it is not clear what connection this phenomenon has with the current conception of Moonshine. See **[Nor]** for further discussion.

(2.2) Quilts of T-systems. Let T be the T-system of a pair of elements (α, β) of G. As a transitive permutation representation of $\mathbf{B}_3$, T determines a unique quilt Q (Theorem 1), which we call the quilt of T. Naturally, Q contains complete information on the action of $\mathbf{B}_3$ on T, much as any coset diagram would. For instance, if M is the modulus of Q, then M is the smallest power of cba which commutes with both α and β. However, because quilts were designed specifically for encoding information about T-systems, by labelling Q suitably, we may use Q to study T in a much more interesting manner.

First, recall that a choice of a given quilt diagram in the homology class of Q is the same as the choice of a representative for each $\langle Z \rangle$-orbit in T, or in other words, a pair in each $\langle Z \rangle$-orbit. Also, recall that each $\langle Z \rangle$-orbit of T corresponds to a seam of Q. Therefore, our first step in labelling Q is to write the representative for each $\langle Z \rangle$-orbit on the corresponding seam. For instance, since (α, β) is certainly in T, by choosing new representatives, if necessary, we know that one seam of Q may be labelled as in Figure 8.

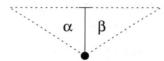

Figure 8. Labelling a seam with a pair

Now, we may extend the definition of quilt consistently by using not only whole boundaries of seams to define the homology class of a quilt diagram, but also boundaries of left and right halves of seams. In fact, this extended definition of quilt is compatible with our previous labelling in the following manner. For t an integer $\bmod M$ and α an element of a pair in T, define α_t to be α conjugated by $(cba)^t$, and consider a seam labelled as in Figure 8. If we replace α with α_t (resp. β with β_t) each time we add a left (resp. right) counterclockwise boundary of $+t$ to this seam, we see that our labelling remains consistent, since $(\alpha, \beta)Z^t = (\alpha_t, \beta_t)$. Therefore, we will consider quilts and their labellings in this extended sense from now on.

Two examples of this extended notion of labelled quilt are shown in Figures 9 and 10. Figure 9 shows the effect of adding a left boundary of $+t$ to the seam in Figure 8. Figure 10 shows another typical seam which might appear in the extended labelled quilt of T; such a seam is interpreted as saying that (γ_{t+u}, δ_u) is in T for all u $(\bmod M)$.

Remark. We remark that in **[CH95]**, quilts are first defined in terms of the labelled quilt of a T-system. Seen from the viewpoint taken here, the main result of **[CH95]** is the fact that quilts may be interpreted as coset diagrams of subgroups of $\mathbf{B}_3$.

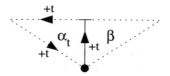

Figure 9. Effect of a left boundary

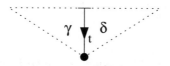

Figure 10. Interpreting arbitrary seams

(2.3) Drawing the quilt of a T-system. Let (α, β) be a pair of elements of a finite group G, let T be its T-system, and let Q be the quilt of T. Using the extended notion of quilt developed above, we have the following set of rules for drawing Q, based on the assumptions that there is no collapse in Q, and that Q has genus 0. Of course, many quilts of T-systems have nonzero genus, and many quilts of T-systems have some kind of collapsed feature (edge, vertex, or patch). In such cases, these rules only produce a cover of the quilt of T, and more work is required to find all of the identifications that occur. However, as we will see, these rules often suffice for the cases in which we are most interested.

First, as observed in (2.2), we may choose the quilt diagram representing Q such that one seam is labelled as in Figure 8. Therefore, we begin by drawing this seam.

The rest of the process consists of repeating the use of the basic properties of quilts (the edge and vertex rules, trivalence) and two principles specific to quilts of T-systems. The first principle is called the *multiplication principle*, and it is described as follows. Suppose we already know that the seam in Figure 8 is in Q. Then, since

$$(\alpha, \beta)E = (\beta_1, \alpha),$$
$$(\alpha, \beta)L = (\alpha, \beta)EV^{-1} = (\alpha, \alpha\beta),$$
$$(\alpha, \beta)R = (\alpha, \beta)E^{-1}V = (\alpha\beta, \beta),$$

it follows that we may "create" the seams of the $\langle Z \rangle$-orbits of (β_1, α), $(\alpha, \alpha\beta)$, and $(\alpha\beta, \beta)$, and expand the seam in Figure 8 to the configuration shown in Figure 11. In other words, we expand Figure 8 to a "directed edge" ending in a trivalent vertex, and place the product of α and β in the unlabelled part of this vertex.

Note that Figure 11 also highlights the following conventions for drawing quilts.
- Arrow flows of 0 are omitted, and unnumbered arrow flows are assumed to be flows of 1.
- If the left or right half of a seam is left unlabelled, then it is assumed to be labelled with the same group element as any other region that can be reached from it by crossing only dotted lines with arrow flows of 0.
- As in the modular case (1.1), dotted lines with arrow flows of 0 may be omitted.

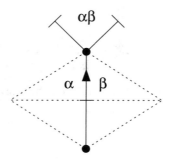

Figure 11. Generic edge and vertex

Our other quilt-drawing principle, known as the *patch principle*, comes from special-ization of the patch observations made in (1.3). Specifically, if n is the size of the patch containing the left side of a seam labelled (α, β), then n divides the order of α, since $(\alpha, \beta)L^i = (\alpha, \alpha^i \beta)$. In particular, if there is no collapse, n is equal to the order of α.

Remark. We can also use the patch principle to recover the following observation of Norton [**Nor87**]. Let (a, b, c) be a triple of 6-transpositions, and let Q be the quilt of the T-system of (ab, bc). Then the patch principle implies that Q is a trivalent polyhedron made of polygons with 6 or fewer sides. Now, in a trivalent polyhedron, we may think of polygons with 5 or fewer sides as "positively curved", and hexagons as "flat". The combinatorial Gauss–Bonnet theorem (or the Riemann–Hurwitz formula, or Euler's formula) then implies that Q must have genus 0 or 1, and furthermore, that Q has genus 1 if and only if it is made entirely of hexagons, with no collapse. This "genus 0" property of 6-transposition groups, with its possible link to Generalized Moonshine, was one of the motivations behind the study of quilts.

(2.4) An example from A_6. We conclude with an example coming from A_6. Let $a = (1\ 4)(2\ 3)$, $b = (2\ 4)(5\ 6)$, $c = (1\ 3)(4\ 5)$, $\alpha = ab = (1\ 2\ 3\ 4)(5\ 6)$, and $\beta = bc = (1\ 3)(2\ 5\ 6\ 4)$. To find the orbit shape of (a, b, c) under the monodromy action, we find the shape of the T-system of (α, β) by drawing its quilt Q.

First, with the benefit of hindsight, we list all of the calculations we need.

$$cba = (1\ 2)(3\ 4\ 6\ 5)$$

$$\alpha\beta = (1\ 5\ 4\ 3\ 2) \qquad\qquad \alpha^{-1}\beta = (1\ 2\ 3\ 5\ 4)$$
$$\alpha^2\beta = (4\ 5\ 6) \qquad\qquad\quad \alpha\beta^2 = (1\ 6\ 4)(2\ 3\ 5)$$

Note that since no nonidentity power of cba commutes with either α or β, the modulus of Q is the order of cba, that is, $M = 4$.

We can now proceed to draw Q, using the following steps.

1. First, we draw the seam in Figure 8 (as discussed in (2.3)). Then, since both α and β have order 4, we apply the patch principle twice to obtain the first picture in Figure 12.

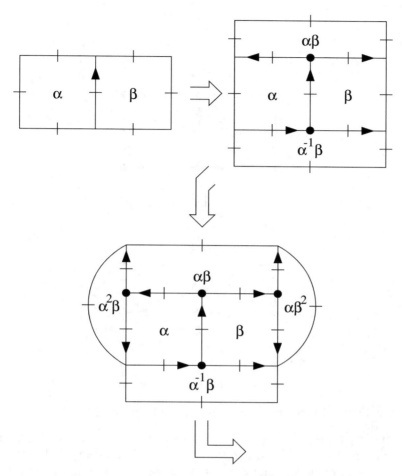

Figure 12. Steps (1)–(3) in producing the quilt in (2.4)

2. Applying the multiplication principle twice, we see that the patches labelled with the elements $\alpha\beta$ and $\alpha^{-1}\beta$ must have the positions shown in the second picture in Figure 12. Furthermore, since both $\alpha\beta$ and $\alpha^{-1}\beta$ have order 5, two applications of the patch principle yield the second picture in Figure 12.

3. Since $\alpha^2\beta$ and $\alpha\beta^2$ both have order 3, two more applications of both the multiplication and patch principles give the third picture in Figure 12.

4. Now, using the fact that the solid lines of Q form a "trivalent graph", we can seal the diagram up into a surface (in this case, a sphere), which means that we have obtained the correct modular structure of Q. The result is the first picture in Figure 13.

5. To obtain the complete structure of Q, we only need to fill in the arrow flows, and since we are assuming no collapse, it is not hard to see that we only need to do so on the solid lines. In fact, we have already filled in many of these arrow flows in the

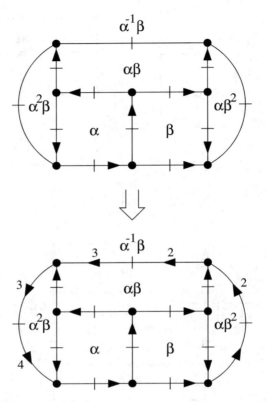

Figure 13. Steps (4) and (5) in producing the quilt in (2.4)

previous steps, when we applied the multiplication principle. The remaining arrow flows may be deduced from the edge rule (outflow from an edge is 1) and vertex rule (inflow into a vertex is 1), and so we arrive at our final answer: the second picture in Figure 13.

As mentioned in (2.3), this result is only guaranteed to be a cover of Q, and not necessarily Q itself. However, it is not hard to check that no pairs from two different seams of Q are in the same $\langle Z \rangle$-orbit, which means that our "guess" is actually the quilt we want.

Remark. Note that at the vertex in the lower left-hand corner of the second picture of Figure 13, the inflow is $5 \equiv 1 \pmod{M}$. Inspection of our quilt-drawing procedure shows that we can arrive at this conclusion without actually computing M. In other words, the rest of the calculation, which is "horizontal" in nature, implies the "vertical" fact that M divides 4. In fact, since Q has 24 seams, this "horizontal implies vertical" effect is a consequence of the first statement of Theorem 2 and the lack of collapse in Q.

References

[Bir74]   J. S. Birman, Braids, links, and mapping class groups, Princeton Univ. Press, Princeton, 1974.

[CH95]   J. H. Conway, T. M. Hsu, Quilts and T-systems, J. Algebra **174** (1995), 856–908.

[Hsu94]   T. M. Hsu, Quilts, T-systems, and the combinatorics of Fuchsian groups, Ph.D. thesis, Princeton Univ., 1994.

[Hsu96]   ———, Some quilts for the Mathieu groups, in: Moonshine, the monster, and related topics, C. Dong, G. Mason (Ed.), Contemp. Math. **193**, Amer. Math. Soc., Providence, RI, 1996, 113–122.

[Nor]   S. P. Norton, Netting the Monster, these proceedings.

[Nor87]   ———, Generalized moonshine, in: The Arcata conference on representations of finite groups, Part 1, Proc. Sympos. Pure Math. **47**, Amer. Math. Soc, Providence 1987, 208–209.

[Nor96a]   ———, Free transposition groups, Comm. Algebra **24** (1996), 425–432.

[Nor96b]   ———, Computing in the monster, submitted to J. Symb. Comp., Proc. 1996 MAGMA conference, 1996.

[Nor96c]   ———, A string of nets, preprint, 1996.

[Sch94]   L. Schneps (ed.), The Grothendieck theory of dessins d'enfants, London Math. Soc. Lecture Note Ser. **200**, Cambridge Univ. Press, Cambridge 1994.

Department of Mathematics
University of Michigan
Ann Arbor, MI 48109, U.S.A.
E-mail: timhsu@math.lsa.umich.edu

The moonshine VOA and a tensor product of Ising models

Masahiko Miyamoto

Abstract.   It is known that $V^\natural$ contains a copy of the tensor product $T = \otimes_{i=1}^{48} L\left(\frac{1}{2}, 0\right)$ of 48 Ising models. We determine the T-module structure of $V^\natural$ explicitly for some T. We also give a new proof of that the full automorphism group of $V^\natural$ is finite.

Section 1. Introduction

The most interesting example of vertex operator algebra (VOA) is the moonshine module $V^\natural = \sum_{i=0}^{\infty} V_i^\natural$. It has many interesting properties, but its structure and the automorphism group are very difficult to see, (see **[FLM]**). On the other hand, the simplest example is one of the minimal series of rational Virasoro VOA $L\left(\frac{1}{2}, 0\right)$ with central charge $\frac{1}{2}$. It has only three inequivalent irreducible modules $L\left(\frac{1}{2}, 0\right)$, $L\left(\frac{1}{2}, \frac{1}{2}\right)$, and $L\left(\frac{1}{2}, \frac{1}{16}\right)$, where the first entry denotes the central charge and the second denotes the highest weights. Its fusion rules are:

(1) $\left(\frac{1}{2}, 0\right)$ is identity,

(2) $L\left(\frac{1}{2}, \frac{1}{2}\right) \times L\left(\frac{1}{2}, \frac{1}{2}\right) = L\left(\frac{1}{2}, 0\right)$,

(3) $L\left(\frac{1}{2}, \frac{1}{2}\right) \times L\left(\frac{1}{2}, \frac{1}{16}\right) = L\left(\frac{1}{2}, \frac{1}{16}\right)$,

(4) $L\left(\frac{1}{2}, \frac{1}{16}\right) \times L\left(\frac{1}{2}, \frac{1}{16}\right) = L\left(\frac{1}{2}, 0\right) + L\left(\frac{1}{2}, \frac{1}{2}\right)$.

It is shown by Dong, Mason, and Zhu in **[DMZ]** that the moonshine VOA $V^\natural$ contains 48 mutually orthogonal conformal vectors e^i with the central charge $\frac{1}{2}$ such that the sum is the Virasoro element of $V^\natural$. This means that $V^\natural$ contains a vertex operator subalgebra $T = \otimes_{i=1}^{48} L\left(\frac{1}{2}, 0\right)$, which is the tensor product of 48 copies of $L\left(\frac{1}{2}, 0\right)$ and so we can see $V^\natural$ as a module of T. It is also proved by them that every irreducible module of a tensor product VOA is isomorphic to a tensor product of each irreducible modules of the corresponding factor VOA's. Namely, every irreducible T-module L can be expressed in the form $L = \otimes_{i=1}^{48} L\left(\frac{1}{2}, h^i\right)$.

Definition 1. We will call the above 48-tuple $(h^1, \ldots, h^{48})$ of highest weights "the highest weight row" of L. Since $L\left(\frac{1}{2}, 0\right)$ is rational and the highest weights of $\otimes L\left(\frac{1}{2}, h^i\right)$ are limited, $V^\natural$ is a finite direct sum of such irreducible T-modules. Let hwr$(T, V^\natural)$ denote the set of all highest weight rows of irreducible T-submodules of $V^\natural$ with multiplicities. For an irreducible T-module W with a highest weight row $h = (h^1, \ldots, h^{48})$, we assign a word $\tilde{h} = (\tilde{h}^1, \ldots, \tilde{h}^{48})$ by defining $\tilde{h}^i = 1$ if $h^i = \frac{1}{16}$ and $\tilde{h}^i = 0$ otherwise. We will call this word "a word of W of $\frac{1}{16}$-positions" and denote it by $\tilde{h}(W)$. Even if W is not irreducible, we can use the same notation $\tilde{h}(W)$ whenever all irreducible T-submodules of W have the same word.

Our purpose in this paper is to determine $\mathrm{hwr}(T, V^\natural)$ explicitly. It is proved by the author in [M1] that if a vertex operator algebra V contains a rational conformal vector e with central charge $\frac{1}{2}$, that is, if V contains $L(\frac{1}{2}, 0)$, then we have an automorphism τ_e of VOA V by defining the endomorphism τ_e on every $\langle e \rangle$-submodule V by

$$\tau_e : \begin{cases} 1 & \text{on } U \cong L(\frac{1}{2}, 0) \text{ or } L(\frac{1}{2}, \frac{1}{2}) \\ -1 & \text{on } U \cong L(\frac{1}{2}, \frac{1}{16}). \end{cases}$$

So in the case of $V^\natural$, we have an elementary Abelian 2-group P generated by mutually commutative 48 automorphisms $\{\tau_{e_i} : i = 1, \ldots, 48\}$. The fixed point set $(V^\natural)^P$ is a vertex operator subalgebra of $V^\natural$ and it is generated by all irreducible T-modules whose word of $\frac{1}{16}$-positions is the zero word (0^{48}). The author showed that such a VOA is isomorphic to a code VOA M_D for some even binary linear code D of length 48, see [M3]. We will call this code "the moonshine generator code" and denote it by MGC. We will explicitly construct MGC and its vertex operator algebra M_{MGC} in Sec. 3. An advantage of MGC is that it contains a lot of Hamming codes and so we can apply the representation theory of Hamming code VOA, see [M3]. For any irreducible character χ of P, the eigenspace

$$V_\chi^\natural = \{v \in V^\natural : gv = \chi(g)v \text{ for } g \in P\}$$

is a nontrivial irreducible $(V^\natural)^P$-module and we have a decomposition

$$V^\natural = \oplus_{\chi \in \mathrm{Irr}(P)} V_\chi^\natural,$$

(see [DM]). Since every irreducible character χ of P is given by a map

$$\chi : \tau_{e_i} \to (-1)^{h_\chi^i} \quad (h_\chi^i = 0, 1),$$

each character χ corresponds to a binary word $h_\chi = (h_\chi^i)$ and the component $V_\chi^\natural$ is generated by all irreducible T-submodules whose codewords of $\frac{1}{16}$-positions are h_χ. Set $S = \{h_\chi : \chi \in \mathrm{Irr}(P)\}$. The results in [DM] and the fusion rules (1.1) imply that S is a binary linear code. One of our purposes is to show that $S = \mathrm{MGC}^\perp$.

In order to avoid the misleading, we have to note that there are probably many sets of unconjugate mutually orthogonal forty eight conformal vectors in $V^\natural$, but our choice of mutually orthogonal conformal vectors is some specific set, which is given by Miracle Octad Generator. We will explain it in Sec. 4.

In Sec. 3, we will introduce Moonshine Generator Code MGC. In Sec. 4, we will investigate the structure of VOA V_Λ of Leech lattice Λ. In Sec. 5, we will show a new proof of that $\mathrm{Aut}(V^\natural)$ is finite by using the fact that the Virasoro element is a sum of conformal vectors with central charge $\frac{1}{2}$. In Sec. 6, we will show the list the multiplicity of irreducible T-submodules.

Section 2. Notation

α	Codewords.
$\lvert\alpha\rvert$	The weight of binary word $\alpha = (a_i)$, that is $\lvert\{i : a_i = 1\}\rvert$.
Bl, Gr, Rd	The three colored blocks in MGC.
C_{24}	The Golay code.
Δ	$= \{1, 2, \ldots, 48\}$.
e	A conformal vector with central charge $\frac{1}{2}$.
$e^{x_i \pm x_j}$	A vector of V_Λ, see [FLM].
$h = (h^i)$	A highest weight row.
$\mathrm{hwr}(T, V)$	The set of all highest weight rows, see Def. 1.
Λ	The Leech lattice.
M_D	A VOA constructed by an even binary linear code D.
MGC	The Moonshine generator code, see Subsec. 3.1.
$\mathrm{MGC}^\perp$	The orthogonal complement of MGC.
Ω	$= \{1, 2, \ldots, 24\}$.
P	$= \langle \tau_{e_i} : i = 1, \ldots, 48 \rangle$.
r_i	$r_i = (0, \cdots, 0, 1, \underset{i}{0}, \cdots 0) = \{i\}$.
T	$= \otimes_{i=1}^{48} L(\frac{1}{2}, 0)$, see Sec. 1.
τ_e	An automorphism of VOA given by e, see [M1].
θ	An automorphism of V_Λ induced by -1 on Λ.
U_i	The i-th upper space.
$\{x_i : i = 1, \ldots, 24\}$	An orthogonal basis of $\mathbb{R} \otimes \Lambda$ with $\langle x_i, x_j \rangle = 2\delta_{ij}$.

Section 3. Moonshine generator code

3.1 Definition of MGC. We will introduce a binary linear code "Moonshine Generator Code" MGC of length 48 in this section. It is generated by 3 colored blocks and 4 orthogonal reflections. Let $\mathbb{Z}_2^{48}$ be a 48-dimensional vector space over $\mathbb{Z}_2$ and Δ the set of 48 coordinates. Identify $\mathbb{Z}_2^{48}$ as the set of $2 \times 2 \times 2 \times 2 \times 3$-matrices. Divide Δ into three $2 \times 2 \times 2 \times 2$-matrices and call them "Red", "Green", and "Blue" blocks and denote them by Rd, Gr, and Bl, respectively. Namely, we have a presentation $\Delta = \{(\pm 1, \pm 1, \pm 1, \pm 1, \text{colors})\}$. Let U_i be the subset of Δ consisting of the elements whose i-th entry is $+1$ for $i = 1, \ldots, 4$ and call it the i-th upper space. For example, $U_2 = \{(\pm 1, +1, \pm 1, \pm 1, \text{colors})\}$.

Definition 2. For i, the reflection ϕ_i denotes the permutation on Δ which changes the sign of i-th entry. We view ϕ_i as a linear map of $\mathbb{Z}_2^{48}$.

For each $i = 1, \ldots, 4$, take any even subset α of i-th upper space U_i, take the image $\phi_i(\alpha)$ of α and get the union $\alpha \cup \phi_i(\alpha)$. We often identify the subsets of Δ with the binary words. Collect all such unions and then MGC is defined as the binary linear code generated by them.

Definition 3. $\mathrm{MGC} = \langle \alpha \cup \phi_i(\alpha) : \alpha \subset U_i, \ |\alpha| \ \text{even}, \ i = 1, \dots, 4 \rangle.$

We note that if the intersection of MGC and one colored block is isomorphic to Reed-Muller code $R(2,4)$ which has minimum distance 4, (see [Mu], [Re]). It is easy to check the following:

Lemma 3.1. MGC *has dimension* 41 *and the minimal length* 4.

Proof. Clearly, MGC is an even linear code. Since the dimension of the set $E(\mathbb{Z}_2^n)$ of all even words in $\mathbb{Z}_2^n$ is $n-1$, the dimension of $E(U_1)$ is 23, that of $E(U_1 \cap U_2)$ is 11, that of $E(U_1 \cap U_2 \cap U_3)$ is 5, and that of $E(U_1 \cap U_2 \cap U_3 \cap U_4)$ is 2. Let $\{\alpha_i : i \in I_1\}$, $\{\beta_i : i \in I_2\}$, $\{\gamma_i : i \in I_3\}$ and $\{\delta_i : i \in I_4\}$ be bases of $E(U_1)$, $E(U_1 \cap U_2)$, $E(U_1 \cap U_2 \cap U_3)$ and $E(U_1 \cap U_2 \cap U_3 \cap U_4)$, respectively. It is easy to see that

$$\{\phi_1(\alpha_h) \cup \alpha_h, \ \phi_2(\beta_i) \cup \beta_i, \ \phi_3(\gamma_j) \cup \gamma_j, \ \phi_4(\delta_k) \cup \delta_k : h \in I_1, \ i \in I_2, \ j \in I_3, \ k \in I_4\}$$

is a linearly independent set. Totally, the dimension of MGC is greater than or equal to $23 + 11 + 5 + 2 = 41$. On the other hand, the orthogonal complement of MGC contains the three colored blocks and four upper spaces, which generate a subspace of dimension 7. Hence, we have $\dim \mathrm{MGC} = 41$. By the definition of MGC, for any $\emptyset \neq \alpha \in \mathrm{MGC}$ and i, $\alpha \cap U_i$ and $\alpha \cap (\Delta - U_i)$ are both even sets. If $\alpha \cap (\Delta - U_4) = \emptyset$, then α is generated by three reflections ϕ_1, ϕ_2, ϕ_3. Repeating them, we can check that there is no word $\alpha \in \mathrm{MGC}$ with $|\alpha| = 2$.

It is easy to see the following lemma.

Lemma 3.2. *The orthogonal complement* $\mathrm{MGC}^\perp$ *of* MGC *has dimension* 7 *and it is spanned by* U_1, U_2, U_3, U_4 *and three colored blocks* Rd, Gr, Bl. *In particular, if* $\alpha \in \mathrm{MGC}^\perp$, *then* $|\alpha| = 0, 16, 32, 48$ *or* 24. *The weight enumerator of* $\mathrm{MGC}^\perp$ *is*

$$x^{48} + 3x^{32}y^{16} + 120x^{24}y^{24} + 3x^{16}y^{32} + y^{48}.$$

We will show in the next section that $\mathrm{MGC}^\perp$ coincides with the set of words of $\frac{1}{16}$-positions associated to the highest weight rows $\mathrm{hwr}(T, V^\natural)$.

In order to see a relation with the Golay code constructed by Miracle Octad Generator, it is useful to write MGC as a subspace of the set of 4×12-matrices. Divide $\mathbb{Z}_2^{48}$ into three colored blocks "4×4-matrices". We will introduce 4 axes Ω_i $(i = 1, \dots, 4)$ for the reflections ϕ_i $(i = 1, \dots, 4)$ as follows. I here write only one block, but every colored block has the same axes.

	Ω_4	Ω_1	Ω_4
Ω_3	1 ⋮ 2	9	10
	3 ⋮ 4	11	12
Ω_2	5 ⋮ 6	13	14
Ω_3	7 ⋮ 8	15	16

Each axis divides 48 coordinates into 24 and 24. The 24 coordinate set containing the first coordinate is "i-th upper space" U_i.

Remark 1. An important property of MGC is that it contains a lot of Hamming codes. For example, take $i \neq j$ and two points set $\beta \subseteq U_i \cap U_j$, then $\beta \cup \phi_i(\beta) \in$ MGC and for any even set $\alpha \in \beta \cup \phi_i(\beta)$, we have $\phi_j(\alpha) \cup \alpha \in$ MGC. Hence,

$$\langle \beta \cup \phi_i(\beta), \ \alpha \cup \phi_j(\alpha) : \alpha \in \beta \cup \phi_i(\beta) \ \text{even} \rangle$$

is a $[8, 4, 4]$-Hamming subcode and $\beta \cup \phi_i(\beta) \cup \phi_j(\beta) \cup \phi_i \phi_j(\beta)$ is a support of Hamming subcode. In particular, an intersection of an upper space and a colored block is a support of some Hamming subcode.

Section 4. The structure of V_Λ

In this section, we will investigate the structure of VOA V_Λ of Leech lattice Λ. Comparing to the moonshine VOA, it is not difficult to study its structure.

Let $(V_\Lambda)^\theta$ be the fixed point set of V_Λ by θ, where θ is the automorphism of V_Λ induced from the automorphism -1 on Λ. V_Λ^θ is the untwisted part of the moonshine VOA $V^\natural$.

4.1. MOG. As we told at the end of the introduction, we will choose a specific set of 48 mutually orthogonal conformal vectors in V_Λ. Before that, we will explain the Miracle Octad Generator. As usual, $F_4 = \{0, 1, w, \bar{w}\}$ denotes the finite field consisting of four elements with the relations $1 + w + w^2 = 0$ and $w^2 = \bar{w}$.

The hexacode C_6 is a 3-dimensional code of length 6 over F_4 generated by the codewords

$$(w\bar{w}w\bar{w}w\bar{w}), \quad (w\bar{w}\bar{w}ww\bar{w}), \quad (\bar{w}www\bar{w}w), \quad (\bar{w}ww\bar{w}ww\bar{w})$$

Actually, the first three codewords generate C_6.

We will define the Golay code C_{24} of length 24 as a subset of 4×6 matrices as follows.

First, for each element of F_4, we define an odd interpretation and an even interpretation. Although it is given as a 4×1 column vector in **[CS]**, we will write it by a row-vector here.

$$
\begin{array}{rcllll}
0 & \to & [*, 0|0, 0], & [0, *|*, *], & [0, 0|0, 0], & [*, *|*, *] \\
1 & \to & [0, *|0, 0], & [*, 0|*, *], & [*, *|0, 0], & [0, 0|*, *] \\
w & \to & [0, 0|*, 0], & [*, *|0, *], & [*, 0|*, 0], & [0, *|0, *] \\
\bar{w} & \to & [0, 0|0, *], & [*, *|*, 0], & [*, 0|0, *], & [0, *|*, 0]
\end{array}
$$

The codewords of C_{24} are defined from hexacodewords by either

i) replacing each digit by an odd interpretation, in any way such that the top row becomes odd; or

ii) replacing each digit by an even interpretation, in any way such that the top row becomes even.

We next define the Leech lattice using the above the Golay code C_{24}. Let $\{x_1, \ldots, x_{24}\}$ be an orthogonal basis with $\langle x_i, x_i \rangle = 2$. The Leech lattice Λ is given by

$$\Lambda = \sum_{c \in C_{24}} \mathbb{Z} x_c / 2 + \sum_{i=1}^{24} \mathbb{Z}(x_i - x_\Omega / 4),$$

where $\Omega = \{1, \ldots, 24\}$ and $x_c = \sum_{i \in c} x_i$. For two coordinates x_{2i-1}, x_{2i}, we have four mutually orthogonal conformal vectors

$$e^{4i+1} = \tfrac{1}{16}\{(x_{2i+1} + x_{2i+2})(-1)\}^2 + \tfrac{1}{4}(e^{(x_{2i+1}+x_{2i+2})} + e^{-(x_{2i+1}+x_{2i+2})}),$$
$$e^{4i+2} = \tfrac{1}{16}\{(x_{2i+1} + x_{2i+2})(-1)\}^2 - \tfrac{1}{4}(e^{(x_{2i+1}+x_{2i+2})} + e^{-(x_{2i+1}+x_{2i+2})}),$$
$$e^{4i+3} = \tfrac{1}{16}\{(x_{2i+1} - x_{2i+2})(-1)\}^2 + \tfrac{1}{4}(e^{(x_{2i+1}-x_{2i+2})} + e^{-(x_{2i+1}-x_{2i+2})}),$$
$$e^{4i+4} = \tfrac{1}{16}\{(x_{2i+1} - x_{2i+2})(-1)\}^2 - \tfrac{1}{4}(e^{(x_{2i+1}-x_{2i+2})} + e^{-(x_{2i+1}-x_{2i+2})}),$$

and totally, we have forty eight mutually orthogonal conformal vectors $\{e^1, \ldots, e^{48}\}$ with central charge $\tfrac{1}{2}$, (see [DMZ], [M1].) Set $P = \langle \tau_{e^i} : i = 1, \ldots, 48 \rangle$ and $T = \langle e^1 \rangle \otimes \cdots \otimes \langle e^{48} \rangle \cong \otimes L(\tfrac{1}{2}, 0)$.

It is not difficult to check the following correspondences for $\begin{pmatrix} e^1 & e^2 \\ e^3 & e^4 \end{pmatrix}$, see (4.3) in

[M3]:

$$(x_1 + x_2)(-1) \quad \rightarrow \quad \begin{pmatrix} \frac{1}{2} & \frac{1}{2} \\ 0 & 0 \end{pmatrix}, \qquad (x_1 - x_2)(-1) \quad \rightarrow \quad \begin{pmatrix} 0 & 0 \\ \frac{1}{2} & \frac{1}{2} \end{pmatrix},$$

$$e^{x_1} + e^{-x_1} + e^{x_2} + e^{-x_2} \quad \rightarrow \quad \begin{pmatrix} \frac{1}{2} & 0 \\ \frac{1}{2} & 0 \end{pmatrix}, \qquad e^{x_1} + e^{-x_1} - e^{-x_2} - e^{-x_2} \quad \rightarrow \quad \begin{pmatrix} 0 & \frac{1}{2} \\ 0 & \frac{1}{2} \end{pmatrix},$$

$$e^{(x_1+x_2)/2} + e^{-(x_1+x_2)/2} \quad \rightarrow \quad \begin{pmatrix} \frac{1}{2} & 0 \\ 0 & 0 \end{pmatrix}, \qquad e^{(x_1-x_2)/2} + e^{(-x_1+x_2)/2} \quad \rightarrow \quad \begin{pmatrix} 0 & 0 \\ \frac{1}{2} & 0 \end{pmatrix},$$

$$e^{(x_1+x_2)/2} - e^{-(x_1+x_2)/2} \quad \rightarrow \quad \begin{pmatrix} 0 & \frac{1}{2} \\ 0 & 0 \end{pmatrix}, \qquad e^{(x_1-x_2)/2} - e^{(-x_1+x_2)/2} \quad \rightarrow \quad \begin{pmatrix} 0 & 0 \\ 0 & \frac{1}{2} \end{pmatrix},$$

$$e^{x_1}/2 \quad \rightarrow \quad \begin{pmatrix} \frac{1}{16} & \frac{1}{16} \\ \frac{1}{16} & \frac{1}{16} \end{pmatrix}, \qquad e^{x_2}/2 \quad \rightarrow \quad \begin{pmatrix} \frac{1}{16} & \frac{1}{16} \\ \frac{1}{16} & \frac{1}{16} \end{pmatrix},$$

$$e^{(x_1+x_2)/4} \quad \rightarrow \quad \begin{pmatrix} \frac{1}{16} & \frac{1}{16} \\ 0 & 0 \end{pmatrix}, \qquad e^{(x_1-x_2)/4} \quad \rightarrow \quad \begin{pmatrix} 0 & 0 \\ \frac{1}{16} & \frac{1}{16} \end{pmatrix},$$

where $v \rightarrow \begin{pmatrix} h_1 & h_2 \\ h_3 & h_4 \end{pmatrix}$ means that the $\langle e^1 \rangle \otimes \cdots \otimes \langle e^4 \rangle$-subspace generated by v contains $L(\tfrac{1}{2}, h_1) \otimes \cdots \otimes L(\tfrac{1}{2}, h_4)$ as $\langle e^1 \rangle \otimes \cdots \otimes \langle e^4 \rangle$-modules.

Hence, $(x_{2k-1} \pm x_{2k})(-1)$ gives a word of type $\alpha \cup \phi_4(\alpha)$ and so $x_j(-1)x_i(-1)$ defines an element $\alpha \cup \phi_4(\alpha)$ with even set α in $U_4 = \{1, 3, 5, 7, \ldots, 47\}$. Moreover, e^{x_i} gives a word of type $\alpha \cup \phi_3(\alpha)$ and $e^{x_i-x_j}$ defines $\alpha \cup \phi_3(\alpha)$ for even set $\alpha \subseteq U_3 = \{1, 2, 5, 6, 9, 10, \ldots, 45, 46\}$. The properties of MGC define the other two reflections as we will see.

In V_Λ^θ, let H be the set of all highest weight rows $h = (h^1, \ldots, h^{48})$ without $\frac{1}{16}$. We will see it as binary code by replacing $h^i = \frac{1}{2}$ by 1. The fusion rules and Lemma 3.1 in [DMZ] implies that H is closed by the sum, that is, H is a linear code. We have the following proposition.

Theorem 4.1. H is isomorphic to MGC and $(V_\Lambda^\theta)^P \cong M_{\mathrm{MGC}}$.

Proof. As we showed, $\alpha \cup \phi_4(\alpha) \in H$ for any even subset $\alpha \subset U_4 = \{1, 3, 5, \ldots, 47\}$ and $\beta \cup \phi_3(\beta) \in H$ for any even subset $\beta \subset U_3 = \{1, 2, 5, 6, 9, 10, \ldots, 45, 46\}$. Here

$$\phi_4 : 2i - 1 \leftrightarrow 2i$$

and

$$\phi_3 : 4i - 3 \leftrightarrow 4i - 1 \qquad 4i - 2 \leftrightarrow 4i.$$

A hexacodeword (000000) and its even interpretation $\begin{pmatrix} * & 0 & * & 0 & 0 & 0 \\ * & 0 & * & 0 & 0 & 0 \\ * & 0 & * & 0 & 0 & 0 \\ * & 0 & * & 0 & 0 & 0 \end{pmatrix}$ gives

$(1, 5, 9, 13) \in H$. By adding $(5, 7, 13, 15) = \phi_3(5, 13) \cup (5, 13)$, we have $(1, 7, 9, 15) = \phi_2(1, 9) \cup (1, 9)$. Similarly, we have $\alpha \cap \phi_2(\alpha) \in H$ for any even set $\alpha \subset U_2 = \{16n + 1, 2, 3, 4, 9, 10, 11, 12 : n = 0, 1, 2\}$, where

$$\phi_2 : 8n + 1 \leftrightarrow 8n + 7, \quad 8n + 2 \leftrightarrow 8n + 8, \quad 8n + 3 \leftrightarrow 8n + 5, \quad 8n + 4 \leftrightarrow 8n + 6.$$

A hexacodeword (111100) and its even interpretation $\begin{pmatrix} 0 & 0 & * & * & 0 & 0 \\ 0 & 0 & * & * & 0 & 0 \\ * & * & 0 & 0 & 0 & 0 \\ * & * & 0 & 0 & 0 & 0 \end{pmatrix}$ gives

$(5, 13, 17, 25) \in H$. By adding $(13, 14, 25, 26) = \phi_4(13, 25) \cup (13, 25)$, we have $\phi_1(5, 17) \cup (5, 17)$. Similarly, we have $\beta \cap \phi_1(\beta) \in H$ for any even set $\beta \subset U_1 = \{16n + 1, 2, 3, 4, 5, 6, 7, 8 : n = 0, 1, 2\}$, where

$$\phi_1 : 16n + 2i - 1 \leftrightarrow 16n + 9 + 2i - 1, \quad 16n + 2i \leftrightarrow 16 + 7 + 2i.$$

Since $(V_\Lambda^\theta)^P$ is generated by θ-invariants in

$$\langle x_i(-1)x_j(-1),\ e^{x_i},\ e^{(x_{2i-1} \pm x_{2i})} : i, j = 1, \ldots, 24 \rangle,$$

we have already obtained all elements of H. This completes the proof of Theorem 4.1.

We next study the codewords $\tilde{h}$ of $\frac{1}{16}$-positions of elements $h = (h^i)$ of $\mathrm{hwr}(T, V^\natural)$.

Theorem 4.2. *The set of codewords of $\frac{1}{16}$-positions of elements of $\mathrm{hwr}(T, V_\Lambda^\theta)$ is equal to $\langle U_1, U_2, U_3, \mathrm{Bl}, \mathrm{Gr}, \mathrm{Rd} \rangle$.*

Proof. By the fusion rule of Ising models, we can ignore the action of $(V_\Lambda^\theta)^P$ since the action of $(V_\Lambda^\theta)^P$ does not change the positions of $L(\frac{1}{2}, \frac{1}{16})$. Set

$$\Lambda_1 = \sum_{c \in C_{24}} \mathbb{Z} \frac{x_c}{2} + \sum \mathbb{Z}(x_1 - x_i).$$

Then $V^\theta_{\Lambda_1}$ is a vertex operator subalgebra of V^θ_Λ and we have the decomposition

$$(V_\Lambda)^\theta = (V_{\Lambda_1})^\theta \oplus V^\theta_{\Lambda_1+t}$$

of V^θ_Λ into the direct sum of $V^\theta_{\Lambda_1}$-modules, where $t = \frac{x\Omega}{4} - x_1$ and $V^\theta_{\Lambda_1+t}$ is an irreducible $V^\theta_{\Lambda_1}$-module. So, the set of codewords $\tilde{h}$ of positions of $\frac{1}{16}$ in elements of $\mathrm{hwr}(T, V^\theta_\Lambda)$ is the code generated by that of $\mathrm{hwr}(T, V^\theta_{\Lambda_1})$ and that of $\mathrm{hwr}(T, V_{\Lambda_1+t})$. Clearly, the positions of $L(\frac{1}{2}, \frac{1}{16})$-components in the T-modules generated by $e^{\frac{x\Omega}{4}-x_1} + e^{-\frac{x\Omega}{4}+x_1}$ is $U_3 = \{16n + 1, 2, 5, 6, 9, 10, 13, 14 : n = 0, 1, 2\}$. A hexacodeword $(w\bar{w}0101)$

and its even interpretation $\begin{pmatrix} * & * & 0 & * & 0 & * \\ 0 & 0 & 0 & * & 0 & * \\ * & 0 & 0 & 0 & 0 & 0 \\ 0 & * & 0 & 0 & 0 & 0 \end{pmatrix}$ gives $\mathrm{Bl} = \{1, 2, \ldots, 16\}$. Sim-

ilarly, we have $\mathrm{Gr} = \{17, 18, \ldots, 32\}$ and $\mathrm{Rd} = \{33, 34, \ldots, 48\}$. A hexacodeword

$(w1w1w1)$ and its even interpretation $\begin{pmatrix} * & * & * & * & * & * \\ 0 & * & 0 & * & 0 & * \\ * & 0 & * & 0 & * & 0 \\ 0 & 0 & 0 & 0 & 0 & 0 \end{pmatrix}$ gives $U_1 = \{16n +$

$1, 2, 3, 4, 5, 6, 7, 8 : n = 0, 1, 2\}$. A hexacodeword $(w\bar{w}w\bar{w}w\bar{w})$ and its odd interpreta-

tion $\begin{pmatrix} 0 & 0 & 0 & 0 & 0 & * \\ 0 & 0 & 0 & 0 & 0 & * \\ * & 0 & * & 0 & * & * \\ 0 & * & 0 & * & 0 & 0 \end{pmatrix}$ gives $U_2 = \{16n + 1, 2, 3, 4, 9, 10, 11, 12 : n = 0, 1, 2\}$.

Theorem 4.3. *The set B of codewords of $\frac{1}{16}$-positions of elements of $\mathrm{hwr}(T, V^\natural)$ is equal the orthogonal to $\langle U_1, U_2, U_3, U_4, \mathrm{Bl}, \mathrm{Gr}, \mathrm{Rd}\rangle$.*

Proof. For any $\chi \in \mathrm{Irr}(P)$, since M_{MGC} and $(V^\natural)_\chi$ are both inside of $V^\natural$, the vertex operators of all elements of M_{MGC} on $(V^\natural)_\chi$ have all integer powers. Hence, by Proposition 5.3 in **[M3]**, the words of $(V^\natural)_\chi$ of $\frac{1}{16}$-positions are all orthogonal to MGC. Namely, it is contained in $\mathrm{MGC}^\perp = \langle U_1, U_2, U_3, U_4, \mathrm{Bl}, \mathrm{Gr}, \mathrm{Rd}\rangle$. By the above theorem, B contains $\langle U_1, U_2, U_3, \mathrm{Bl}, \mathrm{Gr}, \mathrm{Rd}\rangle$. Since $V^\natural \neq (V_\Lambda)^\theta$, B should be greater than $\langle U_1, U_2, U_3, \mathrm{Bl}, \mathrm{Gr}, \mathrm{Rd}\rangle$ and so $B = \langle U_1, U_2, U_3, U_4, \mathrm{Bl}, \mathrm{Gr}, \mathrm{Rd}\rangle$.

Section 5. Automorphism group

In this section, we will show a new way to prove that the full automorphism group $V^\natural$ is finite. This result is true in the more general situation, that is, we assume the following

Hypotheses.
(1) $V = \sum_{i=0}^\infty V_i$ is a VOA over the real field $\mathbb{R}$.
(2) $\dim V_0 = 1$.
(3) $V_1 = 0$.
(4) V has a positive invariant bilinear form $\langle\,,\,\rangle$.

(5) The Virasoro element is a sum of mutually orthogonal conformal vectors with central charge $\frac{1}{2}$.

Under the above hypotheses, we have the following result.

Theorem 5.1. *Let* e, f *be two distinct conformal vectors with central charge* $\frac{1}{2}$, *that is,*

$$\langle e, e \rangle = \langle f, f \rangle = \frac{1}{4}.$$

Then we have

$$\langle e, f \rangle \leq \frac{1}{12}$$

and so

$$\langle e - f, e - f \rangle \geq \frac{1}{3}.$$

In particular, there are only finitely many conformal vectors with central charge $\frac{1}{2}$.

Proof. By the product $ab = a_1 b$ and the inner product $\langle a, b \rangle \mathbf{1} = a_3 b$, V_2 becomes an algebra with an invariant bilinear form $(,)$, which is called Griess algebra. Let $\mathbb{R}e \oplus \mathbb{R}e^{\perp}$ be the decomposition of V_2. For another conformal vector f, we have $r \in \mathbb{C}$ and $w \in \mathbb{R}e^{\perp}$ such that

$$f = re + w.$$

Since $\langle ew, e \rangle = \langle w, e^2 \rangle = \langle w, 2e \rangle = 0$, we have $ew \in \mathbb{R}e^{\perp}$. Moreover, we have

$$2re + 2w = 2f = f^2 = \{r^2 2e + w_e^2\} + \{(w^2 - w_e^2) + 2rew\},$$

where w_e^2 denotes the first entry of w^2 in the decomposition $\mathbb{R}e \oplus \mathbb{R}e^{\perp}$. Hence, we have $r^2/2 + \langle e, w_e^2 \rangle = \langle e, 2r^2 e + w_e^2 \rangle = \langle e, f^2 \rangle = \langle e, 2f \rangle = \langle e, 2re \rangle = r/2$.
Hence, $\langle e, w_e^2 \rangle = r(1 - r)/2$. On the other hand, we have

$$\frac{1}{4} = \langle f, f \rangle = r^2 \frac{1}{4} + \langle w, w \rangle,$$

and so $\langle w, w \rangle = \frac{1}{4}(1 - r^2)$. Moreover, since $V_1 = 0$ and dim $V_0 = 1$, if W is an irreducible $\langle e \rangle$-module with $\langle e \rangle \cap W = 0$, then $W \subseteq \oplus_{n=2}^{\infty} V_n$. Since W is isomorphic to $L(\frac{1}{2}, h)$, W is generated by $\{v, e_n v : n \leq 0\}$ and so $W \cap V_2$ is contained in the highest weight space of W. Therefore, the eigenvalues of e on $\mathbb{R}e^{\perp}$ are 0, $\frac{1}{2}$, or $\frac{1}{16}$. Hence, we have

$$r/2 - r^2/2 = \langle e, w_e^2 \rangle = \langle e, w^2 \rangle = \langle we, w \rangle \leq \frac{1}{2} \langle w, w \rangle = \frac{1}{8}(1 - r^2)$$

and so we have $3r^2 - 4r + 1 \geq 0$. This implies $r \geq 1$ or $r \leq \frac{1}{3}$. If $r \geq 1$, then it contradicts to $\langle w, w \rangle \geq 0$. Hence we have $r \leq \frac{1}{3}$ and so $\langle e, f \rangle \leq \frac{1}{12}$. In particular, we have $\langle e - f, e - f \rangle \geq \frac{1}{3}$ and so there are only finitely many conformal vectors with central charge $\frac{1}{2}$ since $(V_2 - \{0\})/(\mathbb{R} - \{0\})$ is a compact space.

Theorem 5.2. *If V satisfies Hypotheses, then* Aut(V) *is finite.*

Proof. Clearly Aut(V) permutes all conformal vectors with central charge $\frac{1}{2}$. By Theorem 5.1, the number of conformal vectors with central charge $\frac{1}{2}$ is finite and so a finite index subgroup G of Aut(V) fixes all conformal vectors. In particular, G fixes the set of coordinate conformal vectors point-wise so that G fixes all components of V into the direct sum of $\otimes L\left(\frac{1}{2}, h^i\right)$. Since the Virasoro element is a sum of mutually orthogonal conformal vectors with central charge $\frac{1}{2}$, V is a finite direct sum of $\otimes L\left(\frac{1}{2}, h^i\right)$. Let $g \in G$. Then g is an identity on T and acts as scalar multiples on $L\left(\frac{1}{2}, \frac{1}{2}\right)$ and $L\left(\frac{1}{2}, \frac{1}{16}\right)$. Using the fusion rules of the Ising model, we can prove that $g^4 = 1$. Then G is compact. Thus G is finite.

Section 6. Multiplicity of irreducible T-modules

Now we want to show the list of the multiplicity of irreducible $\otimes_{i=1}^{48} L\left(\frac{1}{2}, 0\right)$-module L_h with highest weight row $h = (h^i)$. Let L be a T-submodule of $V^\natural$ with word $\tilde{h}$, then $\tilde{h}$ is in MGC$^\perp$ by Theorem 4.3. In particular, the weight $|\tilde{h}|$ is 0, 16, 32, 48, or 24. Set $K_{\tilde{h}} = \{\alpha \in \text{MGC} : \alpha \subseteq \tilde{h}\}$. The important thing is that $K_{\tilde{h}}$ contains a direct sum $E_{\tilde{h}}$ of Hamming codes such that $\tilde{h}$ is a code of $E_{\tilde{h}}$ with the maximum weight. For example, if $|\tilde{h}| = 24$, then $K_{\tilde{h}}$ contains a direct product $H_8 \oplus H_8 \oplus H_8$ of 3 Hamming codes. By Corollary 5.2 in [**M3**], we have:

Lemma 6.1. *Let* $h = (h^i) \in \text{hwr}(T, V^\natural)$ *and let* $\tilde{h}$ *denote the word of* $\frac{1}{16}$*-positions of* h. *Then the multiplicity of* T*-irreducible submodule with highest weight row* h *is*

$$\frac{|K_{\tilde{h}}|}{2^{|\tilde{h}|/2}}.$$

By Lemma 6.1, the investigation in §4, and the fact that the sums $\sum h^i$ are integers, we have the following theorem.

Main Theorem. *If* $|\tilde{h}| = 0$, *then* mult(h) $= 1$.

$$h \in \text{hwr}(T, V^\natural) \Leftrightarrow h = \frac{1}{2}\alpha : \alpha \in \text{MGC}.$$

Here we view a word α as in $\mathbb{Z}^{48}$ and mult(h) denotes the multplicity of irreducible T-module with highest weight row h.

If $|\tilde{h}| = 24$, then $|K_{\tilde{h}}| = 2^{11+5+2}$ and so mult(h) $= 2^6$.

$$h \in \text{hwr}(T, V^\natural) \Leftrightarrow h = \frac{1}{16}\tilde{h} + \frac{1}{2}r_i + \frac{1}{2}\alpha : \alpha \in \text{MGC}$$

where $r_i = (0, \ldots, 0, 1, 0, \ldots 0)$ and $i \notin \tilde{h}$.
If $|\tilde{h}| = 16$, then $|K_{\tilde{h}}| = 2^{7+3+1}$ and so mult(h) $= 2^3$.

$\tilde{h}$ is one of 3 colored blocks, {Bl, Gr, Rd} and

$$h \in \mathrm{hwr}(T, V^{\natural}) \Leftrightarrow h = \frac{1}{16}\tilde{h} + \frac{1}{2}r_i + \frac{1}{2}r_j + \frac{1}{2}\alpha : \alpha \in \mathrm{MGC}.$$

where $i, j \notin S$, i and j are from different colored blocks.
If $|\tilde{h}| = 32$, then $|K_{\tilde{h}}| = 2^{15+7+3+1}$ and so $\mathrm{mult}(h) = 2^{10}$.
$\tilde{h}$ is a union of two colored blocks and

$$h \in \mathrm{hwr}(T, V^{\natural}) \Leftrightarrow h = \frac{1}{16}\tilde{h} + \frac{1}{2}\alpha : \alpha \in \mathrm{MGC}.$$

If $|\tilde{h}| = 48$, then $|K_{\tilde{h}}| = 2^{23+11+5+2}$ and so $\mathrm{mult}(h) = 2^{17}$.
$\tilde{h}$ is the set of all coordinates and

$$h \in \mathrm{hwr}(T, V^{\natural}) \Leftrightarrow h = \frac{1}{16}\tilde{h}.$$

In the above expressions, the rule of sum is given by the fusion rules $0 + \frac{1}{2} = \frac{1}{2}$, $0 + \frac{1}{16} = \frac{1}{16}$, $\frac{1}{2} + \frac{1}{2} = 0$, $\frac{1}{2} + \frac{1}{16} = \frac{1}{16}$ of the irreducible modules of $L(\frac{1}{2}, 0)$.

Acknowledgement. The author would like to thank the Mathematical Department of Ohio State University for its support.

References

[B]   R. E. Borcherds, Vertex algebras, Kac–Moody algebras, and the Monster, Proc. Nat. Acad. Sci. U.S.A. **83** (1986), 3068–3071.

[CS]   J. H. Conway , N. J. A. Sloane, Sphere packings, lattices and groups, Springer-Verlag, New York 1988.

[DM]   C. Dong, G. Mason, On quantum Galois theory, Duke Math. J. **86** (1997), 305–321.

[DMZ]   C. Dong, G. Mason, Y. Zhu, Discrete series of the Virasoro algebra and the moonshine module, in: Algebraic groups and their generalizations, Part 2, Proc. Sympos. Pure. Math. **56** II (1994), 295–316.

[FLM]   I. B. Frenkel, J. Lepowsky, A. Meurman, Vertex operator algebras and the monster, Pure Appl.Math. **134**, Academic Press, Boston 1988.

[FZ]   I. Frenkel , Y. Zhu, Vertex operator algebras associated to representations of affine and Virasoro algebras, Duke Math. J. **66** (1992), 123–168.

[FQS]   D. Friedan, Z. Qiu, S. Shenker, Conformal invariance, unitarity and two dimensional critical exponents, in: Vertex Operators in mathematics and physics, Math. Sci. Res. Inst. Publ. **3**, Springer-Verlag, New York 1985, 419–449.

[M1]   M. Miyamoto, Griess algebras and conformal vectors in vertex operator algebras, J. Algebra **179** (1996), 523–548.

[M2]   M. Miyamoto, Binary codes and vertex operator (super)algebras, J. Algebra **181** (1996), 207–222.

[M3]   M. Miyamoto, Representation theory of code VOAs and construction of VOAs, preprint.

[Mu]   D. E. Muller, Application of Boolean algebra to switching circuit design and to error detection, IEEE Trans. Comput. **3** (1954), 6–12.

[R]   I. S. Reed, A class of multiple-error-correcting codes and the decoding scheme, IEEE Trans. Inform. Theory **4** (1954), 38–49

Institute of Mathematics
University of Tsukuba
Tsukuba-Shi Ibaraki, 305
Japan
E-mail: masahiko@math.tsukuba.ac.jp

Netting the Monster

S. P. Norton

Abstract. We define nets, examine their properties, and summarise the progress so far made with their enumeration.

Introduction

Sections 1–3 describe the history of nets and give basic definitions, conventions and results. The current state of the enumeration of nets is outlined in Sections 4–7, and Section 8 describes some open questions related to nets.

Section 1. The name of the game

We start with two partial definitions.

Definition 1. Let G be a group, and let a, b and c be three involutions of G. Then a *quilt* is a finite geometry associated with the triple (a, b, c) in a manner to be described later. Alternatively, a quilt may be associated with any pair of group elements (g, h). G is called the *underlying group* of the quilt, while $\langle a, b, c \rangle$ is the *quilt group*.

This term "quilt" comes from [**Hsu1; Hsu2; Hsu3**]. To relate the two versions of quilt to each other define $g = ab$ and $h = bc$. In the next definition (and elsewhere in this paper) we denote the Fischer–Griess Monster by $\mathbb{M}$ and use [**ATLAS**] notation for conjugacy classes.

Definition 2. Let a, b and c be three $2A$-involutions of $\mathbb{M}$. Then a *net* is a finite geometry associated with the triple (a, b, c) in the manner applicable to Definition 1. $\langle a, b, c \rangle$ is the *net group*.

Note. The concept of "net" was referred to in [**Gem; Monalg; AM1**]. In the first of these no name was given, but in the other two the word "football", which "net" is intended to supersede, was used.

Definitions 1 and 2 seem remarkably similar. So why use a separate name? To answer this question one must look at the motivation behind the study of nets.

Any quilt gives rise to a subgroup of the modular group Γ (up to conjugation) in a way that will be defined precisely later. We now refer to the Generalized Moonshine Conjecture (GMC) introduced in [**Gem**]. (Note: Condition 2 of the GMC should be changed to assert that for fixed g and any rational number u, the coefficient of q^u in the function associated with (g, h) is, up to a factor of a root of unity, the value of a *projective* character of $C_M(g)$ on h.)

The GMC postulates a correspondence between pairs of commuting elements (g, h) in the Monster and certain modular functions that are Hauptmoduls for genus zero groups. Specifically, if Δ_0 is the subgroup of Γ associated with the quilt corresponding to (g, h), then the Hauptmodul will be fixed by Δ_0, possibly up to multiplication by a root of unity. Call the subgroup of Δ_0 that fixes the Hauptmodul Δ_1. Δ_1 may have genus zero, but it is also possible that the fixing group of the Hauptmodul is bigger then Δ_1, containing elements of $PGL_2(\mathbb{Q})$ that aren't in Γ.

In **[Gem]** the question was asked what happened if g and h did not commute. One property of the Monster is that elements of class $2A$ are *6-transpositions*, i.e., the product of any two of them has order at most 6. It was observed that because of this, non-commuting pairs (g, h) would also lead to groups of genus zero (or possibly 1) provided they were associated as above with a triple of transpositions. This led to the question of whether there is a link between the two occurrences of the "genus zero" motif.

The answer to this question is still unknown, but this paper reveals enough links between nets and Moonshine to suggest that it isn't all a coincidence, and to explain our use of a separate name for quilts linked with the Monster.

The main difference between Generalized Moonshine and nets is that the modular group associated with any *commuting* pair (g, h) is necessarily a congruence group, i.e., for some n it contains $\Gamma(n)$, the kernel of the natural map from Γ to $SL_2(\mathbb{Z}_n)/\{\pm 1\}$. (In fact, n can be taken as the exponent of $\langle g, h \rangle$.) This means that the coefficients of the Fourier expansion of the associated Hauptmodul will be cyclotomic **[Sh**, Section 6.2], as is necessary to satisfy Condition 2 of the GMC (partly stated above). On the other hand, if g does *not* commute with h, a character of $C_M(g)$ doesn't have a value on h. To confuse things even more, it is sometimes possible to assign a meaning to such a character value if h *normalizes* g, and in some cases this appears to lead to a Hauptmodul — but the subgroup of Γ that fixes the Hauptmodul is *not* that associated with the quilt corresponding to (g, h). We shall return to this in the last section (Question 5).

Section 2. The elements of the game

In this section we use Generalized Moonshine to motivate the completion of Definitions 1 and 2. See **[Hsu1; Hsu2; Hsu3]** for further details, also for proofs where not given in Section 3.

Let $\tilde{\Gamma}$ be the 3-string braid group, which has presentations $\langle x, y \mid xyx = yxy \rangle$ and $\langle s, t \mid s^3 = t^2 \rangle$. If we define $s = xy$ and $t = xyx$ we see that the two presentations are equivalent. The element $z = (xy)^3 = s^3 = t^2$ is central; quotienting out z^2 gives rise to $SL_2(\mathbb{Z})$, in which we may take $x = \begin{pmatrix} 1 & -1 \\ 0 & 1 \end{pmatrix}$, $y = \begin{pmatrix} 1 & 0 \\ 1 & 1 \end{pmatrix}$, $s = \begin{pmatrix} 0 & 1 \\ -1 & 1 \end{pmatrix}$, $t = \begin{pmatrix} 0 & 1 \\ 1 & 0 \end{pmatrix}$; and quotienting out z gives rise to the modular group Γ.

In [**Gem**] an action of the $SL_2(\mathbb{Z})$ on commuting pairs (g, h) is defined, namely that $\begin{pmatrix} \alpha & \beta \\ \gamma & \delta \end{pmatrix}$ takes (g, h) to $(g^\alpha h^\gamma, g^\beta h^\delta)$. If we specialize we find that x and y take (g, h) to $(g, g^{-1}h)$ and (gh, h) respectively. We extend these definitions to the non-commuting case. As operations, x and y satisfy $xyx = yxy$, so this certainly gives a representation of the braid group (in which s, t and z can be defined as above). Furthermore, x and y, and hence the entire braid group, preserves $h^{-1}ghg^{-1}$; and the action of z in the commuting case is to invert g and h.

We are now ready to define the geometry of a quilt in the (g, h) formulation. There are three types of element: vertices, edges and faces, which correspond to the orbits of $\langle s \rangle$, $\langle t \rangle$ and $\langle x, z \rangle$ on pairs (g, h). Two of these (of whatever kind) are incident if the relevant orbits intersect. As z is in all three of the above groups, the orbits of $\langle z \rangle$ on pairs correspond to the flags of the geometry (i.e., mutually incident sets containing one of each kind of element). In quilt terminology flags and faces are called *seams* and *patches*.

We now pass to the (a, b, c) formulation mentioned in Section 1 and put $g = ab$ and $h = bc$. To consider how the braiding operations should be extended from (g, h) to (a, b, c), we temporarily consider g and h as generators of a *free* group F_2, so that the braiding operations are actually outer automorphisms. Then the involutions a, b and c defined above will generate the free product Z_2^{*3} of three cyclic groups of order 2. In this group b is the unique involution inverting both g and h, so any autmorphism of F_2 extends in at most one way to Z_2^{*3}. (In fact the outer automorphism groups of Z_2^{*3} and F_2 are *exactly* $\tilde{\Gamma}$ and its double cover respectively.)

We therefore find that x, y, s and t must take the triple (a, b, c) to (b, a^b, c), (a, c, b^c), (b, c, a^{bc}) and (c, b^c, a^{bc}) respectively; that the action of z is to conjugate each of $\{a, b, c\}$ by abc; and that all the above operations preserve abc. Incidentally, in this formulation the correspondence with braiding is intuitive; one can think of three strings labelled a, b and c, with operation x passing a under b and y passing b under c. See [**Bir**, Theorem 1.9].

The action of $\tilde{\Gamma}$ divides the set of all pairs (g, h) or triples (a, b, c) into orbits. Each orbit in turn divides into flags, vertices, edges and faces in such a way as to yield a connected geometry. It is these geometries that we call quilts or nets. We use the term "the quilt (or net) (a, b, c)" to mean the quilt or net corresponding to the orbit of $\tilde{\Gamma}$ containing (a, b, c) itself.

We immerse a quilt in space in such a way that "going round" a vertex, edge or face clockwise corresponds to applying s^{-1}, t or x^{-1} to the triple (a, b, c). The product of the above three operations, in the order stated, is the identity (as it has to be).

Section 3. The terms of the game

We have now defined the concepts of quilt, net, flag (or seam), vertex, edge and face (or patch). Here are definitions for the other net theoretic concepts. Some but not all apply more generally to quilts. It may easily be seen that they are all well defined.

The *class* of the net (a, b, c) is the conjugacy class of abc in $\mathbb{M}$. The class of the face of this net corresponding to this triple is the conjugacy class of ab. The *order* of a net or face is the order of abc or ab respectively.

As the index of $\langle z \rangle$ in $\langle s \rangle$ is 3, a vertex will normally belong to 3 flags. Sometimes, however, it will belong to only one, in which case we say it *collapses*. Similarly an edge collapses if it belongs to 1 rather than 2 flags, and a face collapses if the number of flags it belongs to, its *collapsed order*, is less than its order. In general faces of collapsed order n are called n-gons with the usual specializations (triangles if $n = 3$ etc.).

The smallest positive integer n such that $(abc)^n$ commutes with each of $\{a, b, c\}$ is called the *modulus*. A net is *faithful* if its modulus equals its order.

The *defect* of a vertex, edge or face is its contribution to the genus of the net. It is 0 for an uncollapsed vertex or edge, 4 for a collapsed vertex, 3 for a collapsed edge and $6 - o$ for a face of collapsed order o. If $\mathcal{G}$ is the genus of a net, then Euler's formula shows that the total defect of all its elements is $12(1 - \mathcal{G})$.

By the 6-transposition property, any face of a net has order at most 6. This implies that the defect is non-negative, so the genus is at most 1. A net of genus zero is called a *netball*; a net of genus 1 is called a *honeycomb* as all faces must be hexagons (and there can be no collapses of any kind).

The *exponent* of a vertex, edge or face is essentially the power of abc by which one conjugates when going all the way round it, i.e., when applying to (a, b, c) the operation s^{-1}, t or x^{-1} repeatedly until one first returns to the original flag. This defines an integer modulo m, where m is the modulus of the net. To facilitate the counting arguments of Section 4, we choose the following representatives of the relevant conjugacy classes. The exponent of a vertex is -1 unless it collapses, in which case it is $(m - 1)/3$ or $(2m - 1)/3$, whichever is an integer. The exponent of an edge is 1 if it doesn't collapse and $(m + 1)/2$ if it does. The exponent of a face is 0 if it doesn't collapse and a positive integer between 0 and m if it does.

The *exponent* of a net is the sum of the exponents of all its elements. As any net has more edges than vertices, its exponent is always positive, and it can be seen by homology arguments that it is always a multiple of the modulus. A net is *small* if it is a netball and its exponent equals its modulus, otherwise it is *large*. The significance of these concepts will appear in the next section. A conjugacy class is *perfect* if all nets of that class are small.

The *net rotation group* is the set of orientation preserving automorphisms of the geometry realised by conjugation by elements of $\mathbb{M}$. The geometry may have other automorphisms that correspond to outer automorphisms of the net group not realised in $\mathbb{M}$, or that do not correspond to any group action at all. *Symmetric nets* also have orientation reversing automorphisms, which will invert abc; the *net reflection group* is obtained by adjoining these.

$\mathbb{M}$ is the automorphism group of a 196883-dimensional (non-associative) algebra known as the *Griess algebra*. In [**Con**] Conway constructed a related 196884-dimensional algebra which may be called the *Griess–Conway algebra*. This construction is summarised in [**ATLAS**]. Further properties of the algebra are developed in [**Monalg**] (in which all inner products are doubled to eliminate fractions); in particular, it was stated there that if A, B and C are certain vectors corresponding to $2A$-elements a, b and c, then the triple product $(A - 2, B - 2, C - 2)$ is an invariant of the net. We call this invariant the *net weight*.

From a given net, we may derive other nets by taking one of its triples (a, b, c) and replacing b by another element b' of the dihedral group $\langle a, b \rangle$. If ab has order 5, then the net (a, a^b, c) (which has the same net group, but in general a different geometry) is called a *mate* of the net (a, b, c). More generally the net (a, b', c) is called an *ancestor* of the original. The latter term (with $b' = a^b$) was used in [UM] in a related context. The concept of ancestor is of little intrinsic interest, but has some utility in enumerating nets.

The *modular group* of the net (a, b, c) is the image in Γ of the subgroup of the braid group $\tilde{\Gamma}$ that fixes (a, b, c) up to conjugation by a power of abc. For a given net it is well defined up to conjugation in Γ.

Section 4. The counting game

Let $\mathcal{C}$ be a conjugacy class of $\mathbb{M}$, with typical element k. Let us suppose we wish to enumerate the nets of class $\mathcal{C}$. A lot of information can be obtained purely by counting arguments.

To begin with, the total number of triples (a, b, c) with $abc \in \mathcal{C}$ is the product of the structure constant $(2A, 2A, 2A, \mathcal{C})$ by the order of the Monster. So the *mass* of flags, i.e., the sum over all conjugacy classes of flags of the reciprocal of the centralizer order of the net group, equals the structure constant itself.

By applying such arguments within $G_n = C_{\mathbb{M}}(k^n)$ for various values of n one can work out the mass of triples that correspond to nets of various degrees of unfaithfulness; in particular, the mass of faithful nets of class $\mathcal{C}$.

The next step is to work out the mass of a particular type of element. For example, to work out the mass of faces of class $6A$ (which may either be hexagons or collapse to smaller polygons) one evaluates the structure constant $(2A, 6A, \mathcal{C})$. As a $6A$-element extends uniquely to a dihedral group generated by two transpositions it follows that the mass of $6A$-faces is equal to this structure constant. Nine different classes of face are possible: $1A$, $2A$, $2B$, $3A$, $3C$, $4A$, $4B$, $5A$ and $6A$; to count their masses one must multiply the relevant structure constants by 97239461142009186000, 13571955000, 98280, 306936, 1, 276, 1, 1 and 1 respectively.

Then one repeats these structure constant calculations within G_n. One must, however, note that there can exist $2A$-elements b and c such that bc is in G_n but neither b nor c is. In fact, it is exactly this phenomenon that leads to the collapse of faces. For each of the conjugacy classes of G_n that corresponds to one of the nine types of face, one can work out how many of the relevant extensions to a dihedral group centralize $k^{nn'}$ for every n' dividing $o(k)/n$. This in turn enables one to deduce the mass of faces of each class distributed both by degree of unfaithfulness and degree of collapse.

We now consider how to count collapsed vertices. If the triple (a, b, c) corresponds to a collapsed vertex then there must be an n such that $(abc)^n$ conjugates it to (b, c, a^{bc}), this being another of the triples corresponding to that vertex. One deduces that $(abc)^{3n-1}$ commutes with each of $\{a, b, c\}$, so that m, the modulus of the net, divides $3n - 1$. So 3 doesn't divide the modulus; let us temporarily assume that it doesn't divide the order either. Then $(abc)^n$ may be taken as the unique power of abc, say d, such that $d^3 = abc$. As d conjugates (a, b, c) to (b, c, a^{bc}), we have $a^d = b$, $b^d = c$ and so $(ad^{-1})^3 = abcd^{-3} = 1$. Conversely any a of class $2A$ and d of class $\mathcal{C}$ (or $\mathcal{C}^{-1}$ if it is this class whose elements cube to $\mathcal{C}$), such that ad^{-1}, or equivalently ad, has order 3, gives rise to a triple (a, b, c) corresponding to a collapsed vertex. So here again structure constant calculations will give the answer.

If the order of the net is divisible by 3 the situation is only a little more complicated. This time d cubes to the $3'$-part of abc and ad^{-1} cubes to the 3-part, and everything centralizes the 3-part.

The situation with collapsed edges is similar. If $(abc)^n$ conjugates (a, b, c) to (c, b^c, a^{bc}), then the modulus divides $2n - 1$ and is therefore odd; as before, we start by assuming that the order is too. Let d be the power of abc such that $d^2 = abc$. Then a belongs to class $2A$, d to $\mathcal{C}$ or $\mathcal{C}^{-1}$, and, since $(ad)^2 = ad^2a^d = a.abc.c = b$, ad belongs to $4B$, the unique square root of $2A$. We can count such configurations using structure constants. If the order of abc is even, then d squares to its $2'$-part and ad to the product of a $2A$-element with its 2-part.

The number of uncollapsed vertices is obtained from the number of flags by subtracting the number of collapsed vertices and dividing by 3. Similarly the number of uncollapsed edges is obtained from the number of flags by subtracting the number of collapsed edges and dividing by 2.

We now have a formula, in terms of structure constants within some element centralizer of the Monster, for the frequency of occurrence of every type of element. This enables us to calculate the total defect (which when divided by 12 gives the number of netballs) and the total exponent. One can therefore tell, for example, if the class is perfect; this is so if and only if every netball is small and there can be no honeycombs. Note that both these conditions are required.

All this conceals one difficulty: that character tables for many element centralizers in the Monster are not available. But there are many that are. Perhaps some computing buff may be interested in furthering the study of both GMC and net theory by cataloguing them all?

Section 5. The weighting game

As stated earlier, it was observed in [**Monalg**] that if A, B and C are the axis vectors in the Griess–Conway algebra corresponding to $2A$-elements a, b and c, then the triple product $(A - 2, B - 2, C - 2)$ is invariant under the braiding operations of $\tilde{\Gamma}$. The proof of this is simplicity itself; conjugation by a negates a 96256-dimensional subspace of the algebra and fixes its orthogonal complement; algebra multiplication by A has eigenvalue 2 on the 96256-space, so $A - 2$ annihilates it (and takes the orthogonal complement to itself); so conjugation by a followed by multiplication by $A - 2$ is equivalent to the latter; so, from the total symmetry of the triple product, $(A - 2, B - 2, C - 2) = (A^b - 2, B - 2, C - 2) = (B - 2, A^b - 2, C - 2)$ thus showing invariance under x, and clearly y is similar.

One can evaluate the weight of a net at any face. One uses the formulae of [Con], reprinted in [ATLAS; Monalg], to calculate the algebra product of $A - 2$ and $B - 2$, then takes the inner product with $C - 2$. What happens depends on the class of the face. We go through the various cases below, writing V_0 for the sum of the axis vectors corresponding to the elements of the outer half of the dihedral group $\langle a, b \rangle$, and V_i for the axis vector corresponding to $(ab)^i$ (where this exists). As stated earlier, the inner product is defined as in [Monalg], *not* as in [Con; ATLAS], to make all weights integral.

$1A$: $60(V_0, C) - 488$
$2A$: $(6V_0 - 8V_1, C) - 40$
$2B$: $24 - 2(V_0, C)$
$3A$: $(2V_0 - 3V_1, C) - 2$
$3C$: $16 - (V_0, C)$
$4A$: $(V_0 - V_1, C) + 8$
$4B$: $(V_2 - V_0, C) + 16$
$5A$: The sum of the weights of (a, b, c) and its mate (a, a^b, c) is $24 - (V_0, C)$
$6A$: $14 + (V_2 + V_3 - V_0, C)$

For classes $1A$, $2A$, $2B$ and $3C$ the net weight can be calculated solely from the environment of the face (i.e., the classes of the net and the surrounding faces). This is also true for $3A$ given details of the inner products of axis vectors of types $2A$ and $3A$ given in [Monalg] and summarised below, except for nets of class $6A$ where one also needs to know the group generated by ab and c (or some equivalent information). For classes $4B$ and $6A$ one also needs to know the conjugacy classes of $(ab)^2c$ (for both classes) and $(ab)^3c$ (for class $6A$), i.e., the classes of the ancestral nets. For $4A$ the formula can be used to link the weights of the 276 different nets corresponding to a given ab and c, while for $5A$ it links the weights of a net and its mate.

Let t be a $2A$-element with axis vector T and u a $3A$-element with axis vector U. Then if tu has class $6A$, then (T, U) is 0, 16 or 18 according as $\langle t, u \rangle$ is cyclic of order 6, $2 \times A_4$ or $3 \times S_3$. Otherwise (T, U) depends on the conjugacy class of tu as follows.

$2A$: 90; $3A$: 40; $4A$: 26; $4B$: 10; $5A$: 20; $6B$: 17; $6C$: 8; $6D$: 9; $7A$: 15; $8A$: 14; $8B$: 10; $8C$: 10; $9A$: 13; $10A$: 12; $11A$: 12; $12A$: 12; $12C$: 10; $14A$: 11; $15A$: 11.

Clearly the weight of a net is the same whichever face one evaluates it at. When classifying the nets of a given class this can be of considerable value in restricting possibilities.

Section 6. Scoring the game

At the time of writing, all nets have been classified up to order 7. The full list is in [Net2] and can be obtained by email from the author (possibly updated). Here, to save space, we just print a double page (Table 1), which shows all nets of order up to 3 except for one of the three of class $3C$. This allows us to explain the notation used.

Column 1 (repeated as Column 8) shows the class of the net, and Column 2 (repeated as Column 9) shows a reference number to distinguish nets of the same class. Column 3 shows the centralizer of the net group, which is itself shown in Column 10. Added to Column 3, in square brackets, is the net rotation group. A star is appended here when the net isn't symmetric, and a dollar sign if the structure of the net (as defined in column 11) has symmetries that aren't realised in the Monster. (In fact neither of these phenomena holds for the nets in Table 1.) Column 4 shows the net weight.

Columns 5–7 show sample Monster elements that generate the net in question. (Note: to avoid confusion with Column 11, we use bold face for the elements **a**, **b** and **c**.) We use the *projective plane* formulation introduced in [Y555] and developed in [PTM; CTM]. In this notation there are 26 involutions, corresponding to the points and lines of a projective plane of order 3, which generate the Bimonster $M \wr 2$. These are denoted by a, f, and $z_i, a_i, \ldots, g_i$ for $i = 1, 2, 3$, where the "even" letters z, b, d, f correspond to points and the "odd" letters a, c, e, g to lines. These involutions commute unless one is a point and the other an incident line, in which case their product has order 3. To clarify which involutions commute with which in the above notation, we describe the incidence graph of the projective plane we are considering.

Draw a cube on which (z, a, b, c, d, e, f, g) is a Hamiltonian path linking opposite vertices. There is essentially a unique way to do this. Then the only joins between nodes with suffices occur when the letters correspond to *adjacent* vertices on the cube. If the letters are consecutive in the alphabet (where z is considered to come just before a) or if one of them is b or e, then nodes with the *same* suffix are joined; in other cases nodes are joined when they have *different* suffices. a and f are joined to each other and also to the b_i and e_i respectively.

It is easily checked that this is a regular bipartite graph of valence 4 and girth 6 on 26 nodes, which implies that it is a projective plane.

Within this plane the nodes a, b_i, c_i, d_i, e_i and f_i form a Y_{555} diagram. If we restrict to $i = 1$ or 2, we get an A_{11} diagram corresponding to the symmetric group S_{12} acting, say, on the numbers 1 to 12. We abbreviate 10, 11, 12 to X, E, T and define $f_1 = (12), e_1 = (23), \ldots, f_2 = (ET)$.

The other notations used in Table 1 are as follows. If p and l are a non-incident point and line, then $|pl|$ denotes the unique projective plane automorphism of order 2 that fixes all the lines containing p and all the points on l. Note that $|b_3 e_3| = (1T)(2E)(3X)(49)(58)(67)$. Leech lattice vectors refer to elements of the O_2-subgroup of the $2B$-centralizer $2^{1+24}.Co_1$; any minimal vector gives rise to two $2A$-elements and it doesn't matter which we use. And permutations on the letters $A, \ldots, G$ refer to elements of the $3S_7$ that centralizes a dihedral group D_{14} containing a $7A$.

We now come to Column 11. The faces of the net are labelled $a, b, c, \ldots$ as far as necessary. For each face we give its class followed by the labels of the surrounding faces in (say) clockwise order, and the classes of $(ab)^2c$ (for faces of order ≥ 4) and $(ab)^3c$ (for faces of order 6) with the reference numbers of corresponding ancestors or mates. For orders up to 7 the reference numbers of [**Net2**] are used (which are the same as those used here for orders up to 3); for higher orders question marks are used since no reference numbers have been assigned. For honeycombs the notation is similar except that "hc$(n \times n')$" is used at the beginning, where n and n' are the dimensions of the honeycomb (i.e., $n \times n'$ is the structure of the full translation group of the net geometry, including elements not in the net rotation group), and letters refer not to faces but to *orbits* of the net rotation group on faces.

Section 7. The results of the game

As stated above, all nets have been classified up to order 7. Some of the computational background is given in [**CIM**]. In addition, nets of various orders divisible by a large prime (≥ 17) have been *counted*, i.e., the methods of Section 4 have been applied; to be specific, this has been done for all orders except 17, 34 and multiples of 23. Calculations have also been made in some isolated cases. The results show the following:

1. Most nets are small. In particular all nets corresponding to classes with Monstrous Moonshine [**MM**] symbol $p+$ (p prime) are small. This is also true in the other "large prime" cases except for order 38 when large nets exist. For order up to 3 the only large nets are $2B7$, $3B1$ (a honeycomb) and $3C3$ (the net omitted from Table 1).

2. Honeycombs are only known to exist for classes with Moonshine symbol $n-$ or $n|h-$. For order up to 7 the known honeycombs occur just for the $n-$'s with $n > 2$. Indeed for $6-$ there are no less than 6 honeycombs, one of which is non-symmetric and therefore occurs in two forms related by a reflection. Another known honeycomb, where the net group is the whole Monster, occurs for class $13-$.

If the proportion of hexagons is exceptionally large this is a signal that honeycombs are likely to exist. For order 8, $8|4+$, a potential counterexample to the above observation, is a borderline case, but the structure constants lead one to expect honeycombs for $8|2-$, $8-$ and $8|4-$. For larger orders likely looking classes are $9-$, $10-$, $12-$, $15|3-$, $16-$, $18-$, $21+3$, $24|6+$, $24|4+2$ (the last three would also be counterexamples) and $24|12-$.

The observation can be "explained" as follows. Assume that the net rotation group is as large as possible, i.e., that it has a subgroup of index 6 consisting of translations and transitive on the faces. Then the Monster contains elements conjugation by which translates the honeycomb by the width of a hexagon in any of 6 directions. Take two such directions at an angle of 60°. Then it can be shown that the commutator of the Monster elements that realise the corresponding translations can be taken as abc (or its inverse), and that they both commute with this element. This means that abc is the centre of an extraspecial-like configuration which tends to be seen only in " $-$ " like classes.

Table 1

abc	#	$C_M\langle \mathbf{a}, \mathbf{b}, \mathbf{c}\rangle$	wt	a	b	c
1A	1	$2^{2.2}E_6(2)[.S_3]$	−1704	12.34	13.24	14.23
2A	1	$2.B$	14872	12.34	12.34	12.34
2A	2	$2^{2.2}E_6(2)$	1432	12.34	12.34	13.24
2A	3	$2^{2+22}.Co_2$	−488	12.34	12.34	56.78
2A	4	Fi_{23}	292	12.34	12.34	12.45
2A	5	Th	−248	$|b_1 e_1|$	$|b_1 e_1|$	$|b_2 e_2|$
2A	6	$2^{1+22}.McL$	−8	12.34	12.34	$|b_3 e_3|$
2A	7	$2.F_4(2)$	−248	12.34	12.34	15.26
2A	8	$HN[.2]$	−128	14.23	14.23	25.34
2A	9	$2.Fi_{22}$	−188	12.34	12.34	45.67
2A	10	$2^{3+20}.U_6(2)[.2]$	−104	12.34	12.56	56.78
2A	11	$2^{3+10+20}.M_{22}.2[.S_3]$	24	12.34	56.78	$9X.ET$
2A	12	$O_8^+(3)[.A_4]$	−194	12.45	12.56	23.56
2A	13	$2^{2+20}.U_4(3)[.2^2]$	−200	12.34	$12.9X$	$|b_3 e_3|$
2B	1	$2^{3+20}.U_6(2)[.S_3]$	344	12.34	12.56	12.78
2B	2	$2^{3+7+16}.S_6(2)[.2]$	−40	12.34	12.56	$78.9X$
2B	3	$2^{3+4+12+8}.A_8[.S_3]$	24	$b_1 d_1$	$b_2 d_2$	$b_3 d_3$
2B	4	$2^{1+21}.M_{22}[.2^2]$	56	$4^2, 0^{22}$	$-3, 1^{23}$	$1, -3, 1^{22}$
2B	5	$2^{1+21}.2^4.A_7[.2^2]$	−8	$-3, 1^{23}$	$-2^8, 0^{16}$	$3, -1^7, 1^{16}$
2B	6	$2^{2+14}.S_6(2)[.2]$	8	12.34	12.56	17.28
2B	7	$2^2.U_6(2)[.S_3]$	20	12.45	12.67	13.89
3A	1	$2.Fi_{22}[.2]$	78	12.34	12.56	34.67
3A	2	$S_8(2)$	−42	12.34	12.56	23.56
3A	3	$A_{12}[.2]$	−72	12.34	13.24	23.45
3A	4	$O_7(3)[.S_3]$	−84	12.45	23.78	56.78
3A	5	$2^{11}.M_{22}[.2]$	−66	12.78	17.29.38. 4X.5E.6T	18.27.3X. 49.5E.6T
3A	6	$2^{6+8}.A_7[.S_3]$	−74	12.34	17.28.39. 4X.5E.6T	18.29.37. 4X.5E.6T
3B	1	$3^5.O_5(3)[.3^2.6]$	−3	12.56	23.78	45.89
3C	1	$^3D_4(2).3$	−24	$b_1 b_2$	$|b_1 e_1|$	$|b_2 e_2|$
3C	2	$U_3(8).3[.2]$	0	$AB.CD$	$CD.EF$	$BC.DE$

Nets of order up to 3

abc	#	$\langle a, b, c\rangle$	net
1A	1	2^2	$a2A:bc;\ b2A:ca;\ c2A:ab$
2A	1	2	$a1A:aST'$
2A	2	2^2	$a1A:b;\ b2A:abT'$
2A	3	2^2	$a1A:b;\ b2B:abT'$
2A	4	S_3	$a1A:b;\ b3A:abSb$
2A	5	S_3	$a1A:b;\ b3C:abSb$
2A	6	D_8	$a1A:c;\ b2B':c;\ c4A:acbc(2A3|6)$
2A	7	D_8	$a1A:c;\ b2A':c;\ c4B:acbc(2A2|7)$
2A	8	D_{10}	$a1A:c;\ b1A:d;\ c5A:dcacd(2A8);\ d5A:cdbdc$
2A	9	$2 \times S_3$	$a1A:d;\ b2A:dc;\ c3A:dbd;\ d6A:dadcbc(2A4|9,\ 2A2|9^2)$
2A	10	2^3	$a2B:bc;\ b2A:ca;\ c2A:ab$
2A	11	2^3	$a2B:bc;\ b2B:ca;\ c2B:ab$
2A	12	$3^2.2$	$a3A:bcd;\ b3A:adc;\ c3A:dab;\ d3A:cba$
2A	13	$2 \times D_8$	$a2A:cd;\ b2A:cd;\ c4A:adbd(2A6|10);\ d4A:acbc$
2B	1	2^3	$a2A:bc;\ b2A:ca;\ c2A:ab$
2B	2	2^3	$a2A:bc;\ b2B:ca;\ c2B:ab$
2B	3	2^3	$a2B:bc;\ b2B:ca;\ c2B:ab$
2B	4	$2 \times D_8$	$a2A:cd;\ b2A:cd;\ c4A:adbd(2A6|9);\ d4A:acbc$
2B	5	$2 \times D_8$	$a2B:cd;\ b2B:cd;\ c4A:adbd(2A6|10);\ d4A:acbc$
2B	6	$2 \times D_8$	$a2A:cd;\ b2B:cd;\ c4B:adbd(2A7|9);\ d4B:acbc$
2B	7	$2^2 \times S_3$	$a2A:ef;\ b2A:fd;\ c2A:de;\ d6A:ecefbf(2A9^2,\ 2B1|7^2);$ $e6A:fafdcd;\ f6A:dbdeae$
3A	1	$2 \times S_3$	$a2A:bc;\ b2A:ac;\ c6A':ab(6A1^2,\ 1A1^3)$
3A	2	S_4	$a2A:bc;\ b3A:cac;\ c4B:abcTb(4B1|2)$
3A	3	A_5	$a2A:de;\ b3A:ecd;\ c3A:dbe;\ d5A:ceaeb(5A1);\ e5A:bdadc$
3A	4	$S_3 \times S_3$	$a2A:de;\ b2A:de;\ c2A:de;$ $d6A:aebece(6A1|9,\ 3A1^3);\ e6A:cdbdad$
3A	5	$2 \times S_4$	$a2A:ce;\ b2A:de;\ c4A:aede(4A3|5);$ $d4A:bece;\ e6A:cacdbd(6A4^2,\ 2B1|4^2)$
3A	6	$2^4.S_3$	$a3A:dec;\ b3A:ced;\ c4A:aebd(4A4|8);$ $d4A:beac;\ e4A:adbc$
3B	1	$3^3.2^2$	$hc(3 \times 3):a^9:a^6(6A9^2,\ 3A4^3)$
3C	1	S_4	$a2A:bc;\ b3C:cac;\ c4B:abcTb(4B1|3)$
3C	2	A_5	$a2A:de;\ b3C:ecd;\ c3C:dbe;\ d5A:ceaeb(5A2);\ e5A:bdadc$

3. The prism phenomenon. We call a net a $4A$-*prism* if for all integers i, $ab.(bc)^i$ is in class $4A$. This implies that the net is prism-shaped and that all the "side" faces have class $4A$, hence the name. A $4A$-prism is called *special* if in addition both the end faces have the same class as the net, i.e., ab, ab^c and abc are conjugate.

Let a be a $2A$-element and v a $4A$-element such that av is the product of two transpositions. Then there exist transpositions b and c such that $bc = v$ and (a, b, c) is a special $4A$-prism. Furthermore this can be done uniquely up to conjugation by an element of $C_M\langle a, v \rangle$. There are 22 conjugacy classes (including a pair related by inverting v) satisfying the hypothesis and the above is true for each of them.

4. Collapses in nets tend to occur only when they have to, that is, if a collapse is forced by the orders of the faces. The only known exception is $6A8$ (in the notation of [**Net2**]), which is generated by 12.45, 23.45 and 46.57.

A summary of some results.   The number of nets (up to reflection) for each class of order up to 7 is as follows.

$1A$: 1; $2A$: 13; $2B$: 7; $3A$: 6; $3B$: 1; $3C$: 3; $4A$: 22; $4B$: 25; $4C$: 13; $4D$: 2; $5A$: 18; $5B$: 1; $6A$: 38; $6B$: 14; $6C$: 21; $6D$: 8; $6E$: 7; $6F$: 8; $7A$: 28; $7B$: 2.

Note how the number tends to decline as one moves through the classes of each order. The classes are arranged in order of decreasing centralizer order and, more likely to be relevant, the later classes have fewer Atkin-Lehner involutions in their Monstrous Moonshine symbol.

We describe some of the more interesting nets. Two of the "cubes" in class $4A$ have net group $5^2.4.2^2$ and $3^4.4.2^2$, centralizing $U_3(5)$ and $3^4.S_6$ respectively. The former is non-symmetric and therefore comes in two forms.

The honeycombs for classes $4C$ and $5B$ generate groups $2^8.3^2.2$ and $U_3(5)$ respectively. The former consists of even combinations of the 9 points of an affine plane inside our projective plane, together with its translations and half turns.

One of the nets of class $6B$ has two triangles and six pentagons. Yet finding its generators turned out to be exceedingly difficult. The process is described in [**CIM**], where acknowledgment is given to Eamonn O'Brien for computing the order of the net group.

Three nets of class $6B$ have four squares and four pentagons (with the same geometry). In one case the net group is a non-split $2^6.U_4(2)$, while in the other two it is $2 \times {}^2F_4(2)'$. One of the latter two is non-symmetric. Note that in the Tits group the product of *any* two 6-transpositions has order 2, 4 or 5. The relevant elements (a, b, c) are given in [**CIM; Net2**].

As stated earlier, there are seven conjugacy classes of honeycomb of class $6E$, including two related by a reflection. These two and one other have net group $O_8^+(2)$. The other net groups are $2^2 \times 3^{1+2}.2$, $2^2 \times 2^6.3^{1+2}.2$ (the only known honeycomb whose net rotation group is not transitive on the faces), $2^{1+8}.3^3.2^2$ and $O_8^+(3)$.

Possible net groups for class $7A$ include $2.Fi_{22}$ (two ways), HN and $2^{2.2}E_6(2)$ (two ways). The two for $2.Fi_{22}$ have the same geometry (with four squares, four pentagons and a hexagon) as do the other three (with three squares and six pentagons). For the last two, determination of the class of the mate ($26A$ and $38A$) is due to Robert Wilson.

Finally, the honeycomb of class $7B$ generates the Held group.

Section 8. The quiz game

In this last section we list some questions that seem worth asking.

1. Is a complete enumeration of nets feasible? The Monster has 1400384 orbits on triples of $2A$-elements. This is a big number, even though there can be quite a lot in a single net (for example, six of the nets of order 119 have 238 flags and the other four have 357).

2. Is the weight calculus of any use in limiting the geometric structures of nets which are beyond computational reach, e.g., where the net group is the Monster itself? (See [CIM] for a discussion on whether computing in $\mathbb{M}$ might become practical.)

3. Are there similar geometric structures corresponding to other braid groups? For example, what about quadruples of 4-transpositions in a group such as the Baby Monster or sextuples of 3-transpositions in a group such as Fi_{24}? (For an analysis of *quintuples* of transpositions in Fi_{24}, see [FTG].) We suspect that the prism phenomenon noted above may be linked to the action of the 5-string spherical braid group on conjugacy classes of quintuples of $2A$-elements of $\mathbb{M}$ whose product is the identity.

4. Do the classes and net groups of honeycombs show any pattern? One may also ask, more generally, what groups have *quilts* that are honeycombs. It is of course possible to define a free honeycomb group, and the methods of [FTG] may be used to define a "free honeycomb 6-transposition group", where quotes are used because the latter is not itself actually a 6-transposition group.

5. Does Normalizer Moonshine have any significance? This question may not have any relation to net theory but we include it here because both deal with the extension of Generalized Moonshine to non-commuting pairs of elements of the Monster.

Let g be an element of the Monster with Moonshine symbol $p+$ for p an odd prime. Then GMC asserts that the coefficients of certain modular functions form characters of $C_{\mathbb{M}}(g)$. In general these characters will have irrationalities involving $\sqrt{\pm p}$ and will not extend to the full normalizer of g. However, they *will* extend as p-modular characters, and one may therefore be able to define a modular function for a pair (g, h) with h normalizing g and of order not divisible by p.

We may note that if h acts as an outer involution on $C_{\mathbb{M}}(g)/\langle g \rangle$ (and, if $p = 3$, we require this to be of class $2D$, rather than $2C$, in Fi_{24}) then the first coefficient can be taken to be the same as that of $2p|2+$ or $2p|2+p'$, the "square root(s)" of $p+$ with appropriate constant term. (See [FMN] for a list of functions in which these are included.) Furthermore, functions of these types exist *exactly* when g is fully normalized in $\mathbb{M}$. This can hardly be a coincidence.

However, there are many cases where there is no obvious candidate for a modular function. For example, for $2C$ in Fi_{24}, the first coefficient will be 277. Is there a "good" modular function whose q-expansion begins $q^{-1} + 277q + \cdots$? Furthermore,

$2p|2+$ and $2p|2+p'$ are *not* related to the modular groups obtained by stabilizing (g, h) in $\tilde{\Gamma}$ and taking the homomorphic image in Γ, as a naive attempt to extend Generalized Moonshine might suggest.

The idea of using modular characters comes from [**MM1**], and the question of Normalizer Moonshine was first raised in [**MM2**, Question 6].

6. The weight of a net is defined by applying algebra multiplication and inner producting to axis vectors within the Griess–Conway algebra. But this is just one of a whole host of operations that can be applied in the vertex operator algebra on the Moonshine Module (see [**FLM**]). Can one define and compute an infinite series of weights, and, if so, are they the coefficients of a modular function? If so, this might provide the link between Generalized Moonshine and net theory that we are seeking.

Acknowledgment. The work described here was carried out while the author was visiting Lehrstuhl D für Mathematik at Aachen (supported by the European Community's HCM project on Computational Group Theory), the Mathematics Department at the University of Chicago (who supported the visit) and the Centre Interuniversitaire en Calcul Mathématique Algébrique at Concordia University, Montreal (supported by the Natural Sciences and Engineering Research Council (Canada) and Fonds pour la Formation de Chercheurs et l'Aide à la Recherche (Québec)). The author wishes to thank these institutions for their hospitality. All computations used the GAP system developed at Aachen [**GAP**].

References

[Bir]   J. S. Birman, Braids, links and mapping class groups, Princeton Univ. Press, Princeton 1974.

[MM2]   R. E. Borcherds, A. J. Ryba, Modular moonshine II, Duke Math. J. **83** (1996), 435–459.

[Con]   J. H. Conway, A simple construction for the Fischer–Griess monster Group, Invent. Math. **79** (1985), 513–540.

[ATLAS] J. H. Conway, R. T. Curtis, S. P. Norton, R. A. Parker, R. A. Wilson, Atlas of finite groups, Oxford University Press, Oxford 1985.

[Hsu1]   J. H. Conway, T. M. Hsu, Quilts and T-systems, J. Algebra **174** (1995), 856–903.

[MM]   J. H. Conway, S. P. Norton, Monstrous moonshine, Bull. Lond. Math. Soc. **11** (1979), 308–339.

[Y555]   J. H. Conway, S. P. Norton, L. Soicher, The Bimonster, the group Y_{555}, and the projective plane of order 3, in: Computers in algebra, ed. M. C. Tangora, Lecture Notes in Pure and Appl. Math. **111**, Marcel Dekker, New York 1988, 27–50.

[FMN]   D. Ford, J. McKay, S. P. Norton, More on replicable functions, Comm. Algebra **22** (1994), 5175–5193.

[FLM]   I. B. Frenkel, J. Lepowsky, A. Meurman, Vertex operator algebras and the monster, Pure Appl. Math. **134**, Academic Press, Boston 1988.

[Hsu2]   T. M. Hsu, Quilts, T-systems, and the combinatorics of Fuchsian groups, Ph. D. Thesis, Princeton 1994.

[Hsu3]   T. M. Hsu, Quilts, the 3-string braid group, and braid actions on finite groups: an introduction, these proceedings.

[UM]   S. P. Norton, The uniqueness of the Fischer–Griess monster, in: Finite groups — coming of age, ed. J. McKay, Contemp. Math. **45**, AMS, Providence 1985, 271–285.

[Gem]   S. P. Norton, Generalized moonshine, Proc. Symp. Pure Math. **47** (1987), 208–209.

[PTM]   S. P. Norton, Presenting the monster?, Bull. Soc. Math. Belg. Ser. A 42 (1990), 595–605.

[CTM]   S. P. Norton, Constructing the monster, in: Groups, combinatorics and geometry, ed. M. Liebeck, J. Saxl, Proceedings of the LMS Symposium, Durham 1990, London Math. Soc. Lecture Note Ser. **165**, Cambridge 1992, 63–76.

[Monalg] S. P. Norton, The monster algebra: some new formulae, in: Moonshine, the monster, and related topics, ed. C. Dong, G. Mason, Contemp. Math. **193**, AMS, Providence 1996, 433–441.

[FTG]   S. P. Norton, Free transposition groups, Comm. Algebra **24** (1996), 425–432.

[AM1]   S. P. Norton, Anatomy of the monster: I, in: The atlas of finite groups ten years on, ed. R. T. Curtis, R. A. Wilson, Proceedings of the atlas ten years on conference, Birmingham 1995, London Math. Soc. Lecture Note Ser. **249**, Cambridge 1998, 198–214.

[CIM]   S. P. Norton, Computing in the monster, submitted to J. Symb. Comp. for the Proceedings of the 1996 MAGMA conference in Milwaukee.

[Net2]   S. P. Norton, A string of nets, preprint.

[MM1]   A. J. Ryba, Modular moonshine, in: Moonshine, the monster, and related topics, ed. C. Dong, G. Mason, Contemp. Math. **193**, AMS, Providence 1996, 307–335.

[GAP]   M. Schönert et al., GAP — Groups, algorithms and programming, Lehrstuhl D für Mathematik, Rheinisch Westfälische Technische Hochschule, Aachen, Germany, third edition, 1993.

[Sh]   G. Shimura, Introduction to the arithmetic theory of automorphic functions, Publ. Math. Soc. Jap. **11**, Princeton Univ. Press, 1971.

DPMMS
16 Mill Lane
Cambridge CB2 1SB
England
E-mail: simon@dpmms.cam.ac.uk

Monster roots

*Christopher S. Simons**

1. Introduction

It is now known [**Nor**], [**Iva**] that the *Bimonster*, or wreathed square $M \wr 2$ of the Monster group $\mathbb{M}$, is presented by the single relation

$$(ab_1c_1ab_2c_2ab_3c_3)^{10} = 1 \tag{1}$$

in addition to the Coxeter relations of the $\mathbb{M}_{666}$ diagram (Figure 1).

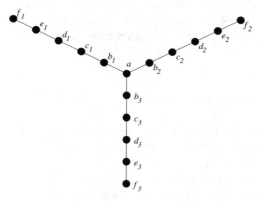

Figure 1. $\mathbb{M}_{666}$ **diagram**

In the Atlas [**Co1**] and previous papers on this subject the group generated by a together with

the first p of b_1, c_1, d_1, e_1, f_1

the first q of b_2, c_2, d_2, e_2, f_2

the first r of b_3, c_3, d_3, e_3, f_3

is called Y_{pqr}. However, since this group appears to be more naturally associated with the parameters $p + 1, q + 1, r + 1$, we introduce the alternate notation $\mathbb{M}_{p+1,q+1,r+1}$.

The Coxeter group $c\mathbb{M}_{666}$ defined by the $\mathbb{M}_{666}$ diagram is a hyperbolic reflection group (a group generated by reflections in hyperplanes containing the origin in a

* Supported in part by an NSERC 1967 graduate fellowship. The author strongly acknowledges the frequent help provided by John H. Conway.

Lorentzian space), so the Norton–Ivanov theorem **[Nor]**, **[Iva]** shows that the Bimonster is a homomorphic image of this very concretely geometrical group.

Each element of the Bimonster has the form

$$(x, y) \text{ or } (x, y)\tau = \tau(y, x) \text{ with } \tau^2 = 1$$

where x, y range over $\mathbb{M}$ and τ is the wreathing involution. The conjugacy class of τ consists of all the elements $(x, x^{-1})\tau = \tau_x$ and so has size $|\mathbb{M}|$. We call the elements of this class the *reflection elements* of $\mathbb{M} \wr 2$, since they are the images of the reflections in the Coxeter group.

In the geometric representation, the generators of the $\mathbb{M}_{666}$ diagram correspond to reflections in certain vectors $r_1, \ldots, r_{16}$. Following **[Co2]** we call these the *fundamental Monster roots* and use the term *Monster roots* for their images under the Coxeter group, $c\mathbb{M}_{666}$ they determine. A Monster root serves as a name for the appropriate reflection element of $\mathbb{M}_{666}$. Conversely, each reflection element has infinitely many such names.

The aim of this paper is to explore this naming relationship: When do two Monster root vectors r, s name the same element of $\mathbb{M}_{666}$? A complete solution to this question, which may be within reach, would provide a very simple way to compute with the Monster. As a partial solution, tables of some of these relations appear in the appendices. These tables have proven to be very useful for further investigations of the Bimonster and the Monster.

2. Coordinate system and roots

In order to enumerate the Monster roots and their equivalences we need coordinate systems. We will use two different coordinate systems, both taken from **[Co2]**. The setting of this section is $c\mathbb{M}_{666}$.

Often we will use $+$ to denote 1 and $-$ to denote -1.

In *System* 1 we have a space of 19 coordinates

$$\begin{array}{cccccc} a & b & c & d & e & f \\ g & h & i & j & k & l & \quad t \text{ with quadratic form } a^2 + \cdots + r^2 - t^2. \\ m & n & o & p & q & r \end{array} \qquad (2)$$

In this system the fundamental Monster roots are as indicated in Figure 2. All the vectors satisfy the following relations.

$$a + b + c + d + e + f = t$$
$$g + h + i + j + k + l = t$$
$$m + n + o + p + q + r = t$$

Thus t is redundant, and we shall sometimes omit it. We call it the *type*. On occasion we use the notation

$$a \quad b \quad c \quad d \quad e \quad f \mid g \quad h \quad i \quad j \quad k \quad l \mid m \quad n \quad o \quad p \quad q \quad r.$$

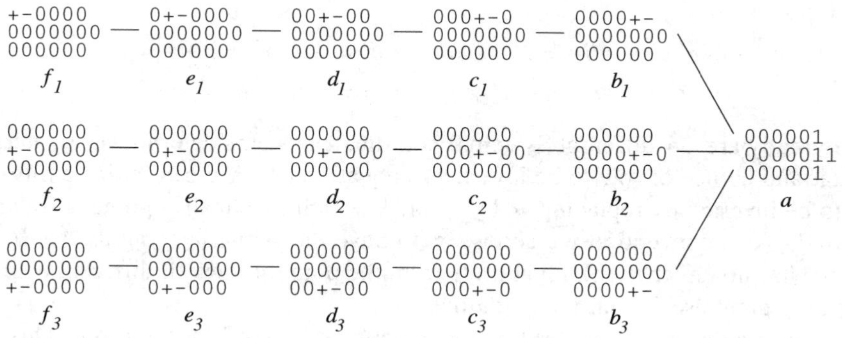

```
+-0000      0+-000      00+-00      000+-0      0000+-
0000000  —  0000000  —  0000000  —  0000000  —  0000000  ⟍
000000      000000      000000      000000      000000
  f₁          e₁          d₁          c₁          b₁


000000      000000      000000      000000      000000        000001
+-00000  —  0+-0000  —  00+-000  —  000+-00  —  0000+-0  ——  0000011
000000      000000      000000      000000      000000        000001
  f₂          e₂          d₂          c₂          b₂           a


000000      000000      000000      000000      000000
0000000  —  0000000  —  0000000  —  0000000  —  0000000  ⟋
+-0000      0+-000      00+-00      000+-0      0000+-
  f₃          e₃          d₃          c₃          b₃
```

Figure 2. The fundamental Monster roots in System 1

For *System* 2 we have a space of 18 coordinates

$$
\begin{array}{cccccc}
a & b & c & d & e & f \\
g & h & i & j & k & l \quad s \text{ with quadratic form } a^2 + \cdots + q^2 - 5s^2. \\
m & n & o & p & q & *
\end{array}
\tag{3}
$$

In this system the fundamental Monster roots are as indicated in Figure 3. All the vectors satisfy the following relations.

$$a + b + c + d + e + f + g + h + i + j + k + l = 6s$$
$$m + n + o + p + q = s$$

Thus s is redundant, and we shall sometimes omit it. We call s the *size*. On occasion we use the notation

$$a \quad b \quad c \quad d \quad e \quad f \quad l \quad k \quad j \quad i \quad h \quad g \mid m \quad n \quad o \quad p \quad q.$$

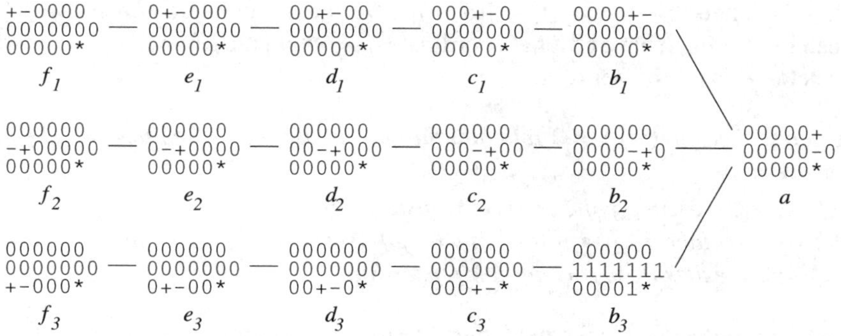

```
+-0000      0+-000      00+-00      000+-0      0000+-
0000000  —  0000000  —  0000000  —  0000000  —  0000000  ⟍
00000*      00000*      00000*      00000*      00000*
  f₁          e₁          d₁          c₁          b₁


000000      000000      000000      000000      000000        00000+
-+00000  —  0-+0000  —  00-+000  —  000-+00  —  0000-+0  ——  00000-0
00000*      00000*      00000*      00000*      00000*        00000*
  f₂          e₂          d₂          c₂          b₂           a


000000      000000      000000      000000      000000
0000000  —  0000000  —  0000000  —  0000000  —  1111111  ⟋
+-000*      0+-00*      00+-0*      000+-*      00001*
  f₃          e₃          d₃          c₃          b₃
```

Figure 3. The fundamental Monster roots in System 2

The two systems are linearly equivalent. Just let

$$
\begin{array}{cccccc}
a & b & c & d & e & f \\
g & h & i & j & k & l \quad t = \\
m & n & o & p & q & r
\end{array}
\begin{array}{cccccc}
a & b & c & d & e & f \\
g' & h' & i' & j' & k' & l' \quad s \\
m & n & o & p & q & *
\end{array}
\tag{4}
$$

where

$$g + g' = h + h' = i + i' = j + j' = k + k' = l + l' = s$$
$$m + n + o + p + q = s = t - r.$$

There really are 36 different System 2 coordinate systems determined by placement of the star and choice of row to be inverted in equation (4). We can indicate this choice of row to be inverted by replacing $*$ by $\uparrow$ or $\downarrow$ which (cyclically) points at the row to be inverted. As a convention we choose $*$ to have the same meaning as $\uparrow$. In $\mathbb{M}_{666}$ this choice has no effect. Looking at roots from the point of view of different coordinate systems is of great use for our computations.

We are now able to consider the (practical) enumeration of Monster roots. This theory is briefly discussed in [Co2], but we shall provide a more detailed treatment of it.

We start with the fundamental Monster roots $r_1, \ldots, r_{16}$. These are such that $(r_i, r_i) = 2$ for all i and $(r_i, r_j) = 0$ or -1 for all $i \neq j$. Since every Monster root r has $(r, r) = 2$ it follows that the formula for a reflection in r is $x \to x - (x, r)r = x^r$. This implies that every Monster root has integral coordinates in both systems.

By the theory of Coxeter groups [**Hum**, Section 5.13] we know that the reflecting hyperplanes of $c\mathbb{M}_{666}$ divide the hyperbolic space into copies of what is called the fundamental region, which we take to be the intersection of the halfspaces defined by $(r_i, x) < 0$. We now choose a point w inside the fundamental region, so that $(w, w) < 0$ and $(r_i, w) < 0$ for all i. In hyperbolic space [**Vin**] the distance d between w and $r^\perp$ can be defined such that

$$\sinh d = \frac{|(w, r)|}{\sqrt{|(w, w)||(r, r)|}}. \tag{5}$$

Therefore we can use (5) as an indicator of distance between $r^\perp$ and w. If a vector r is such that $(r, w) < 0$ while for some i $(r, r_i) > 0$, then $r^{r_i \perp}$ is necessarily closer to w than $r^\perp$ is. Therefore $|(w, r^{r_i})| < |(w, r)|$. We can use induction on $|(w, r)|$ (noting that w can be chosen to have rational coordinates) to obtain the following test for whether a given vector r is a Monster root.

Test 2.1. *First check if $(r, r) = 2$ and that the coordinates of r are integral. Then repeat the following steps:*

1. *If $r \in \{\pm r_1, \ldots, \pm r_{16}\}$ then r is a Monster root.*
2. *Otherwise replace r by r^{r_i} for i such that (r, r_i) and (r, w) have opposite signs. If no such i exists then r is not a Monster root.*

In System 1, reflections in the fundamental Monster roots other than a just interchange two coordinates and therefore generate the *coordinate permutation group* $S_6 \times S_6 \times S_6$ where each S_6 permutes the coordinates within a row. The 18 coordinates $a, \ldots, r$ are divided into 3 *blocks* of 6 coordinates each. Since Monster roots are integral and of norm 2, the Monster roots of type $t = 0$ or $t = 1$ are precisely those of form

$$0^4 + -|0^6|0^6 \text{ or } 0^5 1|0^5 1|0^5 1.$$

To test an integral norm 2 vector v of type $t > 1$ we first check that it satifies the relations that follow equation (2). Taking for example w close to $1^6|1^6|1^6$, we find that if r is a type 1 Monster root then $(w, r) < 0$. So the above procedure enables us to replace v by its reflection in any such r with $(v, r) > 0$. This inner product cannot be $\geq t$, since supposing, for example, that

$$
v = \begin{matrix} a & b & c & d & e & f \\ g & h & i & j & k & l \\ m & n & o & p & q & r \end{matrix} \; t \text{ and } r = \begin{matrix} 0 & 0 & 1 & 0 & 0 & 0 \\ 0 & 0 & 0 & 0 & 1 & 0 & 1 \\ 0 & 1 & 0 & 0 & 0 & 0 \end{matrix}
$$

then by the following argument we have that the inner product $c + k + n - t < t$. We know that $c^2 + k^2 + n^2 - t^2 \leq 2$, and since all coordinates are integers $t \geq 2$. The equality $c^2 + k^2 + n^2 - t^2 = 2$ is inconsistent with our other assumptions, therefore we have $c^2 + k^2 + n^2 - t^2 \leq 1$. This implies $3c^2 + 3k^2 + 3n^2 - 3t^2 \leq 3$, so that $(c + k + n)^2 < 4t^2$. Taking square roots we get $c + k + n < 2t$ and this is equivalent to the desired inequality. Therefore replacing v by v^r reduces to a case of smaller positive type. Monster roots of negative type are precisely the negatives of those of positive type. As a bonus we can inductively show that in System 1 for nonzero type Monster roots, all coordinates have the same sign as their type.

Similar results hold in System 2. The reflections in the fundamental Monster roots other than b_3 just interchange two coordinates and therefore generate the *coordinate permutation group* $S_{12} \times S_5$ where the S_{12} acts on the coordinates in the first two rows and the S_5 acts on the coordinates of the starred third row. The 17 coordinates $a, \ldots, q$ are divided into 2 *blocks*: one of 12 coordinates and the other of 5 coordinates. The Monster roots of size $s = 0$ or $s = 1$ are precisely of form

$$
0^{10} + -|0^5 \quad \text{or} \quad 0^{12}|0^3 + - \quad \text{or} \quad 0^6 1^6|0^4 1
$$

(where the | separates the two coordinate blocks). As in System 1, to test an integral norm 2 vector v of size $s > 1$ we first check that that the relations following equation (3) are satisfied. We then simply replace v by its reflection in any root r of size $s = 1$ with which it has positive inner product. It is easy to prove that this inner product is $< s$, so this test works and in System 2 for nonzero size Monster roots, all coordinates have the same sign as their type. Monster roots of negative size are precisely the negatives of those of positive size.

These facts allow us to enumerate the Monster roots by type in System 1 and by size in System 2. We define the *ancestor* of a Monster root r to be its reflection in a type or size 1 root vector v having maximal inner product (v, r). The ancestor of a root is unique up to the coordinate permutation group. Therefore, after sorting by the coordinate permutation group, the Monster roots acquire tree structures for System 1 and System 2. These structures allow us to eliminate duplication from our enumeration algorithm. Appendix A contains an initial segment of the (infinite) enumerations of the Monster roots in both System 1 and System 2. Naming conventions are also introduced.

3. Axiomatics

In [**Co2**] $\mathbb{M}_{666}$ was defined to be the minimal group other than S_{17} that possesses an S_5-subgroup S whose centralizer is a subgroup S_{12} in which a 7-point stabilizer is conjugate to S. We call this the $S_{5,12}$ axiom. It has the appearance of being a very powerful and difficult axiom. We shall give a simpler, more geometric, axiom.

Let δ_i be the central involution of the Coxeter group $W(D_8)$ generated by a, b_1, b_2, b_3, c_i, d_i, e_i and f_i. On the space spanned by its D_8 roots, δ_i acts as negation.

Axiom 3.1 (D_8). $\mathbb{M}_{666}$ is the Coxeter group determined by the $\mathbb{M}_{666}$ diagram (Figure 1) with the added relations: $1 = \delta_1 = \delta_2 = \delta_3$.

In fact, by a result of Soicher [**Soi**] and some of the following results, it is known that the relation $1 = \delta_1$ implies the other two. We assume that $\mathbb{M}_{666}$ has order > 2.

Theorem 3.1. *The $S_{5,12}$ axiom implies the D_8 axiom.*

Proof. This proof is drawn in Figure 4. Assume the $S_{5,12}$ axiom. Conway and Pritchard [**Co2**] showed that the group obtained is necessarily a quotient of $c\mathbb{M}_{666}$. Without loss of generality take δ to be δ_1. Then δ centralizes the S_5 generated by f_1, e_1, d_1, c_1 so lies in the S_{12} generated by f_2, e_2, d_2, c_2, b_2, a, b_3, c_3, d_3, e_3, f_3. But δ centralizes the 3 S_4 subgroups $\langle d_2, e_2, f_2 \rangle$, $\langle b_2, a, b_3 \rangle$ and $\langle d_3, e_3, f_3 \rangle$. It follows that δ is the identity.   □

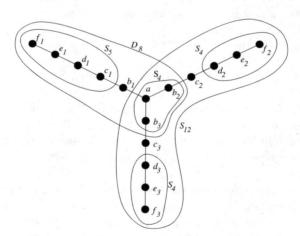

Figure 4. Proof of Theorem 3.1

We now deduce some further relations from the D_8 axiom.

Let δ be δ_1 and ε be the central involution of $W(E_8)$, where E_8 is as indicated in Figure 5. We note that both live in the Weyl group of an extended E_8, $\tilde{E}_8$ and that ε acts as negation on the space spanned by the E_8 roots.

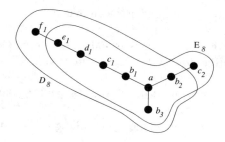

Figure 5. $\tilde{E}_8$

We consider the element $\delta\varepsilon$.

Using the general theory of (affine) Coxeter groups [**Hum**] we introduce the following euclidean coordinates.

f_1	$\frac{1}{2}$	$-\frac{1}{2}$	$-\frac{1}{2}$	$-\frac{1}{2}$	$-\frac{1}{2}$	$-\frac{1}{2}$	$-\frac{1}{2}$	$-\frac{1}{2}$
e_1	$-$	$+$	0	0	0	0	0	0
d_1	0	$-$	$+$	0	0	0	0	0
c_1	0	0	$-$	$+$	0	0	0	0
b_1	0	0	0	$-$	$+$	0	0	0
a	0	0	0	0	$-$	$+$	0	0
b_2	0	0	0	0	0	$-$	$+$	0
c_2	0	0	0	0	0	0	$-$	$+$
b_3	$\frac{1}{2}$	$\frac{1}{2}$	$\frac{1}{2}$	$\frac{1}{2}$	$\frac{1}{2}$	$-\frac{1}{2}$	$-\frac{1}{2}$	$-\frac{1}{2}$

The element $\delta\varepsilon$ clearly fixes the space spanned by the 7 roots in the generating sets of both D_8 and E_8. How does $\delta\varepsilon$ act on the remaining fundamental E_8 root c_2? Since the 7 other fundamental roots are fixed, c_2 must go to a vector of the form

$$a \quad a \quad a \quad a \quad a \quad a \quad (a-1) \quad (3a+1).$$

However the norm must be 2 and $a \in \frac{1}{2}\mathbb{Z}$, so $a = 0$ and c_2 must be fixed. Recall that $W(\tilde{E}_8) \cong \mathbb{Z}^8 : W(E_8)$. Thus we have seen that the action of $\delta\varepsilon$ in $\mathbb{Z}^8 : W(E_8)/\mathbb{Z}^8$ is trivial. Therefore $\delta\varepsilon$ is a translation.

Looking at the fundamental chamber of the affine $\tilde{E}_8$ we see that this translation is a translation by twice r where r is such that $(r, f_1) = 1$ and all inner products with other fundamental D_8 roots are 0. It follows that r is $-\frac{1}{4}^7 - \frac{3}{4}$. Therefore $\delta\varepsilon$ is a translation by a norm 4 vector. We have proven the following lemma.

Lemma 3.1. *We have $\delta\varepsilon = t_{v_4}$ where t_{v_4} is a translation by a norm 4 vector, v_4, of E_8.*

Theorem 3.2. *The D_8 axiom implies that the E_8 central involution and all translations in E_8 are trivial in $\mathbb{M}_{666}$.*

Proof. From the D_8 axiom we have $\delta = 1$ and so $\varepsilon = t_{v_4}$. But the norm 4 vectors of E_8 are all conjugate [**Co3**]. So for any norm 4 vector, v, of E_8 we have $\varepsilon = t_v$.

Choose norm 4 E_8 vectors v, v', and v'' such that $v + v' + v'' = 0$. We have $\varepsilon = t_v$, $\varepsilon = t_{v'}$ and $\varepsilon = t_{v''}$, so $\varepsilon^3 = t_v t_{v'} t_{v''} = 1$. Hence $\varepsilon = 1$. This also implies that the translation by any norm 4 E_8 vector is trivial. The norm 4 vectors of E_8 span E_8 [Co3]. Therefore all translations by E_8 vectors are trivial.   □

Obviously the above results hold for all the D_8 and E_8 subdiagrams of the $\mathbb{M}_{666}$ diagram.

Using the general theory of affine and hyperbolic groups and since

$$\omega = \begin{array}{cccccc} 1 & 1 & 1 & 1 & 1 & 1 \\ 0 & 0 & 0 & 2 & 2 & 2 & 6 \\ 0 & 0 & 0 & 0 & 3 & 3 \end{array} \tag{6}$$

spans the radical, $\tilde{E}_8^{\perp}$, of its $\tilde{E}_8$ lattice, we have the following theorem.

Theorem 3.3. *If r is any root vector of the $\tilde{E}_8$ corresponding to ω then $r \equiv r + m\omega \equiv -r$ for all m in $\mathbb{Z}$. In particular, $r \equiv \omega - r$.*

As an example we have

$$\begin{array}{cccccc} 1 & 1 & 1 & 0 & 0 & 0 \\ 0 & 0 & 0 & 1 & 1 & 1 & 3 \\ 0 & 0 & 0 & 0 & 1 & 2 \end{array} \equiv \begin{array}{cccccc} 0 & 0 & 0 & 1 & 1 & 1 \\ 0 & 0 & 0 & 1 & 1 & 1 & 3 \\ 0 & 0 & 0 & 0 & 2 & 1 \end{array}. \tag{7}$$

In fact this relation (together with the five other ones corresponding to different choices of $\tilde{E}_8$) suffices for the computations in [Co2].

4. Conjugacy classes

We briefly discuss conjugacy in $\mathbb{M} \wr 2$ and $\mathbb{M}_{666}$.

$\mathbb{M} \wr 2$.   We reduce the problem of conjugacy in $\mathbb{M} \wr 2$ to that of conjugacy in $\mathbb{M}$. It is easy to see that

$$(x, y)\tau \sim (x', y')\tau \text{ just if } xy \sim x'y'$$
$$(x, y) \sim (x', y') \text{ just if } x \sim x', \ y \sim y' \text{ or } x \sim y', \ y \sim x'.$$

It follows that the conjugacy classes of $\mathbb{M} \wr 2$ are determined by

1.   a single conjugacy class of $\mathbb{M}$ or
2.   an unordered pair of conjugacy classes of $\mathbb{M}$.

In the first case (the *odd* case) we may take an element of the form $(c, 1)\tau$ as representative. In the second case (the *even* case) we may take an element of the form (c_1, c_2) as representative.

M_{666}.   We identify some conjugacy classes of M_{666} which we call $1A$, $2A$, $3A$ and $4A$.

We define $1A$ to be the class of the identity element. $2A$ is the class of the product of the two reflection elements

$$
\begin{array}{cccccccccccc}
+ & - & 0 & 0 & 0 & 0 & & 0 & 0 & 0 & 0 & 0 & 1 \\
0 & 0 & 0 & 0 & 0 & 0 & 0 \cdot & 0 & 0 & 0 & 0 & 0 & 1 & 1 \,. \\
0 & 0 & 0 & 0 & 0 & 0 & & 0 & 0 & 0 & 0 & 0 & 1
\end{array}
\tag{8}
$$

$3A$ is the class of the product of the two reflection elements

$$
\begin{array}{cccccccccccc}
0 & 0 & 0 & 0 & + & - & & 0 & 0 & 0 & 0 & 0 & 1 \\
0 & 0 & 0 & 0 & 0 & 0 & 0 \cdot & 0 & 0 & 0 & 0 & 0 & 1 & 1 \,. \\
0 & 0 & 0 & 0 & 0 & 0 & & 0 & 0 & 0 & 0 & 0 & 1
\end{array}
\tag{9}
$$

$4A$ is the class of the product of the two reflection elements

$$
\begin{array}{cccccccccccc}
0 & 0 & 0 & 0 & 1 & 1 & & 1 & 1 & 0 & 0 & 0 & 0 \\
0 & 0 & 0 & 0 & 1 & 1 & 2 \cdot & 1 & 1 & 0 & 0 & 0 & 0 & 2 \,. \\
0 & 0 & 0 & 0 & 1 & 1 & & 1 & 1 & 0 & 0 & 0 & 0
\end{array}
\tag{10}
$$

$5A$ is the class of the product of the two reflection elements

$$
\begin{array}{cccccccccccc}
0 & 0 & 0 & 1 & 1 & 2 & & 0 & 2 & 1 & 1 & 0 & 0 \\
0 & 0 & 0 & 1 & 1 & 2 & 4 \cdot & 0 & 2 & 1 & 1 & 0 & 0 & 4 \,. \\
0 & 0 & 0 & 1 & 1 & 2 & & 0 & 0 & 1 & 0 & 2 & 1
\end{array}
$$

$6A$ is the class of the product of the two reflection elements

$$
\begin{array}{cccccccccccc}
0 & 0 & 0 & 1 & 1 & 2 & & 0 & 2 & 1 & 1 & 0 & 0 \\
0 & 0 & 0 & 1 & 1 & 2 & 4 \cdot & 0 & 2 & 1 & 1 & 0 & 0 & 4 \,. \\
0 & 0 & 0 & 1 & 1 & 2 & & 0 & 0 & 0 & 2 & 1 & 1
\end{array}
$$

In fact only $1A$, $2A$, $3A$ and $4A$ are required for the purposes of this paper, but $5A$ and $6A$ are defined here for general reference.

We note immediately that any product of two orthogonal fundamental reflection elements is in $2A$ and any product of any two nonorthogonal fundamental reflection elements is in $3A$. This follows since in cM_{666} any two fundamental roots act as transpositions in some S_{12}. We extend this to all pairs of roots.

Theorem 4.1. *In cM_{666} any two roots with inner product ± 1 or 0 are roots of the same fundamental region.*

Corollary 4.1. *If two roots have inner product 0 then the product of their reflections is in $2A$. If two roots have inner product ± 1 then the product of their reflections is in $3A$.*

Proof of theorem. Let r_1 and r_2 be the two roots and w_1 and w_2 the corresponding walls. The two walls meet in a space of codimension 2 which is a facet of some fundamental region. The angle between w_1 and w_2 is filled with (say) m copies of the fundamental region. By our hypothesis this angle must be $\pi/3$ or $\pi/2$. But any two walls of the fundamental region meet at one of these two angles, so $m = 1$. $\square$

This proves that the elements of $2A$ have order 2 and that the elements of $3A$ have order 3. It is possible to show, using the root and alias tables (Appendices A and B) with a few elementary calculations, that $4A$ is of order 4. We are using our assumption that $\mathbb{M}_{666}$ is of order > 2.

We intend to use the Atlas [Co1] notation, of the form nX, for the conjugacy classes of $\mathbb{M}$ and also for the corresponding classes (nX, nX^{-1}) of $\mathbb{M} \wr 2$. Using the fact that $\mathbb{M}_{666} = \mathbb{M} \wr 2$, we can verify that $2A$ and $3A$ are appropriately named. We do this by using the Monster power maps [Co1] to map conjugacy classes of $\frac{1}{2}(S_{12} \times S_5)$ to conjugacy classes of $\mathbb{M}$.

5. Alias groups

In equation (7) we saw two root vectors whose reflections are equivalent in $\mathbb{M}_{666}$,

$$
\begin{matrix}
1 & 1 & 1 & 0 & 0 & 0 & & 0 & 0 & 0 & 1 & 1 & 1 \\
0 & 0 & 0 & 1 & 1 & 1 & 3 \equiv 0 & 0 & 0 & 1 & 1 & 1 & 3. \\
0 & 0 & 0 & 0 & 1 & 2 & & 0 & 0 & 0 & 0 & 2 & 1
\end{matrix}
$$

We say that each of these roots is an *alias* for the other. This equivalence can be conjugated by elements of the coordinate permutation group $S_6 \times S_6 \times S_6$ to obtain further equivalences, for example we have

$$
\begin{matrix}
1 & 0 & 1 & 1 & 0 & 0 & & 0 & 1 & 0 & 0 & 1 & 1 \\
1 & 1 & 0 & 0 & 0 & 1 & 3 \equiv 1 & 1 & 0 & 0 & 0 & 1 & 3. \\
0 & 2 & 0 & 1 & 0 & 0 & & 0 & 1 & 0 & 2 & 0 & 0
\end{matrix}
$$

This family of equivalences can be summarized as $0^3 1^3 | 0^3 1^3 | 0^4 12 \equiv 1^3 0^3 | 0^3 1^3 | 0^4 21$. We express this by saying that the digit permutation $(01)| \sim |(12)$ preserves the root vector.

For Appendix B one needs to compute similar equivalences for other roots, such as $0^2 12^2 4^2 6^2 78^2 | 0^3 4^2$. These are the interchange of 1 and 7, and the alternating group A_6 on $0, 2, 4, 6, 8, \{1, 7\}$. As an example, an equivalent root vector is $2^2 80^2 4^2 6^2 817 | 0^3 4^2$.

It is therefore natural to define a *digit* of a root to be the equivalence class of coordinate points in the same block with the property that any pair can be interchanged without changing the corresponding reflection element of $\mathbb{M}_{666}$. We define the *(small) alias group* of the root to be the group of block preserving digit permutations which preserve the root in $\mathbb{M}_{666}$. The *large* alias group is the small alias group extended by those permutations of the blocks which preserve the root in $\mathbb{M}_{666}$. In System 2 the large alias group coincides with the small alias group. We will often just use the term *alias group*.

As a final example of the terminology we state that the alias group of $0^{10} + -|0^5$ is of order 1 and that $+$ and $-$ are the same digit. The negation of a root is necessarily an alias of the root.

We shall quote [Co2] for the alias groups that appear there (they are for roots of $\mathbb{M}_{663}$ in System 2). However it was not there proved that the alias groups could not be larger that those stated. We need a method to establish such assertions.

We illustrate by computing the alias group of

$$r = \begin{array}{cccccc} 0 & 0 & 0 & 1 & 1 & 1 \\ 0 & 0 & 0 & 1 & 1 & 1 \\ 0 & 0 & 0 & 0 & 1 & 2 \end{array} \; 3 \, .$$

The relations of the form of equation (7) show that the group is at least

$$\langle \, (01)| \sim |(12), \; \sim |(01)|(12) \, \rangle.$$

Can it be larger?

Might we not, for instance, have

$$r = \begin{array}{cccccc} 0 & 0 & 0 & 1 & 1 & 1 \\ 0 & 0 & 0 & 1 & 1 & 1 \\ 0 & 0 & 0 & 0 & 1 & 2 \end{array} \; 3 \equiv \begin{array}{cccccc} 1 & 0 & 0 & 2 & 0 & 0 \\ 1 & 1 & 0 & 0 & 1 & 0 \\ 0 & 1 & 1 & 0 & 0 & 1 \end{array} \; 3 = r' \; ?$$

No, r commutes with f_3 but r' does not.

In general, if s is the difference of any 2 coordinate vectors in the same block, then s is a root and if $(r, s) = 0$ then $rs \in 2A$ while if $(r, s) = \pm 1$ then $rs \in 3A$ as in Section 4.

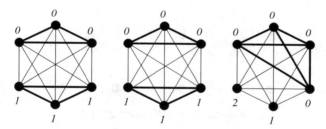

Figure 6. Graph for r. Thick lines for $2A$, thin lines for $3A$

We use these facts to associate a coloured graph with r (Figure 6). The nodes of this graph are the 18 coordinate vectors. Two nodes from the same block are joined by an edge whose colour indicates the conjugacy class of rs, where $s = x - y$. Most of the edges are the determined by easy rules:

1.   Two equal coordinates are joined by a $2A$ edge.
2.   Two coordinates that differ by 1 are joined by a $3A$ edge.

There is no simple rule to find the colour of the edges between coordinates that differ by 2 or more. However, in our example, it is easy to see that the 4 edges between 0 and 2 are all $3A$, since after applying the alias element $(01)| \sim |(12)$ the corresponding differences are replaced by 1's.

Plainly all the elements of the alias group must preserve the graph. The automorphism group of this graph consists of

$$(S_3 \wr S_2) \times (S_3 \wr S_2) \times (S_4 \times S_2)$$

(preserving the blocks) together with the interchange of the first two blocks.

This gives us an upper bound for the alias group, namely

$$
\begin{array}{ccccc}
S_2 & \times & S_2 & \times & S_2 \\
\{0, 1\} & & \{0, 1\} & & \{1, 2\}
\end{array}
$$

(together with the trivial interchange of the first two blocks). We call this the *possible group*. However, what we have seen (the *visible group*) is only of index 2 in this. Can there be more? No! Coset representatives for the visible group in the possible group are the identity element $\sim | \sim | \sim$ and the element $\sim | \sim |(12)$. But $\sim | \sim |(12)$ transforms r into its reflection under

$$
s = \begin{array}{ccccccc}
0 & 0 & 0 & 0 & 0 & 0 & \\
0 & 0 & 0 & 0 & 0 & 0 & 0, \\
0 & 0 & 0 & 0 & + & - &
\end{array}
$$

but we know that r and s do not commute. The alias group is therefore the group $\langle (01)| \sim |(12), \ \sim |(01)|(12) \rangle$ of order 4.

In practice we work with a collapsed graph obtained by identifying all nodes joined by $1A$ edges or $2A$ edges as in Figure 7. Since $3A$ edges are then the most common type, we usually omit them (or draw them in 'invisible ink'). This makes the graph for our example very simple indeed (Figure 8).

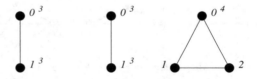

Figure 7. A collapsed graph

Figure 8. A very collapsed graph

The above example is done in System 1. We can, of course, make similar alias computations in System 2.

We have identified the Monster roots of small type or size with their associated reflection elements. The alias groups of these reflection elements have also been computed. These are done simultaneously for System 1 and System 2 using an inductive process. To find the elements of the (visible) alias group of a root r, our main tools are conjugations by reflections in fundamental Monster roots r_i where r^{r_i} has known alias group and change of coordinate systems such that r then has known alias group. Similar techniques determine the reductions of Monster roots to reflection elements. The methods of this section are used to verify that the visible alias groups are the entire alias groups. The results of these computations are presented in Appendices A and B.

Appendix A: Monster root reduction tables

The first column of these tables names a family of roots equivalent under the coordinate permutation group. The second gives the standard root vector of this family. The third column gives the root vector's ancestor. When there is a simpler root vector that yields the same reflection element of M_{666}, it is described in the fourth column. The root vectors of smallest type or size associated with a reflection element are called the *reduced root vectors*. The fifth column gives the *reduced family name*, which takes account of such equivalences.

Family names (first column) have the form n_{sm} where n is the type or size of the root vector and m of the reduced root vector, while s is a tag letter (lower case for System 1 and upper case for System 2). To make the equivalences more visible, we sometimes use letters a, b, c, ... for the distinct digits of certain reduced root vectors. When $n = m$, we say that that n_{sm} is *irreducible*. To make the system explicit, we can prefix the reduced family names by $+$ for System 1 and $-$ for System 2. We define the *type* or *size* of a reflection element to be the type or size of one of its reduced root vectors.

System 1.   We list all Monster roots of type ≤ 8. For type 9, only those roots reducing to reflection elements of type ≥ 5 are listed. For types 10, 11, 12 and 13 we list the irreducible roots. This table is complete up to (and including) type 11, after which there is the (remote) possibility of omission.

0_a	000000|000000|0000+−			0
1_a	000001|000001|000001			1
2_a	000011|000011|000011	1_a		2
3_a	000111|000111|000012	2_a		3
4_a	000112|000112|000112	2_a		4
4_{a2}	001111|000112|000022	3_a	110000|000110|000011	2
4_{b2}	001111|001111|000013	3_a	110000|110000|000011	2
5_{a4}	001112|000113|000122	3_a	110002|000112|000211	4
5_{b4}	000122|000122|000122	4_a	000211|000211|000211	4
5_{a3}	001112|001112|000023	3_a	110001|001110|000012	3
5_{a1}	011111|000122|000023	4_{a2}	100000|000100|000010	1
5_{b1}	011111|011111|000014	4_{b2}	100000|100000|000010	1
6_a	001122|000123|000123	4_a		6
6_b	001122|001122|000114	4_a		6′
6_{a4}	001113|001113|000123	3_a	110002|110002|000121	4
6_{a3}	001113|000123|000222	4_{a2}	110001|000102|000111	3
6_{b3}	011112|000114|000222	4_{a2}	100002|000111|000111	3

6_{c3}	000222|000222|000123	5_{b4}	000111|000111|000210	3
6_{a2}	011112|001113|000033	4_{b2}	100001|110000|000011	2
6_{b2}	011112|001122|000024	4_{a2}	100001|001100|000011	2
6_{c2}	001122|001122|000033	5_{a3}	001100|001100|000011	2
6_{a0}	011112|000222|000033	5_{a1}	+0000-|000000|000000	0
6_{b0}	111111|000123|000033	5_{a1}	000000|000+0-|000000	0
6_{c0}	111111|000222|000024	5_{a1}	000000|000000|0000+-	0
6_{d0}	111111|111111|000015	5_{b1}	000000|000000|0000+-	0
7_{a}	001123|001123|000124	4_{a}		7
7_{a4}	001114|001123|000223	4_{a2}	110002|001120|000112	4
7_{b4}	001123|000133|000223	5_{a4}	001120|000211|000112	4
7_{c4}	011122|000133|000124	5_{a4}	200011|000211|000121	4
7_{d4}	001114|000133|001222	5_{a4}	110002|000211|112000	4
7_{e4}	001222|000223|000124	5_{b4}	112000|000112|000121	4
7_{f4}	011122|000115|001222	5_{a4}	200011|000112|112000	4
7_{g4}	001222|000133|000133	6_{a}	112000|000211|000211	4
7_{a3}	011113|001114|000133	4_{b2}	100002|110001|000111	3
7_{b3}	011122|001123|000034	5_{a3}	011100|001101|000012	3
7_{c3}	011122|011122|000025	5_{a3}	100011|011100|000012	3
7_{a2}	011113|000124|000223	4_{a2}	100001|000101|000110	2
7_{b2}	000223|000223|000223	5_{b4}	000110|000110|000110	2
7_{a1}	011113|011113|000034	4_{b2}	100000|100000|000010	1
7_{b1}	111112|000223|000034	5_{a1}	000001|000001|000001	1
7_{c1}	011113|001222|000034	5_{a1}	100000|001000|000001	1
7_{d1}	111112|001222|000025	5_{a1}	000001|001000|000010	1
7_{e1}	001222|001222|000034	6_{c2}	001000|001000|000010	1
8_{a}	001124|001124|001124	4_{a}		8
8_{a7}	001124|000134|001223	5_{a4}	110023|000124|002113	7
8_{b7}	001223|001223|000125	5_{b4}	002113|002113|000214	7
8_{a6}	011123|001133|000125	5_{a4}	100023|001122|000213	6
8_{b6}	001124|001133|000224	5_{a4}	110022|001122|000114	6'
8_{c6}	001115|001133|001223	5_{a4}	110004|001122|002112	6'
8_{d6}	001133|001133|000134	6_{a4}	110022|110022|000114	6'
8_{a4}	011123|000134|000224	5_{a4}	100021|000121|000112	4
8_{b4}	001124|001124|000233	5_{a3}	001102|001102|000211	4
8_{c4}	001223|000224|000224	5_{b4}	002110|000112|000112	4
8_{d4}	011123|001115|000233	5_{a3}	100012|110002|000211	4
8_{e4}	001223|000233|000134	6_{a}	002110|000211|000121	4
8_{f4}	001124|000233|000233	6_{a}	110020|000211|000211	4
8_{g4}	011222|000233|000125	6_{a}	211000|000211|000211	4
8_{h4}	011222|011222|000116	6_{b}	211000|211000|000112	4

8_{i4}	011222|000134|000134	6_a	211000|000121|000121	4
8_{a3}	011114|001124|000224	4_{a2}	100002|001101|000111	3
8_{a2}	011114|011114|000134	4_{b2}	100001|100001|000110	2
8_{b2}	011123|011123|000035	5_{a3}	100010|100010|000011	2
8_{c2}	001133|000233|000224	6_{a3}	001100|000011|000110	2
8_{d2}	011123|001223|000044	6_{b2}	100010|001001|000011	2
8_{e2}	001133|000125|002222	6_{a3}	001100|000101|110000	2
8_{f2}	001115|000233|002222	6_{a3}	110000|000011|110000	2
8_{g2}	111122|001133|000035	6_{a2}	000011|110000|000011	2
8_{h2}	011222|001223|000035	6_{c2}	011000|001001|000011	2
8_{i2}	111122|011222|000026	6_{b2}	000011|011000|000011	2
8_{j2}	111122|000116|002222	6_{b3}	000011|000110|110000	2
8_{k2}	011114|011222|000044	6_{b2}	100001|011000|000011	2
8_{l2}	111122|001124|000044	6_{b2}	000011|001100|000011	2
8_{m2}	002222|000224|000134	6_{c3}	110000|000110|000101	2
8_{n2}	011222|001133|000044	7_{b3}	011000|110000|000011	2
8_{o2}	111122|000233|000044	7_{b1}	000011|000011|000011	2
8_{a1}	011114|000224|000233	5_{a1}	100000|000001|000100	1
8_{b1}	111113|000125|000233	5_{a1}	000001|000100|000100	1
8_{c1}	000233|000233|000233	7_{b2}	000100|000100|000100	1
8_{a0}	111113|011114|000044	5_{b1}	000000|+0000−|000000	0
8_{b0}	111113|001223|000035	5_{a1}	000000|00+00−|000000	0
8_{c0}	011123|002222|000035	6_{a0}	+000−0|000000|000000	0
8_{d0}	002222|001223|000044	7_{e1}	000000|00+00−|000000	0
8_{e0}	002222|002222|000035	7_{e1}	000000|000000|0000+−	0
9_a	001224|001224|000144	6_b	$aabccd$|$aabccd$|000144	9
9_b	011223|001125|000144	6_b	$baaccd$|$aaccbd$|000144	9
9_c	001233|001134|000144	7_a	$aabdcc$|$ccaabd$|000144	9
9_{a8}	001125|001134|001224	5_{a4}	110024|001124|002114	8
9_{a7}	011124|001134|001224	5_{a4}	100024|001123|002113	7
9_{b7}	001224|001233|000135	6_a	112003|003211|000124	7
9_{c7}	001125|001233|000234	6_a	001123|003211|000124	7
9_{d7}	011223|001233|000126	6_a	200113|002311|000124	7
9_{a6}	011124|000135|001224	5_{a4}	100023|000123|002112	6
9_{b6}	001224|000234|000234	6_a	002112|000132|000132	6
9_{c6}	011223|000234|000135	6_a	200112|000312|000132	6
9_{d6}	001224|001224|000225	5_{b4}	002112|002112|000114	6′
10_a	001225|001225|001144	6_b	$aabccd$|$aabccd$|001144	10
10_b	011224|001126|001144	6_b	$baaccd$|$aaccbd$|001144	10
10_c	001126|001234|001234	6_a		10′
10_d	001234|001234|000145	7_a		10″
10_e	001144|001144|001144	8_a		10‴

11_a	001235|001136|001244	7_a	$aab23c\|0011de\|00fg44$	11
11_b	011234|001127|001244	7_a	$baa23c\|0011de\|00gf44$	11
11_c	001244|001145|001145	8_a		11'
12_a	002244|002217|000444	9_a	$aabbcc\|ddeeff\|000444$	12
12_b	001155|112206|000444	9_b	$aabbcc\|ddeeff\|000444$	12
12_c	112233|001128|000444	9_b	$aabbcc\|ddeeff\|000444$	12
12_d	110046|003315|000444	9_c	$aabbcc\|ddeeff\|000444$	12
12_e	003324|110037|000444	9_c	$aabbcc\|ddeeff\|000444$	12
12_f	001155|002415|000444		$aabbcc\|ddeeff\|000444$	12
12_g	001236|001236|001245	7_a		12'
12_h	001245|001245|001146	8_a		12''
13_a	001246|001246|000445	9_a	$00abcd\|00abcd\|000445$	13
13_b	001237|001345|000445	9_c	$00cabd\|00bcad\|000445$	13
13_c	001246|001246|001246	8_a		13'

System 2.   We list all Monster roots of size ≤ 4. For size 5, only those roots reducing to reflection elements of size ≥ 3 are listed. For sizes 6, 7, 8 and 9 we list the irreducible roots. This table is complete up to (and including) size 6, after which there is the possibility of omission.

0_A	0000000000+−|00000			0
0_B	000000000000|000+−			0'
1_A	000000111111|00001			1
2_A	000011112222|00011	1_A		2
2_{A1}	000111111222|00002	1_A	000111111000|00001	1
3_A	000111222333|00012	1_A		3
3_{A2}	000111123333|00111	2_A	000111102222|11000	2
3_{B2}	000012222333|00111	2_A	000021111222|11000	2
3_{A1}	001111222233|00003	1_A	001111000011|00001	1
3_{B1}	001111113333|00012	2_A	001111110000|00010	1
3_{C1}	000022222233|00012	2_A	000011111100|00010	1
3_{A0}	011111112333|00003	2_{A1}	+0000000−000|00000	0
3_{B0}	000122222223|00003	2_{A1}	000+0000000−|00000	0
4_A	000112233444|00112	2_A		4
4_{A3}	001112223444|00013	2_A	110002221333|00012	3
4_{B3}	000122233344|00013	2_A	000211133322|00012	3
4_{A2}	001111333344|00013	1_A	112222000011|00011	2
4_{B2}	001111224444|00112	2_A	110000112222|00110	2
4_{C2}	000022333344|00112	2_A	000011222200|00110	2
4_{D2}	001111233444|00022	2_{A1}	110000211222|00011	2
4_{E2}	000112333344|00022	2_{A1}	000110222211|00011	2

4_{F2}	000122223444|00022	3_A	000211110222|00011	2		
4_{A1}	111111222444|00004	2_{A1}	000000111111|00001	1		
4_{B1}	011112223344|00004	2_{A1}	011110001100|00001	1		
4_{C1}	001122233334|00004	2_{A1}	001100011110|00001	1		
4_{D1}	000222222444|00013	3_A	000111111000|00010	1		
4_{E1}	000112224444|01111	3_{A2}	000110001111|10000	1		
4_{F1}	000022233444|01111	3_{B2}	000011100111|10000	1		
4_{G1}	000013333344|01111	3_{B2}	000011111100|10000	1		
4_{A0}	011111333334|00004	1_A	+0000000000−|00000	0		
4_{B0}	001222222344|00004	3_{A1}	00+000000−00|00000	0		
4_{C0}	011111124444|00022	3_{B1}	+000000−0000|00000	0		
4_{D0}	000023333334|00022	3_{C1}	0000+000000−|00000	0		
4_{E0}	111111114444|00013	3_{B1}	000000000000|000+−	0'		
4_{F0}	000033333333|00013	3_{C1}	000000000000|000+−	0'		
5_A	001122334455|00014	2_A		5		
5_{A4}	001112334555|00113	2_A	110002332444|00112	4		
5_{B4}	000122344455|00113	2_A	000211244433|00112	4		
5_{C4}	001112244555|00122	3_A	110002233444|00211	4		
5_{D4}	000122334555|00122	3_A	000211332444|00211	4		
5_{E4}	000113344455|00122	3_A	000112244433|00211	4		
5_{A3}	001112344455|00023	2_{A1}	001110233322|00021	3		
5_{B3}	001122234555|00023	3_A	001122201333|00012	3		
5_{C3}	000123334455|00023	3_A	000123331122|00012	3		
5_{D3}	000113334555|01112	3_{A2}	000112221333|10002	3		
5_{E3}	000122244555|01112	3_{B2}	000211122333|10002	3		
6_A	001122345666|00123	3_A	*aabbccdefggg*|00123	6		
6_B	000123445566|00123	3_A	*gggfedccbbaa*|00123	6		
6_C	001122445566|00114	2_A		6'		
7_A	001123456677|00124	3_A		7		
8_A	001224466788|00044		*aabccddeebff*|00044	8		
8_B	001125566778|00044		*aaccbddeeffb*|00044	8		
8_C	011223367788|00044		*bffeeddbccaa*|00044	8		
8_D	001134556788|00044		*ccaabdeedbff*|00044	8		
8_E	001233457788|00044		*ffbdeedbaacc*|00044	8		
8_F	001124467788|01124		*aabbc44deeff*|*ghh2i*	8'		
8_G	001224457788|00125		*aacbb44deeff*|*hhg2i*	8'		
8_H	001134466788|00125		*ffeed44bbcaa*|*hhg2i*	8'		
8_I	001234456788|00116			8''		
9_A	001224577899|00144		*aabccdeffghh*|00144	9		
9_B	011223478899|00144		*baaccdegff99*|00144	9		
9_C	001125677889|00144		*aaccbdeffhhg*|00144	9		
9_D	001233568899|00144		*aabdcceghhff*|00144	9		

9_E	001134667899|00144	*ccaabdffeghh*|00144	9
9_F	001144558899|00144		9'
9_G	001234567899|00045		9''
9_H	001234567899|00126		9'''

Appendix B: Alias tables

In the following alias tables we describe the alias groups of the Monster roots appearing in appendix A. The first column is the reduced family name. The second column gives the reduced root vector. The third column describes the associated (small) alias group, often by listing the group generators. The fourth column is the order of the small alias group. The large alias group is obvious, except that for 12 of System 1 we must adjoin $(ad)(be)(cf)|(04)$.

When describing the groups we write $\{ab\ldots c\}$ to mean the symmetric group on the digits a, b, $\ldots$, c. We use $\frac{1}{2}(G)$ to designate the even part of G. The dihedral group of order $2n$ on the letters a_0, a_1, $\ldots$, a_{n-1} in this cyclic order is denoted by $D_{2*n}(a_0 a_1 \ldots a_{n-1})$, and $\mathrm{PGL}_2(n)(a_0 a_1 \ldots a_n)$ is the 2-dimensional projective linear group over $\mathbb{F}_n$ where a_0, a_1, $\ldots$, a_n are acted upon as 0, 1, $\ldots$, $n-1$, ∞.

System 1.   We list alias groups for System 1.

0	000000|000000|0000+−	$\sim\mid\sim\mid\sim$	1
1	000001|000001|000001	$\sim\mid\sim\mid\sim$	1
2	000011|000011|000011	$\sim\mid\sim\mid\sim$	1
3	000111|000111|000012	$01\mid\sim\mid12,\ \sim\mid01\mid12$	4
4	000112|000112|000112	$\sim\mid\sim\mid\sim$	1
6	001122|000123|000123	$210\|123\mid\sim,\ 12\|12\|12,$ $210\mid\sim\mid123$	18
6'	001122|001122|000114	$012\|210\mid\sim,\ 02\|02\mid\sim$	6
7	001123|001123|000124	$01\|23\|12,\ 23\|01\|12$	4
8	001124|001124|001124	$\sim\mid\sim\mid\sim$	1
9	*aabccd*|*aabccd*|000144	$bd\|bd.ac\mid\sim,\ ac\|ac\mid\sim$	4
10	*aabccd*|*aabccd*|001144	$ac\|ac\mid\sim,\ bd\|bd\mid\sim,\ bd\|ac\|04$	8
10'	001126|001234|001234	$01\|23\|12,\ 01\|12\|23$	6
10''	001234|001234|000145	$1234\|4321\mid\sim,\ 12.34\|13\|45$	8
10'''	001144|001144|001144	$\{04\}\|\{04\}\|\{04\}$	8
11	*aab23c*|0011*de*|00*fg*44	$bc\|01\|04,\ 23\|01\|fg$	4
11'	001244|001145|001145	$12\|01.45\mid\sim,\ 12\mid\sim\mid01.45,\ 04\mid\sim\mid\sim$	8
12	*aabbcc*|*ddeeff*|000444	$\frac{1}{2}(\{abc\}\|\{def\}\|\{04\})$	36
12'	001236|001236|001245	$12\|12\|12.45$	2

12″	001245|001245|001146	$15\|15\| \sim$	2
13	00*abcd*|00*abcd*|000445	$ad.bc\|ab.cd\| \sim,\ ab.cd\|ad.bc\| \sim,$ $bcd\|dcb\| \sim$	12
13′	001246|001246|001246	$\sim \| \sim \| \sim$	1

System 2.   We list alias groups for System 2.

0	0000000000+−|00000	$\sim \| \sim \| \sim$	1
0′	000000000000|000+−	$\sim \| \sim \| \sim$	1
1	000000111111|00001	$01\| \sim$	2
2	000011112222|00011	$\{012\}\| \sim$	6
3	000111222333|00012	$\frac{1}{2}(\{0123\}\|\{12\})$	24
4	000112233444|00112	$\frac{1}{2}(\{04\}\{123\}\|\{01\})$	12
5	001122334455|00014	$\text{PGL}_2(5)(023451) \equiv$ $01.34\| \sim,\ 12.45\| \sim,\ 01.23.45\| \sim$	120
6	*aabbccdefggg*|00123	$bc.ef\|12,\ fed\|123,\ cba\|123$	18
6′	001122445566|00114	$012.456\| \sim,\ 01.45\| \sim,\ 04\| \sim$	48
7	001123456677|00124	$01.45\|12,\ 07.16.25.34\| \sim$	8
8	*aabbccddeeff*|00044	$\frac{1}{2}\{abcdef\}$	360
8′	*aabbc44deeff*|*ghh*2*i*	$\frac{1}{2}(D_{2*5}(ab4ef)\{cd\}\|\{04\})$	20
8″	001234456788|00116	$12.56\|01,\ 15.48\|01,\ 08.17.26.35\| \sim$	144
9	*aabccdeffghh*|00144	$bd.eg.fh\| \sim,\ ah.bg.cf.de\| \sim,\ be.fh\|04$	32
9′	001144558899|00144	$\{048\}\{159\}\|\{04\},\ 09.18.45\| \sim$	144
9″	001234567899|00045	$\frac{1}{2}(\text{PGL}_2(7)(12634578)\|\{45\}),\ 09\| \sim$	672
9‴	001234567899|00126	$12.45.78\|12,\ 123.678\| \sim,\ 09.18.27.36.45\| \sim$	12

Bibliography

[Co1] J. H. Conway, R. T. Curtis, S. P. Norton, R. A. Parker, R. A. Wilson, Atlas of finite groups, Oxford University Press, 1985.

[Co2] J. H. Conway, A. D. Pritchard, Hyperbolic reflections for the bimonster and $3Fi_{24}$, in: Groups, combinatorics and geometry , ed. M. Liebeck, J. Saxl, Proceedings of the LMS Symposium, Durham 1990, London Math. Soc. Lecture Note Ser. **165**, Cambridge 1992, 24–45.

[Co3] J. H. Conway, N. J. Sloane, Sphere packings, lattices and groups, Springer-Verlag, second edition, 1993.

[Hum] J. E. Humphreys, Reflection groups and Coxeter groups, Cambridge Stud. Adv. Math. **29**, Cambridge University Press, Cambridge 1990.

[Iva] A. A. Ivanov, A geometric chracterization of the monster, in: Groups, combinatorics and geometry, ed. M. Liebeck, J. Saxl, Proceedings of the LMS Symposium, Durham 1990, London Math. Soc. Lecture Note Ser. **165**, Cambridge 1992, 46–62.

[Nor]   S. P. Norton, Constructing the monster, in: Groups, combinatorics and geometry, eds. M. Liebeck, J. Saxl, Proceedings of the LMS Symposium, Durham 1990, London Math. Soc. Lecture Note Ser. **165**, Cambridge 1992, 63–76.

[Soi]   L. H. Soicher, More on the group Y_{555} and the projective plane of order 3, J. Algebra **136** (1991), 168–174.

[Vin]   E. B. Vinberg, Some arithmetical discrete groups in Lobacevskii spaces, in: Discrete subgroups of Lie groups and applications to moduli, Oxford University Press, 1975, 323–348.

Department of Mathematics
Princeton University

Current address:

Department of Mathematics & Statistics
Concordia University
1455 de Maisonneuve West
Montreal, Quebec H3G IM8
Canada
E-mail: simons@cicma.concordia.ca

PART II

LIE ALGEBRAS

On graded Lie algebras of characteristic three with classical reductive null component

Georgia Benkart,[1] Thomas Gregory, and Michael I. Kuznetsov[2]

Abstract. We consider the finite-dimensional transitive irreducible graded Lie algebras $L = \bigoplus_{i=-2}^{r} L_i$ of depth two over an algebraically closed field of characteristic three with classical reductive component L_0 and show that the L_0-module L_{-1} in such algebras must be restricted. This result contrasts with the depth-one case, where nonrestricted modules L_{-1} can occur.

1. Preliminaries

Finite-dimensional transitive irreducible graded Lie algebras $L = \bigoplus_{i=-q}^{r} L_i$ of depth $q \geq 1$ with classical reductive component L_0 play a crucial role in the classification of finite-dimensional simple Lie algebras over algebraically closed fields F of characteristic $p > 0$. In the case when L_{-1} is a restricted irreducible L_0-module, the description of such algebras for $p > 5$ is given by the Recognition Theorem of Kac [K1] (see also [BGP]), and for $p = 3$, $q = 1$ by the Kostrikin–Ostrik Theorem [KO]. In [BG] it is shown that for $p > 5$, L_{-1} must be a restricted L_0-module (the assertion is also true for $p = 5$ (see [BGP])). In the case $p = 3$ the situation is more complicated. For $q = 1$, all simple graded Lie algebras of characteristic three with nonrestricted L_0-module L_{-1} were determined in [BKK]. In the present paper we show that when $p = 3$ and $q = 2$, the L_0-module L_{-1} must be restricted. More exactly, we prove the following theorem:

Theorem 1.1. *Let* $L = L_{-2} \oplus L_{-1} \oplus L_0 \oplus L_1 \oplus \cdots \oplus L_r$ *be a finite-dimensional graded Lie algebra over an algebraically closed field* F *of characteristic* $p = 3$ *such that the following hold:*
(A) *L_0 is classical reductive;*
(B) *L_{-1} is an irreducible L_0-module;*
(C) *for all $j \geq 0$, if $x \in L_j$ and $[x, L_{-1}] = (0)$, then $x = 0$ (i.e. L is transitive);*

1 The main results of this paper were obtained while the first and third authors visited the Ohio State University at Columbus. They would like to thank the Ohio State University for its hospitality and support. The first author also gratefully acknowledges support from National Science Foundation Grants #DMS–9300523 and #DMS–9622447, and from the Ellentuck Fund at the Institute for Advanced Study, Princeton.

2 The third author gratefully acknowledges partial support from the Russian Foundation of Basic Research Grant #96-01-01756.

(D) $L_{-i} = [L_{-i+1}, L_{-1}]$ *for all* $i > 1$;
(E) $[L_{-2}, L_1] \neq (0)$.
   Then L_{-1} *is a restricted module for* $L_0' = [L_0, L_0]$ *under the adjoint action.*

Because there are only finitely many irreducible restricted modules for the derived algebra of a classical reductive Lie algebra, Theorem 1.1 reduces what needs to be considered in classifying the simple Lie algebras over algebraically closed fields of characteristic three. It is also the first step towards proving the following conjecture:

Conjecture. *Let* L *be a graded Lie algebra of characteristic three satisfying conditions* (A)–(E) *of Theorem* 1.1. *If* $q > 1$, *then* L_{-1} *is a restricted module for* L_0' *under the adjoint action.*

Our proof of Theorem 1.1 will require some definitions and preliminary results. Recall that *a classical Lie algebra* over a field F of characteristic $p > 0$ is obtained from using a Chevalley basis to construct a $\mathbb{Z}$-form of a complex simple Lie algebra, reducing the scalars modulo p, extending them to F, and then, perhaps, factoring out the center. The Lie algebras $\mathfrak{gl}(pk)$, $\mathfrak{pgl}(pk)$ are also considered to be classical Lie algebras. Thus, a classical Lie algebra $\mathfrak{g}$ may have a nontrivial center $Z(\mathfrak{g})$ such as in the case of the algebras $\mathfrak{gl}(pk)$, $\mathfrak{sl}(pk)$, and E_6 $(p = 3)$, or a noncentral ideal as in the case of the algebras $\mathfrak{gl}(pk)$, $\mathfrak{pgl}(pk)$, and G_2 $(p = 3)$. In characteristic three, G_2 contains an ideal isomorphic to $\mathfrak{psl}(3)$.

A *classical reductive Lie algebra* is a sum of commuting ideals which are classical Lie algebras or is a central extension of such a sum. The extension is not necessarily a split one, and a "bouquet" of classical Lie algebras with a common nonsplit center might be possible. Since by Schur's lemma an element of the center acts as a scalar on a finite-dimensional irreducible module, the classical reductive Lie algebra occurring as the L_0-component in Theorem 1.2 can have at most a one-dimensional center because of transitivity. A classical Lie algebra $\mathfrak{g}$ has a natural p-structure such that $e_\alpha^{[p]} = 0$, $h_i^{[p]} = h_i$ for any Chevalley basis $\{e_\alpha, \alpha \in R, h_i, i = 1, \ldots, \mathrm{rank}(\mathfrak{g})\}$ of $\mathfrak{g}$, where R is the root system of the corresponding complex simple Lie algebra. For the algebras $\mathfrak{g} = \mathfrak{gl}(pk)$, $\mathfrak{pgl}(pk)$, the derived algebra $\mathfrak{g}' = \mathfrak{sl}(pk)$, $\mathfrak{psl}(pk)$ has such a p-structure. A reductive Lie algebra has a p-structure which is the natural p-structure on each classical summand. To obtain a p-structure on a bouquet of classical ideals, each with a one-dimensional center, we factor out a central ideal of codimension one from the p-ideal which is the sum of their centers.

Let $\pi : L \longrightarrow \mathfrak{gl}(V)$ be a finite-dimensional irreducible representation of a restricted Lie algebra L. The *character* χ of π is a linear functional on L such that $\pi(y)^p - \pi(y^{[p]}) - \chi(y)^p I = 0$ for all $y \in L$. The representation is restricted when the character $\chi = 0$.

Lemma 1.2. *(see* [BG, *Lemma* 1]*). Assume* L *is a graded Lie algebra satisfying conditions* (A)–(D) *of Theorem* 1.1. *If* χ *is the character of* L_0 *on* L_{-1}, *then* L_0 *has character* $-j\chi$ *on* L_j *for all* j.

The following theorem of Weisfeiler **[W]** has played a critical role in the study of graded Lie algebras.

Theorem 1.3. *Let* $G = G_{-q} \oplus \cdots \oplus G_{-1} \oplus G_0 \oplus G_1 \oplus \cdots \oplus G_r$ *be a graded Lie algebra such that conditions* (B)–(D) *hold. Let* $M(G)$ *denote the largest ideal of* G *contained in* $G_{-q} \oplus \cdots \oplus G_{-1}$. *Then:*

(i) $G/M(G)$ *is semisimple and contains a unique minimal ideal* $I = S \otimes A(n : \underline{1})$, *where* S *is a simple Lie algebra,* $n \geq 0$, *and* $A(n : \underline{1}) = F[t_1, \ldots, t_n]/(t_1^p, \ldots, t_n^p)$. *The ideal* I *is graded and* $I_i = (G/M(G))_i$ *for all* $i < 0$.

(ii) *If* $I_1 = (0)$, *then for some* κ, $1 \leq \kappa \leq n$, *the algebra* $A(n : \underline{1})$ *is graded by setting* $\deg(t_i) = -1$ *for* $1 \leq i \leq \kappa$ *and* $\deg(t_i) = 0$ *for* $\kappa < i \leq n$. *Then* $I_i = S \otimes A(n : \underline{1})_i$ *for all* i, $G_2 = (0)$, $I_0 = [G_{-1}, G_1]$, *and* $G_1 \subseteq \{D \in 1 \otimes \text{Der } A(n : \underline{1}) | \deg(D) = 1\}$.

(iii) *If* $I_1 \neq (0)$, *then* S *is graded and* $I_i = S_i \otimes A(n : \underline{1})$ *for all* i. *Moreover,* $(0) \neq [G_{-1}, G_1] \subseteq I_0$.

In what follows, all Lie algebras will be finite-dimensional over an algebraically closed field F of characteristic $p = 3$. The ideal $[L, L]$ of a Lie algebra L will be denoted by L'.

2. The case $q = 1$

Here we collect some results from **[Ko]**, **[BKK]** that we will need later. We begin by describing the depth-one simple Lie algebras of characteristic three with a classical reductive component L_0 and a nonrestricted L_0-module L_{-1}.

For an m-tuple $\underline{n} = (n_1, \ldots, n_m)$ of positive integers, let

$$A(m : \underline{n}) = \langle x^{(a)} = x_1^{(a_1)} \ldots x_m^{(a_m)} \mid 0 \leq a_i < p^{n_i}, \ i = 1, \ldots, m \rangle$$

denote the algebra of divided powers over F with $x^{(a)} x^{(b)} = \prod_i \binom{a_i + b_i}{a_i} x^{(a+b)}$. For $i = 1, \ldots, m$, let $\partial_i : x^{(a)} \mapsto x_1^{(a_1)} \ldots x_i^{(a_i - 1)} \ldots x_m^{(a_m)}$ be the ith "partial derivative" on $A(m : \underline{n})$. The Lie algebra $W(m : \underline{n})$ consists of the derivations $D = \sum_i f_i \partial_i$, $f_i \in A(m : \underline{n})$, of the algebra $A(m : \underline{n})$.

The Hamiltonian Lie algebra $L = H(2 : \underline{n}, \omega)$, $\underline{n} = (1, n_2)$, corresponds to the differential form $\omega = (\exp x^{(3)}) dx \wedge dy$ (see **[K2]**, **[Ki]**). This Lie algebra has a realization as the divided power algebra $A(2 : \underline{n})$, $\underline{n} = (1, n_2)$, with the Poisson bracket

$$\{f, g\} = \partial_y f (\partial_x g + g x^{(2)}) - \partial_y g (\partial_x f + f x^{(2)}). \tag{2.1}$$

Relative to the grading of type $(0, 1)$ on the algebra $H(2 : \underline{n}, \omega)$, in which $\deg x^{(i)} y^{(j)} = j - 1$,

$$L_{-1} = \langle 1, x, x^{(2)} \rangle, \quad L_0 = \langle y, xy, x^{(2)} y \rangle, \quad \text{and} \quad L_1 = \langle y^{(2)}, xy^{(2)}, x^{(2)} y^{(2)} \rangle. \tag{2.2}$$

Obviously, $L_0 \cong \mathfrak{sl}(2)$. Set

$$f = \text{ad}_{L_{-1}} y, \quad e = \text{ad}_{L_{-1}} x^{(2)} y, \quad \text{and} \quad h = -\text{ad}_{L_{-1}} xy. \tag{2.3}$$

Then e, f, h satisfy the standard $\mathfrak{sl}(2)$-relations,

$$[e, f] = h, \quad [h, e] = 2e, \quad [h, f] = -2f, \tag{2.4}$$

and in addition,

$$e^3 = 0, \quad h^3 - h = 0, \; f^3 = 1, \quad \text{and} \quad t \overset{\text{def}}{=} (-h+1)^2 + ef = 1 \tag{2.5}$$

on L_{-1}. In **[Ku**, Theorem 3.2] the derivations of the Lie algebra $H(2 : \underline{n}, \omega)$ have been shown to be given by

$$\text{Der } H(2 : \underline{n}, \omega) = CH(2 : \underline{n}, \omega) \overset{\text{def}}{=} \{D \in W(2 : \underline{n}) \mid D\omega = c\omega, \; c \in F\}. \tag{2.6}$$

The contact Lie p-algebra K_3 on three variables x, y, z of characteristic $p = 3$ is identified with the divided power algebra $A(3 : \underline{1}) = \langle x^{(i)} y^{(j)} z^{(k)} \mid 0 \le i, j, k \le 2 \rangle$ under the Jacobi bracket

$$[f, g] = \Delta f \partial_z g - \Delta g \partial_z f + \partial_x f \partial_y g - \partial_y f \partial_x g,$$

where

$$\Delta f = 2f - x \partial_x f - y \partial_y f.$$

For each $\varepsilon \in F^\times = F \setminus \{0\}$, the Lie algebra $L(\varepsilon)$ (see **[Ko]**) is the subalgebra of K_3 given by

$$L(\varepsilon) = L_{-1} + L_0 + L_1,$$

where

$$L_{-1} = \langle 1, x, x^{(2)} \rangle, \quad L_0 = \langle y, xy, z, (1+\varepsilon)x^{(2)}y + \varepsilon xz \rangle \cong \mathfrak{gl}(2) \tag{2.7}$$

and

$$L_1 = \langle y^{(2)}, (1+\varepsilon)xy^{(2)} - \varepsilon yz, \varepsilon(1+\varepsilon)x^{(2)}y^{(2)} + \varepsilon^2 z^{(2)} \rangle.$$

When $\varepsilon = 1$, the algebra $L(\varepsilon)$ is isomorphic to the classical simple Lie algebra of type C_2. For $\varepsilon \neq 1$ set

$$e = \text{ad}_{L_{-1}} y, \quad h = \text{ad}_{L_{-1}} (\varepsilon - 1)^{-1}(xy + \varepsilon z), \tag{2.8}$$

and

$$f = \text{ad}_{L_{-1}} \left(\frac{\varepsilon + 1}{1 - \varepsilon} x^{(2)} y + \frac{\varepsilon}{1 - \varepsilon} xz \right).$$

These operators satisfy the $\mathfrak{sl}(2)$-relations (2.4) along with the relations

$$e^3 = 0, \quad h^3 - h = -\frac{\varepsilon(\varepsilon + 1)}{\varepsilon^3 - 1}, \quad f^3 = 0,$$

and

$$t = (-h+1)^2 + ef = \frac{(\varepsilon + 1)^2}{(\varepsilon - 1)^2}. \tag{2.9}$$

The Lie algebra M (see [**BKK**]) is by definition the 10-dimensional subalgebra of K_3 defined by

$$M = M_{-1} + M_0 + M_1,$$

where

$$M_{-1} = \langle 1, x, x^{(2)} \rangle, \quad M_0 = \langle y + x^{(2)}z, z - xy, z, xz \rangle \cong \mathfrak{gl}(2), \qquad (2.10)$$

and

$$M_1 = \langle y^{(2)} - xz^{(2)}, z^{(2)}, yz + z^{(2)}x^{(2)} \rangle.$$

The elements

$$e = \mathrm{ad}_{M_{-1}} xz, \quad h = \mathrm{ad}_{M_{-1}}(z - xy), \quad f = \mathrm{ad}_{M_{-1}}(y + x^2 z)$$

satisfy the standard $\mathfrak{sl}(2)$-relations, and $M_0' \cong \mathfrak{sl}(2)$. In this case we have

$$e^3 = 0, \quad h^3 - h = 0, \quad f^3 = 1, \quad \text{and} \quad t = (-h+1)^2 + ef = 0. \qquad (2.11)$$

The next proposition will follow from Corollary 2.8 of [**BKK**] and Theorem 1.3.

Proposition 2.12. *Let* $L = L_{-1} \oplus L_0 \oplus L_1 \oplus \cdots \oplus L_r$ *be a graded Lie algebra satisfying conditions* (A), (B), (C) *of Theorem* 1.1, *and suppose that* $L_1 \neq (0)$. *If* L_{-1} *is a nonrestricted* L_0'-module, *then either* L *is isomorphic to one of the algebras* $L(\varepsilon)$ *or* M, *or* L *is a Hamiltonian Lie algebra such that* $H(2 : \underline{n}, \omega) \subseteq L \subseteq CH(2 : \underline{n}, \omega)$, *where* $\underline{n} = (1, n_2)$, $\omega = (\exp x^{(3)})dx \wedge dy$, *and the grading is of type* (0, 1).

Proof. It follows from irreducibility (B) and transitivity (C) that $M(L) = (0)$ (in the notation of Theorem 1.3), and that L acts faithfully on its unique minimal ideal $I = S \otimes A(n : \underline{1})$. Thus, $L \subseteq \mathrm{Der}(S \otimes A(n : \underline{1}))$ for some n. Suppose first that for the ideal $I = S \otimes A(n : \underline{1})$, $I_1 = (0)$. If $\kappa < n$, then $I_0 = S \otimes A(n : \underline{1})_0$ is not a classical reductive ideal of L_0. Thus, $n = \kappa$, $I_0 = S \otimes 1$, $I_{-1} = S \otimes \langle t_1, \ldots, t_n \rangle$. But then $(0) = I_{-2} = S \otimes \langle t_i t_j \mid 1 \leq i, j \leq n \rangle$, which shows that the case $I_1 = (0)$ is impossible. If instead $I_1 \neq (0)$, then because $I_0 = S_0 \otimes A(n : \underline{1})$ must be a classical reductive ideal of L_0, we have that $n = 0$ and $I = S$. Consequently, $L \subseteq \mathrm{Der}\, S$ where $S = S_{-1} \oplus S_0 \oplus S_1 \oplus \cdots \oplus S_k$ is a simple graded Lie algebra and $S_{-1} = L_{-1}$.

Let χ denote the character of L_0 on L_{-1}. If $\chi(S_0') = 0$, then according to [**KO**], the Lie algebra S is of classical or Cartan type. If S is of Cartan type, then $(\mathrm{Der}\, S)_0$ is one of the algebras $\mathfrak{gl}(n)$, $\mathfrak{csp}(n)$. For a classical Lie algebra of depth one, the component $S_0 \neq \mathfrak{psl}(3)$. Therefore, $(\mathrm{Der}\, S)_0$, which is contained in the normalizer of S_0 in $\mathfrak{gl}(S_{-1})$, cannot be the classical Lie algebra G_2. In any case $(\mathrm{Der}\, S)_0' = S_0'$. Since $L_0 \subseteq (\mathrm{Der}\, S)_0$ we have $L_0' = S_0'$. But $\chi(L_0') \neq 0$. Thus, $\chi(S_0') \neq 0$, and we may use the classification of the simple depth-one Lie algebras with nonrestricted S_0'-module S_{-1} obtained in [**BKK**]. These algebras are $H(2 : \underline{n}, \omega)$, $\underline{n} = (1, n_2)$, $\omega = (\exp x^{(3)})dx \wedge dy$, with the (0, 1)-grading, and the Kostrikin algebras $L(\varepsilon)$ and M. According to [**Ko**], any derivation of $L(\varepsilon)$ or M is inner. Since the derivations of the Lie algebra $H(2 : \underline{n}, \omega)$ are as described in (2.6) above, the conclusions hold. $\qquad\square$

Corollary 2.13. *Under the assumptions of Proposition* 2.12, $L_0' \cong \mathfrak{sl}(2)$, L_1 *is an irreducible three-dimensional L_0'-module, and $[L_1, L_1] = 0$. If $[L_{-1}, L_1] \cong \mathfrak{sl}(2)$, then L is a Hamiltonian Lie algebra.*

Proof. The assertions in Corollary 2.13 follow from Proposition 2.12 and the description of the Lie algebras $H(2 : \underline{n}, \omega)$, $L(\varepsilon)$, M given in (2.2), (2.7), and (2.10) above.   □

Lemma 2.14. *Let $L = L_{-1} \oplus L_0 \oplus L_1$ be one of the Lie algebras $L(\varepsilon)$, M, $H(2 : \underline{n}, \omega)$ with $\underline{n} = (1, n_2)$, let χ be the nonzero character of the L_0-module L_{-1}, and let V be an L-module such that $l^3 \cdot V = (0)$ for any $l \in L_{-1} \cup L_1$. Suppose that W is an irreducible L_0-submodule of V with character $\chi_W = \zeta \chi$, $\zeta \in F^\times$, and suppose that $L_1 \cdot W = (0)$. Then $L_{-1}^2 \cdot W \neq (0)$.*

Proof. We consider the case $L = L(\varepsilon)$ in detail. The other cases can be treated in a similar fashion.

When $L = L(\varepsilon)$, $\varepsilon \neq 1$, the subalgebra $L_0 \cong \mathfrak{gl}(2)$ has a p-structure given by $e^{[p]} = f^{[p]} = 0$, $h^{[p]} = h$ for the standard basis $e = x^{(2)}y$, $f = y$, $h = -xy$ of $L_0' \cong \mathfrak{sl}(2)$, and $c^{[p]} = c$ for the central element $c = xy + z$ (see (2.7)). The character χ of L_0 on L_{-1} takes on the following values,

$$\chi(e) = \chi(f) = \chi(c) = 0, \quad \chi(h)^3 = -\frac{\varepsilon(\varepsilon + 1)}{\varepsilon^3 - 1},$$

as we can see from (2.9). Therefore,

$$e_W^3 = f_W^3 = 0, \quad h_W^3 - h_W = \chi_W(h)^3 I = \zeta^3 \chi(h)^3 \neq 0, \quad \text{and} \quad c_W^3 - c_W = 0, \quad (2.15)$$

where y_W is the representing transformation of $y \in L_0$ on W. Since h and c act as commuting semisimple transformations on W, the elements $xy, z \in \langle h, c \rangle$ are also semisimple on W. It follows from (2.10) (see **[BO]**) that there exists a basis $\{w_0, w_1, w_2\}$ of W such that

$$e \cdot w_0 = 0, \quad e \cdot w_1 = \delta w_0, \quad e \cdot w_2 = (-\delta + 1)w_1,$$
$$h \cdot w_i = (\delta + i)w_i, \quad i = 0, 1, 2,$$
$$f^i \cdot w_0 = w_i, \quad i = 1, 2,$$

for some scalar $\delta \in F$. Since c acts as a scalar on W, w_0 is an eigenvector of $xy \in L_0$, say $(xy) \cdot w_0 = \theta w_0$, $\theta \in F$. It is easy to verify that $[xy, f] = -f$ in L. Thus,

$$(xy) \cdot w_1 = (\theta - 1)w_1, \quad (xy) \cdot w_2 = (\theta + 1)w_2.$$

Let us consider the subalgebra $\mathfrak{g}_1 = \langle e_1, f_1, h_1 \rangle \subset L$ where $e_1 = y^{(2)} \in L_1$, $f_1 = -x^{(2)} \in L_{-1}$, $h_1 = xy \in L_0$. Because $L_1 \cdot W = (0)$, all the vectors w_0, w_1, w_2 are e_1-extremal and correspond to different weights θ, $\theta - 1$, $\theta + 1$ of $h_1 = xy$. As $e_1^3 \cdot V = f_1^3 \cdot V = (0)$, there exists $i \in \{0, 1, 2\}$ such that the $\mathfrak{g}_1$-module $U(\mathfrak{g}_1) \cdot w_i = \langle w_i, f_1 \cdot w_i, f_1^2 \cdot w_i \rangle$ is a three-dimensional irreducible $\mathfrak{g}_1$-module. In particular, $f_1^2 \cdot w_i \neq 0$. Since $f_1 \in L_{-1}$, we have $0 \neq f_1 \cdot w_i \in L_{-1}^2 \cdot W$, which is the desired conclusion.

For $L = M$, we can argue similarly with $\mathfrak{g}_1 = \langle e_1, f_1, h_1 \rangle$, where

$$e_1 = y^{(2)} - xz^{(2)} \in L_1, \quad f_1 = -x^{(2)} \in L_{-1}, \quad h_1 = xy \in L_0.$$

$L = H(2 : \underline{n}, \omega)$, we may use the subalgebra $\mathfrak{g}_1 = \langle e_1, f_1, h_1 \rangle$, where

$$e_1 = y^{(2)} \in L_1, \quad f_1 = -x^{(2)} \in L_{-1}, \quad \text{and} \quad h_1 = -xy \in L_0. \qquad \square$$

3. The reduction to simple graded Lie algebras

Proposition 3.1. *Let $L = L_{-2} \oplus L_{-1} \oplus L_0 \oplus L_1 \oplus \cdots \oplus L_r$ be a graded Lie algebra satisfying the conditions of Theorem 1.1. If the representation of L_0' on L_{-1} is not restricted, then in the notation of Theorem 1.3,*
i) $M(L) \subsetneqq L_{-2}$, *and*
ii) $I_1 \neq (0)$ *and* $I = S$ *is a simple ideal of L.*

Proof. (Compare [**BG**, the proof of the Proposition].)
    Part i) follows from conditions (B), (C), (E).
    For ii) suppose that $I_1 = (0)$. When $\kappa < n$ (in the notation of Theorem 1.3), then $I_0 = S \otimes A(n : \underline{1})_0$ is not a classical reductive ideal of L_0. Thus, $n = \kappa$, $I_0 = S \otimes 1$, $I_{-1} = S \otimes \langle t_1, \ldots, t_n \rangle$. By condition (D), $2 = q = (p-1)n = 2n$ (recall $p = 3$ here). Therefore, $n = 1$, $I_{-1} = S \otimes \langle t \rangle = L_{-1}$, and $S \otimes 1 \subseteq L_0 \subseteq (\text{Der } S \otimes 1) + (1 \otimes \langle t \frac{d}{dt} \rangle)$. We have $S \otimes 1 \subseteq L_0' \subseteq (\text{Der } S)' \otimes 1$. Thus, one of the following cases occurs:
a) $L_0' = S \otimes 1 \cong S$, a classical simple Lie algebra or
b) L_0' is the classical Lie algebra G_2 and $S = \mathfrak{psl}(3)$.
    In any event, the representation of L_0' on $L_{-1} = S \otimes \langle t \rangle$ is restricted. Thus, case (ii) in Theorem 1.3 cannot occur for such an L. If instead $I_1 \neq (0)$, then $I_0 = S_0 \otimes A(n : \underline{1})$ is a classical reductive ideal of L_0, which is possible only if $n = 0$ and $I = S$. $\qquad \square$

Assume now that the Lie algebra L satisfies the conditions of Theorem 1.1 and L_{-1} is nonrestricted L_0-module; that is, $\chi(L_0') \neq 0$. As in [**BG**] we consider a Lie algebra G generated by $L_{-1} \oplus L_0 \oplus U_1$, where U_1 is an irreducible L_0-submodule of L_1. By Lemma 1.2, U_1 is nontrivial whenever $\chi \neq 0$. Factoring out $M(G)$, we may suppose that $M(G) = (0)$, and just as in [**BG**, Proof of the Main Theorem], we obtain that $\text{Ann}_{G_-} G_1 = (0)$, where "$\text{Ann}_{G_-}$" denotes the annihilator in $G_- \overset{\text{def}}{=} \oplus_{j \geq 1} G_{-j}$.
    Thus in completing the proof of Theorem 1.1, we may assume that (i) $L = L_{-2} \oplus L_{-1} \oplus L_0 \oplus L_1 \oplus \cdots \oplus L_r$ is a graded Lie algebra satisfying conditions (A)–(E) of Theorem 1.1, (ii) L_{-1} is a nonrestricted L_0'-module, (iii) L_1 is an irreducible L_0-module, and (iv) $L_i = [L_{i-1}, L_1]$ for $i > 1$.
    From Proposition 3.1 we have that $S \subseteq L \subseteq \text{Der } S$ where $S = S_{-2} \oplus S_{-1} \oplus S_0 \oplus S_1 \oplus \cdots \oplus S_k$ is a simple Lie algebra with $S_i = L_i$ for $i < 0$. Since S_1 is an L_0-submodule of L_1 and L_1 is irreducible, we have $S_1 = L_1$. Consequently, $S_i = L_i$ for $i > 0$. Thus, L may differ from S only in the null component. In particular, $k = r$. By symmetry we may assume that $r \geq q = 2$, since the depth-one Lie algebras are known from [**BKK**] and [**KO**] and none of them has $r = 2$.

Therefore as in [**BG**, Sec. 5], we will consider graded Lie algebras $L = L_{-2} \oplus L_{-1} \oplus L_0 \oplus L_1 \oplus \cdots \oplus L_r$ such that:

(i) L satisfies conditions (A)–(E) of Theorem 1.1.

(ii) $L \subseteq \text{Der } S$ where $S = S_{-2} \oplus S_{-1} \oplus S_0 \oplus S_1 \oplus \cdots \oplus S_r$ is a simple graded Lie algebra.

(iii) $L_i = S_i$, $i \neq 0$.

(iv) $L_1 = S_1$ is an irreducible L_0-module.

(v) $L_{i+1} = [L_i, L_1]$ for $i > 0$.

(vi)   If x is a nonzero element in L_{-i} for some $i > 0$, then $[L_1, x] \neq (0)$.

(vii)   The character χ of L_0' on L_{-1} is nonzero.

(viii) $r \geq q = 2$.

The next lemma follows from the simplicity of S.

Lemma 3.2. (See [**BG**, Lemma 9].) S_{-2} *is an irreducible* S_0-*module.*

Lemma 3.3. (See [**BG**, Lemma 15].) $(\text{ad } L_{-2})^2 L \neq (0)$.

Corollary 3.4. $[S_{-2}, [S_{-2}, S_2]] \neq (0)$.

Proof. By Lemma 3.3 there exists a minimal k such that $[S_{-2}, [S_{-2}, L_k]] \neq (0)$. Evidently, $k \geq 2$. If $k > 2$, then

$$(0) \neq [S_{-2}, [S_{-2}, L_k]] = [S_{-2}, [S_{-2}, S_k]]$$

is an L_0-submodule in S_{k-4} and $k - 4 \geq -1$. When $k - 4 = -1$, then $[S_{-2}, [S_{-2}, S_k]] = S_{-1}$ since S_{-1} is an irreducible L_0-module. As a result,

$$(0) \neq S_{-2} = [S_{-1}, [S_{-2}, [S_{-2}, S_k]]] = [S_{-2}, [S_{-2}, [S_{-1}, S_k]]] \subseteq [S_{-2}, [S_{-2}, S_{k-1}]],$$

contrary to the choice of k. When $k - 4 > -1$, then by transitivity,

$$(0) \neq [S_{-1}, [S_{-2}, [S_{-2}, S_k]]] = [S_{-2}, [S_{-2}, [S_{-1}, S_k]]] \subseteq [S_{-2}, [S_{-2}, S_{k-1}]],$$

which again contradicts our choice of k. Thus, $k = 2$ and $[S_{-2}, [S_{-2}, S_2]] \neq (0)$.   □

Lemma 3.5. (See [**G2**, Lemma 9].) $\text{Ann}_{L_i} S_{-2} = \text{Ann}_{S_i} S_{-2} = (0)$ *for any* $i > 0$.

Consider the subalgebra $E = E_{-1} \oplus E_0 \oplus \cdots \oplus E_{\lfloor \frac{r}{2} \rfloor}$ of L consisting of the even gradation spaces $E_i = L_{2i}$. Set $A_0 = \text{Ann}_{E_0} E_{-1} = \text{Ann}_{L_0} S_{-2}$, and for $i = 1, 2, \ldots$, let

$$A_i = \{x \in E_i \mid [x, E_{-1}] \subseteq A_{i-1}\}.$$

Then $A = A_0 \oplus A_1 \oplus \ldots \oplus A_{\lfloor \frac{r}{2} \rfloor}$ is an ideal of E, and the factor algebra $G = E/A = G_{-1} \oplus G_0 \oplus G_1 \oplus \cdots \oplus G_k$ is a transitive depth-one graded Lie algebra (see [**BG**, Lemma 5]). By Corollary 3.4, $[S_{-2}, S_2] = [E_{-1}, E_1] \not\subseteq A_0$, so $G_1 \neq (0)$. Thus, the Lie algebra G satisfies conditions (A)–(C) of Theorem 1.1, and G_{-1} is a nonrestricted G_0-module.

Lemma 3.6.
a) $\mathfrak{sl}(2) \subseteq L_0/A_0 \subseteq \mathfrak{gl}(2)$.
b) $\dim S_{-2} = 3$.
c) $A_0' \subseteq A_\chi$, the maximal ideal of L_0 contained in $\ker \chi$.
d) $L_0' = A_0' \oplus \mathfrak{sl}(2)$, where $\mathfrak{sl}(2) \subseteq [S_{-2}, S_2]$.
e) L_0 does not contain a bouquet of classical ideals.

Proof. According to Proposition 2.12, either $G = E/A$ is one of the algebras $L(\varepsilon)$, M, or $H(2 : \underline{n}, \omega) \subseteq G \subseteq CH(2 : \underline{n}, \omega)$ with $\underline{n} = (1, n_2)$; whence we obtain a) and b). Since $S_{-2} = L_{-2} = [L_{-1}, L_{-1}]$ is an irreducible L_0-module, each element of the center $Z(L_0)$ must act as scalar multiplication on it with value twice that of its action on $L_{-1} = S_{-1}$. As an immediate consequence we have $Z(L_0) \cap A_0 = (0)$. Thus, A_0 is a centerless classical ideal of L_0, which implies A_0' is a p-ideal. Because A_0 annihilates $S_{-2} = L_{-2}$, we have $A_0' \subseteq A_\chi$ as claimed. To prove d) note that $I' = I$ for any noncentral ideal I of L_0'. Thus,

$$A_0 \cap L_0' = (A_0 \cap L_0')' \subseteq A_0' \subseteq A_0 \cap L_0',$$

which shows that $A_0' = A_0 \cap L_0'$. Because $G_0' = (L_0' + A_0)/A_0 \cong L_0'/A_0 \cap L_0' \cong L_0'/A_0' = \mathfrak{sl}(2)$, $L_0' = A_0' \oplus \mathfrak{sl}(2)$. Since $\mathfrak{sl}(2) \subseteq [G_{-1}, G_1]$, we have $\mathfrak{sl}(2) \subseteq [S_{-2}, S_2]$ to give d).

Finally, if L_0 contains a bouquet Σ of classical Lie algebras, then $\Sigma \cap A_0 = (0)$ because $Z(L) \cap A_0 = (0)$. Therefore, G_0 contains an isomorphic copy of Σ, but that contradicts a). $\qquad\square$

According to Lemma 3.6 a), there is a unique classical ideal of L_0' with nonrestricted action on $L_{-1} = S_{-1}$. To distinguish that ideal, which is isomorphic to $\mathfrak{sl}(2)$, we denote it by $\mathfrak{sl}(2)^1$.

4. The case $A_0 \neq (0)$

We continue studying the simple core S of a Lie algebra L satisfying conditions (i) - (viii); in this section we focus on the case where $A_0 = \mathrm{Ann}_{L_0} S_{-2} \neq (0)$.

Lemma 4.1. *If* $A_0 \neq (0)$ *and* $A_0 \cap S_0 = (0)$, *then*
a) $\mathfrak{sl}(2)^1 \subseteq S_0 \subseteq \mathfrak{gl}(2)$;
b) $A_0 \cong \mathfrak{sl}(2)$, *and* $L_0 = \mathfrak{sl}(2)^1 \oplus A_0 \oplus Z(L_0)$;
c) $S_{-1} = U \otimes V$, *where* U *is an irreducible nonrestricted* $\mathfrak{sl}(2)^1$-*module with* $\dim U = 3$, *and* V *is an irreducible* A_0-*module with* $\dim V = 2$.

Proof. Suppose that $A_0 \cap S_0 = (0)$. By Lemma 3.6 d) $S_0' = \mathfrak{sl}(2)^1$. Thus, $\mathfrak{sl}(2)^1 \subseteq S_0 \subseteq \mathfrak{sl}(2)^1 \oplus Z(L_0)$. It follows from $A_0 \cap Z(L_0) = (0)$ and $A_0 \cap S_0 = (0)$ that S_{-1} is reducible as an S_0-module, and thus, as an $\mathfrak{sl}(2)^1$-module. Because L_0 is classical reductive, there exists an ideal B of L_0 such that $L_0 = \mathfrak{sl}(2)^1 \oplus B$, $A_0 \subseteq B$, and $\mathrm{codim}_B A_0 \leq 1$. Therefore, $S_{-1} = U \otimes V$ where U is an irreducible nonrestricted

$\mathfrak{sl}(2)^1$-module, and V is an irreducible B-module which is restricted as A_0-module (see Lemma 3.6 c)). Consequently, $\dim U = 3$ and $\dim V = \ell > 1$.

If $\{v_1, \ldots, v_\ell\}$ is a basis of V then $S_{-1} = \oplus_{i=1}^{\ell} U \otimes \langle v_i \rangle$ is a direct sum of irreducible S_0-modules. According to [G2], the number of summands in this sum cannot be more than two; i.e., $\dim V = 2$, whence $\mathfrak{sl}(2) \subseteq B \subseteq \mathfrak{gl}(2)$ and $A_0 \cong \mathfrak{sl}(2)$. Thus, $L_0 = \mathfrak{sl}(2)^1 \oplus A_0 \oplus Z(L_0)$.   □

To distinguish this second copy of $\mathfrak{sl}(2)$, which is the ideal A_0 from Lemma 4.1, we adopt the notation $\mathfrak{sl}(2)^0$.

It follows from the fact that S_{-2} is an irreducible L_0-module and a trivial A_0-module that it has the structure $Y \otimes F$, where Y is an irreducible nonrestricted S_0-module, and F is a trivial A_0-module.

Let $\{v_\mu, v_{-\mu}\}$ be a basis of V consisting of weight vectors with respect to some standard basis $\{e_0, f_0, h_0\}$ of $A_0 = \mathfrak{sl}(2)^0$. According to (iv), S_1 is an irreducible L_0-module. In the simple graded Lie algebra S, we have $S_0 = [S_{-1}, S_1]$ (see [Ka, Lemma 15]). This implies that if $A_0 \cap S_0 = (0)$ and $[A_0, S_1] = (0)$, then

$$(0) \neq [S_{-1}, S_1] = [[A_0, S_{-1}], S_1] = [A_0, [S_{-1}, S_1]] = [A_0, S_0] \subseteq A_0 \cap S_0 = (0).$$

Thus, $[A_0, S_1] \neq (0)$. As a consequence, $S_1 = \overline{U} \otimes \overline{V}$ where $\overline{U}$ is an irreducible nonrestricted S_0'-module with $\dim \overline{U} = 3$, and $\overline{V}$ is an irreducible restricted module for $A_0 = \mathfrak{sl}(2)^0$ with $\dim \overline{V} = 2$ or 3.

Suppose that $\dim \overline{V} = 3$. Then $\overline{V}$ has a basis $\langle \bar{v}_{2\mu}, \bar{v}_0, \bar{v}_{-2\mu} \rangle$ consisting of weight vectors relative to the Cartan subalgebra $\langle h_0 \rangle$ of $\mathfrak{sl}(2)^0 = A_0$. By Lemma 4.1, $S_{-1} = (U \otimes \langle v_\mu \rangle) \oplus (U \otimes \langle v_{-\mu} \rangle)$. The product $[U \otimes \langle v_\nu \rangle, \overline{U} \otimes \langle \bar{v}_0 \rangle]$, for $\nu = \pm \mu$, has weight ν with respect to $\langle h_0 \rangle$ and $[S_{-1}, S_1] = S_0$ has weight zero. Thus, $[S_{-1}, \overline{U} \otimes \langle \bar{v}_0 \rangle] = (0)$, which contradicts (C). Consequently, the dimension of $\overline{V}$ must be 2, and we have proved the following lemma:

Lemma 4.2. *If $A_0 \neq (0)$ and $A_0 \cap S_0 = (0)$, then $S_1 = \overline{U} \otimes \overline{V}$, where $\overline{U}$ is an irreducible nonrestricted $\mathfrak{sl}(2)^1$-module with $\dim \overline{U} = 3$, and $\overline{V}$ is a restricted module for $A_0 = \mathfrak{sl}(2)^0$ with $\dim \overline{V} = 2$.*   □

Now we are ready to prove that the case $A_0 \neq (0)$ and $A_0 \cap S_0 = (0)$ is impossible.

Lemma 4.3. *If $A_0 \neq (0)$, then $A_0 \cap S_0 \neq (0)$.*

Proof. Suppose that $A_0 \cap S_0 = (0)$. By Lemma 4.1, $S_0' = \mathfrak{sl}(2)^1$, $A_0 = \mathfrak{sl}(2)^0$, $L_0 = \mathfrak{sl}(2)^1 \oplus \mathfrak{sl}(2)^0 \oplus Z(L_0)$, and $S_{-1} = U \otimes V$, where U is an irreducible nonrestricted $\mathfrak{sl}(2)^1$-module with $\dim U = 3$, and V is an irreducible two-dimensional $\mathfrak{sl}(2)^0$-module. Choose a standard basis $\{e_0, f_0, h_0\}$ of $\mathfrak{sl}(2)^0$. Then V has a basis $\{v_\mu, v_{-\mu}\}$ consisting of weight vectors of weights $\pm\mu$ with respect to $\langle h_0 \rangle$. Analogously, by Lemma 4.2, $S_1 = \overline{U} \otimes \overline{V}$ where $\overline{U}$ is an irreducible nonrestricted $\mathfrak{sl}(2)^1$-module, and $\overline{V}$ is an irreducible restricted $\mathfrak{sl}(2)^0$-module, which we can assume has a basis $\{\bar{v}_\mu, \bar{v}_{-\mu}\}$. Note that the roots of $\mathfrak{sl}(2)^0$ relative to $\langle h_0 \rangle$ are $\pm 2\mu = \mp \mu$.

Since $S_0 = [S_{-1}, S_1]$ and $A_0 \cap S_0 = (0)$, we have

$$[U \otimes \langle v_\nu \rangle, \overline{U} \otimes \langle \bar{v}_\nu \rangle] = (0) \quad \text{for } \nu = \pm\mu.$$

Then it follows from the transitivity of S that

$$[U \otimes \langle v_\nu \rangle, \overline{U} \otimes \langle \bar{v}_{-\nu} \rangle] \neq (0) \quad \text{for } \nu = \pm\mu.$$

Evidently, for each $t \in F$, $x(t) = \exp t(\mathrm{ad}\, e_0)$ and $y(t) = \exp t(\mathrm{ad}\, f_0)$ are automorphisms of the local Lie algebra $S_{\mathrm{loc}} = S_{-1} \oplus S_0 \oplus S_1$ which act trivially on S_0. Thus, $w = x(1)y(-1)x(1)$ is an automorphism of S_{loc} corresponding to the nonidentity element of Weyl group of $\mathfrak{sl}(2)^0$, and $w(s) = s$ for any $s \in S_0$. As a result, for any weight ν of $\overline{V}$,

$$[U \otimes \langle v_\nu \rangle, \overline{U} \otimes \langle \bar{v}_{-\nu} \rangle] = w[U \otimes \langle v_\nu \rangle, \overline{U} \otimes \langle \bar{v}_{-\nu} \rangle]$$
$$= [U \otimes \langle w(v_\nu) \rangle, \overline{U} \otimes \langle w(\bar{v}_{-\nu}) \rangle]$$
$$= [U \otimes \langle v_{-\nu} \rangle, \overline{U} \otimes \langle \bar{v}_\nu \rangle].$$

Therefore,

$$[U \otimes \langle v_\mu \rangle, \overline{U} \otimes \langle \bar{v}_{-\mu} \rangle] = [U \otimes \langle v_{-\mu} \rangle, \overline{U} \otimes \langle \bar{v}_\mu \rangle] = [S_{-1}, S_1] = S_0.$$

Consider the component $S_2 = [S_1, S_1]$. Since $[S_{-2}, S_2] \subseteq S_0$ and $[S_0, A_0] \subseteq A_0 \cap S_0 = (0)$, we obtain $[S_{-2}, [S_2, A_0]] = [[S_{-2}, S_2], A_0] = (0)$. Hence, by Lemma 3.5, $[S_2, A_0] = (0)$. Because the subspace $[\overline{U} \otimes \langle \bar{v}_\nu \rangle, \overline{U} \otimes \langle \bar{v}_\nu \rangle]$ of S_2 consists of weight vectors of weight 2ν with respect to $\langle h_0 \rangle$, we have $[\overline{U} \otimes \langle \bar{v}_\nu \rangle, \overline{U} \otimes \langle \bar{v}_\nu \rangle] = (0)$ for $\nu = \pm\mu$. Therefore, $S_2 = [S_1, S_1] = [\overline{U} \otimes \langle \bar{v}_\mu \rangle, \overline{U} \otimes \langle \bar{v}_{-\mu} \rangle]$.

According to **[G2]**, a simple graded Lie algebra $S = S_{-2} \oplus S_{-1} \oplus S_0 \oplus S_1 \oplus \cdots \oplus S_r$ with reducible component $S_{-1} = U \otimes \langle v_\mu \rangle \oplus U \otimes \langle v_{-\mu} \rangle$ admits a grading of depth one: $S = C_{-1} \oplus C_0 \oplus C_1 \oplus \cdots \oplus C_k$ where

$$C_{-1} = S_{-2} \oplus (U \otimes \langle v_\mu \rangle),$$

and

$$C_0 = (U \otimes \langle v_{-\mu} \rangle) \oplus S_0 \oplus [U \otimes \langle v_\mu \rangle, S_2] = (U \otimes \langle v_{-\mu} \rangle) \oplus S_0 \oplus (\overline{U} \otimes \langle \bar{v}_\mu \rangle).$$

Now because $S_0' = \mathfrak{sl}(2)^1 \subseteq [U \otimes \langle v_{-\mu} \rangle, \overline{U} \otimes \langle \bar{v}_\mu \rangle] = S_0$, the subalgebra C_0 is a depth-one transitive graded Lie algebra, $C_0 = D_{-1} \oplus D_0 \oplus D_1$ whose homogeneous components are $D_{-1} = U \otimes \langle v_{-\mu} \rangle$, $D_0 = S_0$, and $D_1 = \overline{U} \otimes \langle \bar{v}_\mu \rangle$. Moreover, D_{-1} is an irreducible nonrestricted D_0'-module. According to Proposition 2.12, either $H(2 : \underline{n}, \omega) \subseteq C_0 \subseteq CH(2 : \underline{n}, \omega)$ for $\underline{n} = (1, n_2)$ or else C_0 is one of the algebras $L(\varepsilon)$, M.

Observe that C_{-1} is an irreducible C_0-module such that $(\mathrm{ad}\, D_1)^3 C_{-1} = (0)$, and $U \otimes \langle v_\mu \rangle$ is a D_0-submodule of C_{-1} with the same (nonzero) character as $D_{-1} = U \otimes \langle v_{-\mu} \rangle$, and it satisfies $\mathrm{ad}\, D_1(U \otimes \langle v_\mu \rangle) = [\overline{U} \otimes \langle \bar{v}_\mu \rangle, U \otimes \langle v_\mu \rangle] = (0)$. According to Lemma 2.14, $(\mathrm{ad}\, D_{-1})^2(U \otimes \langle v_\mu \rangle) \neq (0)$. However,

$$(\mathrm{ad}\, D_{-1})^2(U \otimes \langle v_\mu \rangle) = [U \otimes \langle v_{-\mu} \rangle, [U \otimes \langle v_{-\mu} \rangle, U \otimes \langle v_\mu \rangle]] \subset [S_{-1}, [S_{-1}, S_{-1}]] = (0).$$

This contradiction shows that the case $A_0 \neq (0)$, $A_0 \cap S_0 = (0)$ is impossible. Thus, whenever $A_0 \neq (0)$, then $A_0 \cap S_0 \neq (0)$ as well.   □

By the reductivity of L_0, there exists an ideal B of L_0 such that $L_0 = \mathfrak{sl}(2)^1 \oplus B$, $A_0 \subseteq B$, $\text{codim}_B A_0 \leq 1$, $L_0' = \mathfrak{sl}(2)^1 \oplus B' = \mathfrak{sl}(2)^1 \oplus A_0'$, and $\mathfrak{sl}(2)^1 \subseteq S_0$ (see Lemma 3.6). Since S_{-1}, S_1 are irreducible L_0-modules, we have as before, $S_{-1} = U \otimes V$, $S_1 = \overline{U} \otimes \overline{V}$ where U, $\overline{U}$ are irreducible nonrestricted $\mathfrak{sl}(2)^1$-modules, and V, $\overline{V}$ are irreducible B-modules which are restricted over A_0'.

We suppose here that $A_0 \neq (0)$. Hence, $\dim V > 1$ and $\dim \overline{V} > 1$. Let H be a Cartan subalgebra of L_0, say $H = H_1 \oplus H_B$, where H_1 is a Cartan subalgebra of $\mathfrak{sl}(2)^1$, and H_0 is a Cartan subalgebra of A_0 contained in the Cartan subalgebra H_B of B. Then the root system R of L_0 is the union of the root system R_1 of $\mathfrak{sl}(2)^1$ with the root system $R_B = R_{A_0}$ of B. Choose a basis $\{u_a\}$ of U ($\{\bar{u}_b\}$ of $\overline{U}$) consisting of weight vectors with respect to H_1, and a basis $\{v_\mu\}$ of V ($\{\bar{v}_\nu\}$ of $\overline{V}$) consisting of weight vectors relative to H_B.

Since the space $\{s \in S_{-1} \mid [s, S_1] \subseteq B\}$ is an L_0-submodule of the irreducible L_0-module S_{-1}, and $\mathfrak{sl}(2)^1 \subseteq [S_{-1}, S_1] \not\subseteq B$, it follows for any nontrivial $\mathfrak{sl}(2)^1$-submodule W of S_{-1}, that $([W, S_1] + B)/B \cong \mathfrak{sl}(2)^1$. Therefore, for any weight vector $v_\nu \in V$, there exists a weight vector $\bar{v}_{-\nu} \in \overline{V}$ such that $[U \otimes \langle v_\nu \rangle, \overline{U} \otimes \langle \bar{v}_{-\nu} \rangle] \not\subseteq B$. Since $[U \otimes \langle v_\nu \rangle, \overline{U} \otimes \langle \bar{v}_{-\nu} \rangle]$ is a nontrivial $\mathfrak{sl}(2)^1$-submodule of the centralizer $C_{L_0}(H_B) = \mathfrak{sl}(2)^1 \oplus H_B$, we have

$$\mathfrak{sl}(2)^1 \subseteq [U \otimes \langle v_\nu \rangle, \overline{U} \otimes \langle \bar{v}_{-\nu} \rangle] \tag{4.4}$$

for any root vector $v_\nu \in V$. Thus, there exist weight vectors $u_a \in U$, $\bar{u}_b \in \overline{U}$ such that $[u_a \otimes v_\nu, \bar{u}_b \otimes \bar{v}_{-\nu}] = e_\alpha \in \mathfrak{sl}(2)^1$ where e_α is a root vector of $\mathfrak{sl}(2)^1$. Because U is a nonrestricted $\mathfrak{sl}(2)^1$-module, the Cartan subalgebra H_1 may be chosen in such a way that $e_\alpha|_U^p \neq 0$. Evidently, $a + b = \alpha$. Let v_λ be a weight vector of the B-module V. Then for any weight vector $u_c \in U$, the vector $[u_a \otimes v_\nu, u_c \otimes v_\lambda]$ in S_{-2} has weight $(\nu + \lambda)|_{H_0}$ with respect to H_0. Since $A_0 = \text{Ann}_{L_0} S_{-2}$, the space S_{-2} consists of weight vectors of weight zero relative to H_0. Therefore, $[u_a \otimes v_\nu, u_c \otimes v_\lambda] = 0$ whenever $\lambda|_{H_0} \neq -\nu|_{H_0}$. Consequently,

$$0 \neq (e_\alpha \cdot u_c) \otimes v_\lambda = [[u_a \otimes v_\nu, \bar{u}_b \otimes \bar{v}_{-\nu}], u_c \otimes v_\lambda]$$
$$= [[u_c \otimes v_\lambda, \bar{u}_b \otimes \bar{v}_{-\nu}], u_a \otimes v_\nu]. \tag{4.5}$$

If $\lambda \neq \nu$, then for any weight c, $[u_c \otimes v_\lambda, \bar{u}_b \otimes \bar{v}_{-\nu}] = e_\delta \neq 0$ where e_δ is a root vector of B corresponding to the root $\delta = \lambda - \nu$. Since $A_0' = B'$, we have $e_\delta \in A_0$. Moreover, $c + b = 0$. But U has three different weights with respect to H_1, and c may be chosen to be different from $-b$. Thus, $\lambda = \nu$ must hold whenever $\lambda|_{H_0} \neq -\nu|_{H_0}$.

Suppose that $\lambda = \nu$, and let $v_\lambda = v_\nu'$, a vector linearly independent from v_ν. Then $[u_c \otimes v_\nu', \bar{u}_b \otimes \bar{v}_{-\nu}] \in \mathfrak{sl}(2)^1 \oplus H_B$. Take $c = a$. Then $a + b = \alpha$, and we obtain $[u_a \otimes v_\nu', \bar{u}_b \otimes \bar{v}_{-\nu}] = x e_\alpha$, $x \in F$, $x \neq 0$. It follows from (4.5) that $0 \neq e_\alpha \cdot u_a \otimes v_\nu' = x e_\alpha \cdot u_a \otimes v_\nu$, which contradicts the linear independence of v_ν, v_ν'. Thus, if λ is a weight of V such that $\lambda|_{H_0} \neq -\nu|_{H_0}$, then $\lambda = \nu$ and $\dim V_\nu = 1$. This allows us to conclude that for any weight λ of V either $\lambda|_{H_0} = -\nu|_{H_0}$ or $\lambda = \nu$. Since ν is an arbitrary weight of V we have $\dim V_\nu = 1$ for any weight ν. It follows from Lemma 3.6 a) and the structure of classical Lie algebras that $\mathrm{codim}_{H_B} H_0 = \mathrm{codim}_B A_0 \le 1$. Therefore, the B-module V has two weights with respect to H_B, which are $\pm\mu$ on H_0, and $\dim V_\mu = \dim V_{-\mu} = 1$. Thus, $\dim V = 2$ and $A_0 \cong \mathfrak{sl}(2)$. To distinguish the copy of $\mathfrak{sl}(2)$ which is A_0 from $\mathfrak{sl}(2)^1$, we will adopt the notation $\mathfrak{sl}(2)^0$ for A_0, as we did in treating the $A_0 \cap S_0 = (0)$ case earlier. Then we have that $L_0 = \mathfrak{sl}(2)^1 \oplus \mathfrak{sl}(2)^0 \oplus Z(L_0)$. Thus, when $A_0 \cap S_0 \ne (0)$, then $A_0 \subset S_0$ and $S_0 = \mathfrak{sl}(2)^1 \oplus \mathfrak{sl}(2)^0 \oplus Z(L_0)$.

It follows from Kantor's result that $S_0 = [S_{-1}, S_1]$. This implies that $[u \otimes v_\mu, \bar{u} \otimes \bar{v}_\mu] = f(u, \bar{u})e_{2\mu}$ where $e_{2\mu}$ is a root vector of $\mathfrak{sl}(2)^0$, and $f(u, \bar{u})$ is a nondegenerate pairing of U and $\overline{U}$, which allows us to identify $\overline{U}$ with U^* as $\mathfrak{sl}(2)^1$-modules. Since $\mathfrak{sl}(2)$ has a unique irreducible module of each dimension, we also have $\overline{V} \cong V^* \cong V$. Let us summarize the results we have just established:

Lemma 4.6. *Let* L *be a graded Lie algebra satisfying conditions* (i)–(viii). *If* $A_0 = \mathrm{Ann}_{L_0} S_{-2} \ne (0)$, *then*

a) $A_0 = \mathfrak{sl}(2)^0$, *and* $L_0' = S_0' = \mathfrak{sl}(2)^0 \oplus \mathfrak{sl}(2)^1$;

b) $S_1 \cong S_{-1}^*$ *as* L_0-*modules and* $S_{-1} = U \otimes V$, *where* U *is an irreducible nonrestricted* $\mathfrak{sl}(2)^1$-*module with* $\dim U = 3$, *and* V *is an irreducible* $\mathfrak{sl}(2)^0$-*module with* $\dim V = 2$. □

In what follows we will identify S_1 with $U^* \otimes V$.

Lemma 4.7. *Under the assumptions of Lemma* 4.6,

$$[A_0, S_2] = (0), \quad \text{and} \quad S_2 = [U^* \otimes \langle v_\mu \rangle, U^* \otimes \langle v_{-\mu} \rangle].$$

Proof. According to Lemma 4.6, $S_1 = (U^* \otimes \langle v_\mu \rangle) \oplus (U^* \otimes \langle v_{-\mu} \rangle)$. By (iii), $S_2 = [S_1, S_1]$. Therefore, $S_2 = [U^* \otimes \langle v_\mu \rangle, U^* \otimes \langle v_\mu \rangle] + [U^* \otimes \langle v_\mu \rangle, U^* \otimes \langle v_{-\mu} \rangle] + [U^* \otimes \langle v_{-\mu} \rangle, U^* \otimes \langle v_{-\mu} \rangle]$. We argue that $[U^* \otimes \langle v_\mu \rangle, U^* \otimes \langle v_\mu \rangle] = (0)$.

Consider the subalgebra T of L generated by $T_{-1} = U \otimes \langle v_{-\mu} \rangle \subset S_{-1}$, $T_0 = \mathfrak{sl}(2)^1 \oplus Z(L_0) \oplus H_0 \subset L_0$, and $T_1 = U^* \otimes \langle v_\mu \rangle \subset S_1$. By (4.4), $\mathfrak{sl}(2)^1 \subseteq [U \otimes \langle v_{-\mu} \rangle, U^* \otimes \langle v_\mu \rangle] \subset T_0$. Since S_{-2} has weight zero with respect to H_0,

$$[U \otimes \langle v_{-\mu} \rangle, U \otimes \langle v_{-\mu} \rangle] = [T_{-1}, T_{-1}] = (0).$$

Therefore, T is a depth-one graded Lie algebra with an irreducible nonrestricted T_0'-module T_{-1}. Let $X_0 = \mathrm{Ann}_{T_0} T_{-1}$, and $X_i = \{x \in T_i \mid [x, T_{-1}] \subseteq X_{i-1}\}$, $i \ge 1$. According to [**BG**, Lemma 5], $X = X_0 \oplus X_1 \oplus \cdots$ is an ideal of T, and T/X is a transitive irreducible depth-one graded Lie algebra. Note that $X_1 \ne T_1$, and since T_1 is an irreducible T_0-module, $X_1 = (0)$. The Lie algebra T/X must be one of the exceptional Lie algebras from Proposition 2.12. By Corollary 2.13, $[\overline{T}_1, \overline{T}_1] = (0)$ where $\overline{T}_1 = (T/X)_1$. Thus, $[T_1, T_1] = X_2$. Since $X_1 = (0)$, then $[T_{-1}, [T_1, T_1]] = (0)$, so that

$$[U \otimes \langle v_{-\mu} \rangle, [U^* \otimes \langle v_\mu \rangle, U^* \otimes \langle v_\mu \rangle]] = (0). \tag{4.8}$$

Now the space $[U \otimes \langle v_\mu \rangle, [U^* \otimes \langle v_\mu \rangle, U^* \otimes \langle v_\mu \rangle]] \subset S_1$ consists of vectors of weight $3\mu = 0$ relative to H_0. But $S_1 = U^* \otimes V$ does not have nonzero vectors of weight zero. Therefore,

$$[U \otimes \langle v_\mu \rangle, [U^* \otimes \langle v_\mu \rangle, U^* \otimes \langle v_\mu \rangle]] = (0). \tag{4.9}$$

Equations (4.8) and (4.9) combine to give that $[S_{-1}, [U^* \otimes \langle v_\mu \rangle, U^* \otimes \langle v_\mu \rangle]] = (0)$, which by the transitivity of L implies that $[U^* \otimes \langle v_\mu \rangle, U^* \otimes \langle v_\mu \rangle] = (0)$. Analogously, $[U^* \otimes \langle v_{-\mu} \rangle, U^* \otimes \langle v_{-\mu} \rangle] = (0)$. Hence, $S_2 = [U^* \otimes \langle v_\mu \rangle, U^* \otimes \langle v_{-\mu} \rangle]$. Because $A_0 = \mathfrak{sl}(2)^0$, then $[A_0, S_2] = (0)$. □

Lemma 4.10. *Under the assumptions of Lemma* 4.6,

$$\mathfrak{sl}(2)^1 \subseteq [S_{-2}, S_2] \subseteq \mathfrak{sl}(2)^1 + H_0 + Z(L_0).$$

Proof. By Lemma 3.6 d), $\mathfrak{sl}(2)^1 \subseteq [S_{-2}, S_2]$, and $[S_{-2}, S_2]$ has weight zero with respect to H_0 by Lemma 4.7. Thus, $[S_{-2}, S_2] \subseteq \mathfrak{sl}(2)^1 + H_0 + Z(L_0)$. □

Lemma 4.11. *Under the assumptions of Lemma* 4.6, $S_3 = (0)$ *and* S_2 *is an irreducible* $\mathfrak{sl}(2)^1$-*module.*

Proof. If $S_3 \neq (0)$, then according to **[G1**, the proof of Lemma 8], $S_0 = [S_{-2}, S_2]$. But by Lemma 4.10, $S_0' = [S_{-2}, S_2]' = \mathfrak{sl}(2)^1$ which contradicts Lemma 4.6 a). Thus, $S_3 = (0)$. Now, by the simplicity of S it follows that S_2 is an irreducible S_0-module. By Lemma 4.6 a) and Lemma 4.7, S_2 is an irreducible $\mathfrak{sl}(2)^1$-module. □

At this juncture we know that if L is a graded Lie algebra satisfying conditions (i)–(viii) and $A_0 = \mathrm{Ann}_{L_0} S_{-2} \neq (0)$, then

$$S = S_{-2} \oplus S_{-1} \oplus S_0 \oplus S_1 \oplus S_2,$$
$$S_0 = \mathfrak{sl}(2)^1 \oplus \mathfrak{sl}(2)^0 \oplus Z(S_0),$$
$$S_{-1} = (U \otimes \langle v_\mu \rangle) \oplus (U \otimes \langle v_{-\mu} \rangle),$$
$$S_1 = (U^* \otimes \langle v_\mu \rangle) \oplus (U^* \otimes \langle v_{-\mu} \rangle), \quad \text{and} \quad S_1 \cong S_{-1}^*,$$
$$U \text{ is an irreducible nonrestricted } \mathfrak{sl}(2)^1\text{-module,}$$
$$V = \langle v_\mu, v_{-\mu} \rangle \text{ is an irreducible } \mathfrak{sl}(2)^0\text{-module,}$$

S_{-2} and S_2 are irreducible $\mathfrak{sl}(2)^1$-modules which are trivial as $\mathfrak{sl}(2)^0$-modules.

Moreover,

$$\mathfrak{sl}(2)^1 \subseteq [U \otimes \langle v_\nu \rangle, U^* \otimes \langle v_{-\nu} \rangle] \subseteq \mathfrak{sl}(2)^1 + H_0 + Z(L_0) \text{ for } \nu = \pm \mu, \tag{4.12}$$

$$\mathfrak{sl}(2)^1 \subseteq [S_{-2}, S_2] \subseteq \mathfrak{sl}(2)^1 + H_0 + Z(L_0), \tag{4.13}$$

$$[U \otimes \langle v_\mu \rangle, U \otimes \langle v_{-\mu} \rangle] = S_{-2}, \tag{4.14}$$

$$[U^* \otimes \langle v_\mu \rangle, U^* \otimes \langle v_{-\mu} \rangle] = S_2. \tag{4.15}$$

Evidently,

$$[U \otimes \langle v_\nu \rangle, S_2] = U^* \otimes \langle v_\nu \rangle \quad \text{for } \nu = \pm\mu, \tag{4.16}$$

$$[U^* \otimes \langle v_\nu \rangle, S_{-2}] = U \otimes \langle v_\nu \rangle \quad \text{for } \nu = \pm\mu. \tag{4.17}$$

Choose a standard basis $\{e_0, f_0, h_0\}$ of $\mathfrak{sl}(2)^0$ and suppose e_0, f_0 correspond to the roots $\pm 2\mu$ with respect to $H_0 = \langle h_0 \rangle$. Consider the subspaces

$$N_{-2} = \langle f_0 \rangle,$$

$$N_{-1} = (U \otimes \langle v_{-\mu} \rangle) \oplus (U^* \otimes \langle v_{-\mu} \rangle),$$

$$N_0 = S_{-2} \oplus \mathfrak{sl}(2)^1 \oplus H_0 \oplus Z(L_0) \oplus S_2,$$

$$N_1 = (U \otimes \langle v_\mu \rangle) \oplus (U^* \otimes \langle v_\mu \rangle),$$

$$N_2 = \langle e_0 \rangle.$$

This is just the grading by the weight with respect to H_0. From (4.12)–(4.17) it follows that $N = N_{-2} \oplus N_{-1} \oplus N_0 \oplus N_1 \oplus N_2$ is a graded Lie algebra, N_{-1} is an irreducible N_0-module, $N_{-2} = [N_{-1}, N_{-1}]$, and $N_0 = P \oplus Z(N_0)$, where $P = S_{-2} \oplus [S_{-2}, S_2] \oplus S_2$ is one of the exceptional Lie algebras from Proposition 2.12. Now $P = P_{-1} \oplus P_0 \oplus P_1$, where $P_{-1} = S_{-2}$, $P_0 = [S_{-2}, S_2]$, and $P_1 = S_2$. The subspace $W = U^* \otimes \langle v_{-\mu} \rangle$ of the P-module N_{-1} is a P_0-submodule under the adjoint action. Moreover, $[P_1, W] = [S_2, U^* \otimes \langle v_{-\mu} \rangle] \subseteq [S_2, S_1] = (0)$, and the character $\chi|_W$ of P_0 on W is equal to the character $\chi|_{P_{-1}}$ of P_0 on P_{-1}. Evidently, $(\text{ad } P_i)^2(N_{-1}) = (0)$ for $i = \pm 1$. But by Lemma 2.14, $(\text{ad } P_{-1})^2(W) \neq (0)$. Thus, such an algebra N does not exist. This proves that the case $A_0 \neq (0)$ is impossible.

5. The conclusion of the proof of Theorem 1.1

We show that a Lie algebra L satisfying conditions (i)–(viii) does not exist, thereby proving Theorem 1.1. We know that $A_0 = \text{Ann}_{L_0} S_{-2} = (0)$. Therefore, by Lemma 3.6, $S_0' = \mathfrak{sl}(2)^1 = L_0'$, and $L_0 = \mathfrak{sl}(2)^1 \oplus Z(L_0)$. Since S_{-1} is an irreducible nonrestricted L_0-module, $\dim S_{-1} = 3$. Furthermore, S_{-2} is a nonrestricted irreducible factor of the $\mathfrak{sl}(2)^1$-module $\Lambda^2 S_{-1}$ and $\dim S_{-2} = \dim \Lambda^2 S_{-1} = 3$. Whence $S_{-2} = \Lambda^2 S_{-1}$.

According to Kantor [Ka], S is a subalgebra of the universal graded Lie algebra $Q = S_{-2} \oplus S_{-1} \oplus Q_0 \oplus Q_1 \oplus \cdots$ which has $Q_0 = \mathfrak{gl}(S_{-1})$, $S_{-2} = \Lambda^2 S_{-1}$, and $\dim S_{-1} = 3$. Such a Lie algebra Q must be a classical Lie algebra of type B_3 (see [Ka]), and $Q = S_{-2} \oplus S_{-1} \oplus Q_0 \oplus Q_1 \oplus Q_2$, $Q_1 \cong S_{-1}^*$, and $Q_2 \cong S_{-2}^*$ must hold. Therefore, by the irreducibility of Q_2 as an L_0-module, $L_0 \subseteq Q_0$, $Q_2 = S_2$ and $[S_{-2}, S_2]$ is a noncentral ideal of $\mathfrak{gl}(S_{-1}) = \mathfrak{gl}(3)$. As a result, $\mathfrak{sl}(3) \subseteq [S_{-2}, S_2] = S_0'$. However $S_0' = \mathfrak{sl}(2)^1$. This final contradiction shows that the case that S_{-1} is nonrestricted cannot happen, and it concludes the proof of Theorem 1.1. $\qquad\qquad\square$

References

[BG]   G. M. Benkart, T. B. Gregory, Graded Lie algebras with classical reductive null component, Math. Ann. **285** (1989), 85–98.

[BGP]   G. Benkart, T. B. Gregory, A. Premet, Graded Lie algebras of prime characteristic, in preparation.

[BKK]   G. Benkart, A. I. Kostrikin, M. I. Kuznetsov, The simple graded Lie algebras of characteristic three with classical reductive component L_0, Comm. Algebra **24** (1996), 223–234.

[BO]   G. M. Benkart, J. M. Osborn, Representations of rank one Lie algebras of characteristic p, Lecture Notes in Math. **933**, Springer, Berlin–New-York, 1982, 1–37.

[G1]   T. B. Gregory, A characterization of the contact Lie algebras, Proc. Amer. Math. Soc. **82** (1981), 505–511.

[G2]   T. B. Gregory, On simple reducible Lie algebras of depth two, Proc. Amer. Math. Soc. **83** (1981), 31–35.

[K1]   V. G. Kac, The classification of simple Lie algebras over a field of nonzero characteristic, Izv. Akad. Nauk SSSR Ser. Mat. **34** (1970), 385–408, Russian; English translation, Math. USSR-Izv. **4** (1970), 391–413.

[K2]   V. G. Kac, Description of filtered Lie algebras with which graded Lie algebras of Cartan type are associated, Izv. Akad. Nauk SSSR Ser. Mat. **38** (1974), 800–834; Errata, **40** (1976), 1415, Russian; English translation, Math. USSR-Izv. **8** (1976), 801–835, Errata, **10** (1976), 1339.

[Ka]   I. L. Kantor, Graded Lie algebras, Trudy Sem. Vector Tenzor Anal. **15** (1970), 227–266, Russian.

[Ki]   S. A. Kirillov, Hamiltonian Lie algebras of Cartan type, Preprint # 257 IPF Akad. Nauk SSSR (1989), Russian.

[Ko]   A. I. Kostrikin, A parametric family of simple Lie algebras, Izv. Akad. Nauk SSSR Ser. Mat. **34** (1970), 744–756, Russian; English translation, Math. USSR-Izv. **4** (1970), 751–764.

[KO]   A. I. Kostrikin, V. V. Ostrik, On a recognition theorem for Lie algebras of characteristic 3, Mat. Sb. **186** (1995), 73–88, Russian.

[Ku]   M. I. Kuznetsov, Truncated induced modules over transitive Lie algebras of characteristic p, Izv. Akad. Nauk SSSR Ser. Mat. **53** (1989), 557–589, Russian; English translation, Math. USSR-Izv. **34** (1990), 575–608.

[W]   B. J. Weisfeiler, On the structure of the minimal ideal of some graded Lie algebras in characteristic $p \geq 0$, J. Algebra **53** (1978), 344–361.

Department of Mathematics
University of Wisconsin
Madison, Wisconsin 53706, U.S.A.

Department of Mathematics
The Ohio State University at Mansfield
Mansfield, Ohio 44906, U.S.A.

Department of Mathematics
Nizhny Novgorod State University
Nizhny Novgorod 603000, Russia

Auslander–Reiten theory for restricted Lie algebras

R. Farnsteiner[*]

0. Introduction

This article provides a survey of methods and results concerning that aspect of the representation theory of restricted Lie algebras that is related to the study of finite dimensional indecomposable modules. Although the first results in this direction date back to the late sixties [Pa], [Po], the field lay fallow until Pfautsch's and Voigt's determination [PV] of the finite algebraic groups of finite representation type. This development markedly contrasts with the modular representation theory of finite groups, where the corresponding problems have enjoyed considerable attention since Green's groundbreaking work on vertices and sources [Gr]. Subsequently, two additional tools have been introduced, namely Carlson's support varieties [Ca] and Auslander–Reiten theory. We refer the reader to [Be1], [Be2], [Be3] and [ARS] concerning a comprehensive treatment of these topics. Auslander–Reiten quivers of group algebras were first studied by Webb [We] in the early 80s. Recently, K. Erdmann [Er2] completed the classification of AR-quivers.

In view of the success for modular group algebras it is natural to apply this approach within the context of modular Lie algebras. Thanks to the work by Friedlander–Parshall [FP1], [FP2], [FP3] and Jantzen [Ja1], Carlson's algebro-geometric techniques are available. However, a comprehensive block theory remains elusive.

The purpose of this article is to provide an exposition of the current state of Auslander–Reiten theory for reduced enveloping algebras of restricted Lie algebras. Aside from displaying the major results, we will illustrate the methods by occasionally delineating proofs that we consider particularly illuminating in this regard.

In Section 1 we review methods concerning Auslander–Reiten quivers in the generality that will be needed later. No attempt has been made to give a comprehensive introduction. In particular the functorial point of view (cf. [Ga1]) has been left out of the account altogether. The main goal of Section 2 is to illustrate the importance of periodic modules in the determination of the components of the AR-quiver. The connection is made by means of subadditive functions. These are usually defined via cohomology groups and we provide the relevant results concerning and cohomological support varieties in Section 3. In particular, we illustrate how information on the cohomology rings can be used to study periodic modules. Our first application concerns blocks of finite representation type. In contrast to the modular representation theory of finite groups, such blocks are Nakayama algebras, and their Auslander–Reiten components are well understood. Using subadditive functions one can write down a fairly short list of the possible shapes of AR-components.

* Supported by N.S.A. Grant MDA-96-1-0040

Support varieties are then employed to show that components of type $\mathbb{Z}[A_\infty]$ occur "most often".

The fifth section presents results on particular classes of restricted Lie algebras. As usual, Lie algebras associated to algebraic groups are best understood, primarily since they afford a natural action of the group on the AR-quiver of the underlying Lie algebra. In the concluding section we present applications and a list of open problems.

I would like to thank Tom Gregory and Joe Ferrar for given me the opportunity to talk on the results of this paper at the Conference on the Monster and Lie Algebras, held at Ohio State University.

1. Auslander–Reiten quivers

Throughout we will be working over an algebraically closed field F. All vector spaces are assumed to be finite dimensional. Given an associative F-algebra Λ, we shall be interested in the category $\mod \Lambda$ of (left) Λ-modules. The theorem of Krull–Remak–Schmidt asserts that $\mod \Lambda$ can be understood by studying the indecomposable Λ-modules. From that point of view algebras may be subdivided into three classes. More precisely,

- Λ has *finite representation type* if there are only finitely many isoclasses of indecomposable Λ-modules,
- Λ has *tame representation type* if it is not of finite representation type, and in each dimension all but finitely many indecomposables may be parametrized by a one-dimensional affine variety,
- Λ has *wild representation type* if there exists a functor $\mod F\{X, Y\} \longrightarrow \mod \Lambda$ from the module category of a free algebra in two non-commuting variables to $\mod \Lambda$ that preserves indecomposability and reflects isomorphisms.

In 1977 Drozd (cf. [CB1], [Dr]) showed that every algebra belongs to exactly one of these three types. A complete classification of the indecomposable modules is possible only in the first two cases. Before presenting some examples, we note some necessary (but not sufficient) conditions for an algebra to have a certain representation type. Let S, T be simple Λ-modules and recall that the nonzero elements of the group $\operatorname{Ext}^1_\Lambda(S, T)$ correspond to the equivalence classes of the nonsplit extensions

$$(0) \longrightarrow T \longrightarrow M \longrightarrow S \longrightarrow (0)$$

in the sense of Baer. Accordingly, the middle terms are indecomposables of dimension $\dim_F S + \dim_F T$. Direct calculation shows that the isoclasses of such extensions correspond to the projective variety $\operatorname{Proj}(\operatorname{Ext}^1_\Lambda(S, T))$. Thus, if Λ is of finite or tame type, then the dimension of this variety will be 0 or 1. Consequently, $\dim_F \operatorname{Ext}^1_\Lambda(S, T) \leq 1, 2$ according as Λ has finite or tame representation type. If Λ is local with Jacobson radical J, then the self-extensions of the unique simple module are isomorphic to J/J^2. We are now in a position to understand the following basic examples.

Examples. Let X, Y, Z be indeterminates over F.

(1) The truncated polynomial algebra $F[X]/(X^n)$ has finite representation type. In fact, every indecomposable module is cyclic and hence an image of $F[X]/(X^n)$. It turns out that $F[X]/(X^n)$ is the only local representation-finite algebra of dimension n. Thus, a commutative algebra has finite representation type if and only if it is a direct sum of truncated polynomial algebras.

(2) The *Kronecker algebra* $F[X, Y]/(X^2, Y^2)$ is tame. This requires more effort than (1) and was essentially shown by Kronecker at the end of the last century.

(3) The algebra $F[X, Y, Z]/(X^2, Y^2, Z^2)$ is wild. This follows directly from the fact that J/J^2 has dimension 3.

The foregoing arguments can be significantly refined by considering the so-called *Gabriel quiver* of Λ. This directed graph has the isoclasses of the simple Λ-modules as its set of vertices. There are $\dim_F \operatorname{Ext}^1_\Lambda (S, T)$ arrows from the vertex $[S]$ to the vertex $[T]$. Using covering theory (cf. [**BG**], [**Er1**]) one can produce a list of directed graphs that identify an algebra Λ as wild, whenever its Gabriel quiver contains such a graph. For instance, an algebra Λ having two simple modules S, T such that $\dim_F \operatorname{Ext}^1_\Lambda (S, S) \geq 1$ and $\dim_F \operatorname{Ext}^1_\Lambda (S, T) \geq 2$ is wild.

The preceding examples already suggest that most algebras are wild. Accordingly, it does not seem expedient to attempt to classify indecomposable Λ-modules. Instead we shall, following Ringel, endow the set of isoclasses of indecomposable Λ-modules with the structure of a directed graph. This so-called *Auslander–Reiten quiver* has turned out to be an important invariant of the Morita equivalence class of the algebra Λ. For our purposes it suffices to confine our attention to the case where Λ is a Frobenius algebra. By definition, such an algebra possesses a nondegenerate bilinear form

$$(,) : \Lambda \times \Lambda \longrightarrow F$$

satisfying the associativity condition

$$(ab, c) = (a, bc) \quad \forall\, a, b, c \in \Lambda.$$

The form is not necessarily symmetric; its departure from symmetry is controlled by its *Nakayama automorphism* $\mu : \Lambda \longrightarrow \Lambda$, which is defined by

$$(a, b) = (b, \mu(a)) \quad \forall\, a, b \in \Lambda.$$

Hopf algebras are known to be Frobenius (cf. [**LS**]), and group algebras of finite groups are *symmetric*, that is, μ can be taken to be the identity. The automorphism μ depends of course on the choice of the bilinear form $(,)$. It can be shown, however, that any two Nakayama automorphisms differ by an inner automorphism of Λ.

Given an automorphism $\lambda \in \operatorname{Aut}(\Lambda)$ and a Λ-module M, we let $M^{(\lambda)}$ be the Λ-module with underlying F-space M and action

$$a \cdot m := \lambda^{-1}(a)\, m \quad \forall\, a \in \Lambda,\ m \in M.$$

It follows that the functor $M \mapsto M^{(\mu)}$ does not depend on the choice of μ; it will be referred to as the *Nakayama functor* of Λ.

Frobenius algebras are self-injective, that is, the Λ-module Λ is injective. Consequently, the concepts of injectivity and projectivity of Λ-modules coincide. This fact enables us to define an operator Ω on the set of isoclasses of nonprojective indecomposable Λ-modules in the following fashion. Given a Λ-module M, we consider a projective cover

$$(0) \longrightarrow K_M \longrightarrow P \longrightarrow M \longrightarrow (0).$$

The isoclass of K_M depends only on the isoclass of M, and we thus obtain a map $[M] \mapsto \Omega([M])$. It turns out that Ω actually defines a functor on the stable category $\underline{\mathrm{mod}}\,\Lambda$. More important for our purposes is Heller's observation [He] that Ω is a bijection on the set of isoclasses of nonprojective indecomposable modules. In addition, the Heller operator Ω is an important ingredient of the *stable Auslander–Reiten quiver* $\Gamma_s(\Lambda)$. The vertices of this directed graph are the isoclasses of the nonprojective indecomposable Λ-modules. There is an arrow $[M] \to [N]$ whenever there exists an *irreducible* map $f : M \longrightarrow N$. By definition, f is a nonisomorphism such that any facorization $f = g \circ h$ implies that g is split surjective or h is split injective.

The map Ω is an automorphism of the quiver $\Gamma_s(\Lambda)$. The same is true for the Nakayama functor, and one defines the *Auslander–Reiten translation* $\tau : \Gamma_s(\Lambda) \longrightarrow \Gamma_s(\Lambda)$ by means of $\tau := \Omega^2 \circ {}^{(\mu)}$. The motivation for this somewhat contrived looking definition stems from the theory of almost split sequences, where $\tau(M)$ is the initial term of the almost split sequence terminating in M.

Examples. (1) Let $\Lambda := F[X]/(X^\ell)$ be a truncated polynomial ring and put $M_i := F[X]/(X^i)$, $1 \le i \le \ell - 1$. The $[M_i]$ form the set of vertices of $\Gamma_s(\Lambda)$. Moreover, Λ is a symmetric algebra with $\Omega([M_i]) = [M_{\ell-i}]$, so that we have $\tau = \Omega^2 = id$. The quiver is given by

$$\mathbb{Z}[A_{\ell-1}]/(\tau): \quad M_1 \; \underset{\leftarrow}{\overset{\rightarrow}{}} \; M_2 \; \underset{\leftarrow}{\overset{\rightarrow}{}} \; \cdots \; \underset{\leftarrow}{\overset{\rightarrow}{}} \; M_{\ell-2} \; \underset{\leftarrow}{\overset{\rightarrow}{}} \; M_{\ell-1}$$

(2) Let $\Lambda := F[X, Y]/(X^2, Y^2)$. Here the quiver of the symmetric algebra Λ decomposes into several (τ-invariant) components. More precisely, we have infinitely many components of type

$$\mathbb{Z}[A_\infty]/(\tau): \quad \bullet \; \underset{\leftarrow}{\overset{\rightarrow}{}} \; \bullet \; \underset{\leftarrow}{\overset{\rightarrow}{}} \; \bullet \; \underset{\leftarrow}{\overset{\rightarrow}{}} \; \bullet \; \underset{\leftarrow}{\overset{\rightarrow}{}} \; \bullet \; \cdots,$$

and one component of type

$$\mathbb{Z}[\tilde{A}_{1,2}]: \quad \cdots \; \bullet \; \underset{\rightarrow}{\overset{\rightarrow}{}} \; \bullet \; \underset{\rightarrow}{\overset{\rightarrow}{}} \; \bullet \; \underset{\rightarrow}{\overset{\rightarrow}{}} \; \bullet \; \underset{\rightarrow}{\overset{\rightarrow}{}} \; \bullet \; \cdots.$$

The last example indicates that our definition above did not include the fact that $\Gamma_s(\Lambda)$ is actually a valued quiver. We will not dwell on this technical aspect here and focus instead on the notation that we have used. There is a general theory of stable representation quivers, i.e., directed graphs with an automorphism τ that enjoys certain properties of the Auslander–Reiten translation. The components Q of such quivers can be constructed in the following fashion. We begin with a directed tree, a Dynkin diagram of type A_ℓ, say. The quiver $\mathbb{Z}[A_\ell]$ is obtained from A_ℓ by taking infinitely many copies of A_ℓ (indexed by the integers) and introducing new arrows as indicated below. In the quiver $\mathbb{Z}[A_\ell]$, the automorphism τ is the map that moves every vertex one level down.

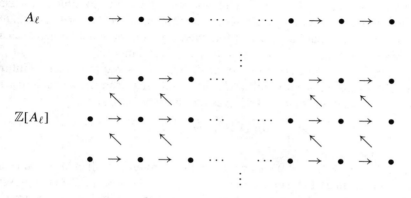

Trees and quivers

We can construct more quivers by considering a subgroup Π of the automorphism group of $\mathbb{Z}[A_\ell]$ (by definition an automorphism is assumed to commute with τ) and by forming (if it makes sense) the orbit graph $\mathbb{Z}[A_\ell]/\Pi$. For instance, if $\Pi = (\tau^k)$, then we obtain a *tube* $\mathbb{Z}[A_\ell]/(\tau^k)$ of rank k. For $k = 1$ we retrieve the quiver of our truncated polynomial ring (1-tube).

This procedure can of course be carried out with any type of tree. The fact that all Auslander–Reiten components can be gotten in this fashion is the contents of Riedtmann's **[Ri1]** important result, which can roughly be stated as follows:

Theorem 1.1 (C. Riedtmann). *Let* $\Theta \subset \Gamma_s(\Lambda)$ *be a connected component of the stable Auslander–Reiten quiver of* Λ. *Then there exists a directed tree* T *and a subgroup* $\Pi \subset \mathrm{Aut}(\mathbb{Z}[T])$ *such that* $\Theta \cong \mathbb{Z}[T]/\Pi$. *Moreover, the undirected tree* T_Θ *is uniquely determined, it is called the* tree class *of* Θ.

Accordingly, one is now interested in the tree classes and groups given by the Auslander–Reiten components.

In general the Auslander–Reiten component provides virtually no information concerning the structure of its vertices. However, there are important parts of the quiver, which are defined by the almost split sequences involving principal indecomposable modules. Given such a module P, we denote its radical and socle by $\mathrm{Rad}(P)$ and $\mathrm{Soc}(P)$, respectively. The *heart* of P is defined by $H(P) := \mathrm{Rad}(P)/\mathrm{Soc}(P)$. The *standard almost split sequence* involving P has the form

$$(0) \longrightarrow \mathrm{Rad}(P) \longrightarrow H(P) \oplus P \longrightarrow P/\mathrm{Soc}(P) \longrightarrow (0).$$

General theory (cf. **[ARS]**) then shows that the indecomposable constituents of $H(P)$ define the successors of $[\mathrm{Rad}(P)]$ and the predecessors of $[P/\mathrm{Soc}(P)]$. Since Ω is a quiver automorphism, this also provides information concerning the component containing the simple vertex $[\mathrm{Soc}(P)] = \Omega([P/\mathrm{Soc}(P)])$.

2. Subadditive functions and periodic modules

In this section we shall indicate how one can obtain control over the possible tree classes. Let T_Θ be the tree class of an Auslander–Reiten component Θ. It turns out that the valuation of Θ induces a valuation of T_Θ. Moreover, since our base field is algebraically closed the matrix corresponding to the valued tree T_Θ is symmetric.

Let $A = (a_{ij})_{i,j \in I}$ be the Cartan matrix of the tree T_Θ. While I can be infinite, the matrix A will always be locally finite (all but finitely many entries in each row and column are zero). A function $d : I \longrightarrow \mathbb{N}$ is said to be *subadditive* if

$$\sum_{j \in I} a_{ij} d(j) \geq 0 \quad \forall\, i \in I.$$

(Here we define $\mathbb{N} := \{1, 2, \ldots\}$.) If equality holds throughout, then d is said to be *additive*. According to an early result by Happel, Preiser and Ringel **[HPR]** the presence of a subadditive function implies that the tree T_Θ is either a finite Dynkin diagram (in which case the finite Cartan matrix is nonsingular and the function d is not additive), an infinite Dynkin diagram of type A_∞, D_∞, A_∞^∞, or one of the Euclidean diagrams of type $\tilde{A}_{12}$, $(\tilde{D}_n)_{n \geq 4}$, $(\tilde{E}_n)_{6 \leq n \leq 8}$.

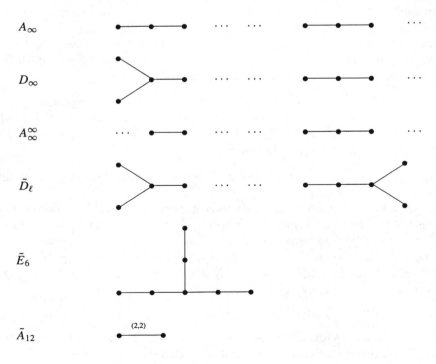

Tree classes

For $\tilde{A}_{12}$ we have provided the valuation, in all other cases the valuation equals $(1, 1)$. There is one more diagram that will appear in the classification: the graph $\tilde{A}_n$ is a circle with $n + 1$ nodes.

If $\Lambda = F[G]$ is the group algebra of a finite group G whose order is divisible by the characteristic $p > 0$ of F, then the existence of subadditive functions was established by Webb [**We**]. His result was refined by Okuyama [**Ok**] who showed that Euclidean components can only occur for $p = 2$. (We shall see later that enveloping algebras of restricted Lie algebras admit Euclidean components in any characteristic.) K. Erdmann and A. Skowroński observed that Okuyama's approach, that was formulated within the context of Green rings, can be rephrased for an arbitrary algebra Λ by using extension functors (cf. [**ES**]). Thus, one first defines certain "subadditive" functions $\delta : \Theta \longrightarrow \mathbb{N}$ on the AR-component Θ and then passes to T_Θ. This passage is possible whenever δ is constant on the τ-orbits of Θ, that is, when $\delta \circ \tau = \delta$ holds. Let M be a Λ-module. Under certain technical conditions on M (see (2.1) below) the functions

$$\delta_M^i : \Theta \longrightarrow \mathbb{N}_0; \quad [X] \mapsto \dim_F \mathrm{Ext}_\Lambda^i(M, X) \quad (i \geq 1)$$

have the requisite properties. If, in addition, $\tau(M) \oplus P \cong M$ for a projective Λ-module P, then $\delta_M^i \circ \tau = \delta_M^i$. Such modules M can be constructed from the so-called τ-periodic modules. By definition, these satisfy the relation $\tau^n(N) \cong M$ for some $n > 0$, and one obtains a suitable module by considering $M := \bigoplus_{i=0}^{n-1} \tau^i(N)$. Recalling that $\tau(N) = \Omega^2(N^{(\mu)})$, we thus have to look for modules satisfying $\Omega^n(N) \cong N$. If the Nakayama automorphism of Λ has finite order, as it does in all relevant cases, this actually suffices for the construction of M. For future reference we summarize the foregoing observations in the following:

Theorem 2.1 (cf. [ES]). *Let $\Theta \subset \Gamma_s(\Lambda)$ be a component of the stable Auslander–Reiten quiver of Λ and suppose that M is a Λ-module such that*
  (a) *$\tau(M) \oplus P \cong M$, with P projective, and*
  (b) *no indecomposable constituent of M belongs to $\Omega^{-n}(\Theta) \cup \Omega^{-n+1}(\Theta)$, and*
  (c) *there is $[X] \in \Theta$ with $\mathrm{Ext}_\Lambda^n(M, X) \neq (0)$.*
  Then the function δ_M^n defines a subadditive function on the tree class T_Θ of Θ, and T_Θ is either a finite Dynkin diagram, an infinite Dynkin diagram or a Euclidean diagram.

There is additional information concerning the subadditive functions that can occur for certain tree classes. A comparison with the behaviour of the extension groups then sometimes allows to rule out some of the tree classes.

3. Cohomology of reduced enveloping algebras

In this section we shall specialize the theory to reduced enveloping algebras of restricted Lie algebras. Henceforth we assume that $char(F) = p > 0$. A *restricted Lie algebra* is a pair $(L, [p])$ consisting of a (finite dimensional) Lie algebra L and an operator $[p] : L \longrightarrow L$ that satisfies the formal properties of a p-power operator of an associative algebra. In fact, every restricted Lie algebra can be realized as a subspace $L \subset \Lambda$ of an associative F-algebra such that

(i) $[L, L] \subset L$, where $[a, b] := ab - ba \ \forall \ a, b \in \Lambda$ is the ordinary commutator product, and

(ii) $x^p \in L \ \forall x \in L$.

We will, however, continue to denote the p-mapping on L by $x \mapsto x^{[p]}$. According to the theorem of Poincaré–Birkhoff–Witt L can be imbedded into its universal enveloping algebra $\mathfrak{U}(L)$. In fact, $L \hookrightarrow \mathfrak{U}(L)$ is a Lie subalgebra of the commutator algebra of $\mathfrak{U}(L)$. Given a linear form $\chi \in L^*$, we consider the reduced enveloping algebra

$$u(L, \chi) := \mathfrak{U}(L)/(\{x^p - x^{[p]} - \chi(x)^p 1; \ x \in L\}).$$

The representation theory of the family $(u(L, \chi))_{\chi \in L^*}$ has been studied from various points of view. The algebras $u(L, \chi)$, whose definition was motivated by Veisfeiler's and Kac' study [VK] of simple L-modules, are in many respects similar to group algebras of finite groups. We shall list a few relevant properties.

(1) L is a subalgebra of the commutator algebra of $u(L, \chi)$.

(2)   If $\dim_F L = n$, then $\dim_F u(L, \chi) = p^n$.

(3) $u(L, \chi)$ is a Frobenius algebra whose Nakayama automorphism is uniquely determined by $\mu(x) = x - \text{tr}(\text{ad } x)1 \ \forall \ x \in L$. Thus, in contrast to group algebras, $u(L, \chi)$ is not necessary symmetric. However, the Nakayama automorphism has order $r \in \{1, p\}$.

The interested reader can find a very nice introduction to the subject in J. Feldvoss' survey article [Fe2], where more properties of $u(L, \chi)$ are given.

The case where $\chi = 0$ is of importance for technical and other reasons. First of all $u(L) := u(L, 0)$ is a Hopf algebra, and second, if $L = \text{Lie}(G)$ is the Lie algebra of an algebraic group G, then the differential $d(\varrho) : L \longrightarrow \text{gl}(M)$ of a rational representation $\varrho : G \longrightarrow \text{GL}(M)$ uniquely extends to a representation $d(\varrho) : u(L) \longrightarrow \text{End}_F(M)$.

Different choices of χ do not necessarily give different algebras. In fact, the group $\text{Aut}(L)$ of automorphisms of the restricted Lie algebra $(L, [p])$ acts contragrediently on L^*, i.e.

$$g \cdot \chi := \chi \circ g^{-1} \ \ \forall \ g \in \text{Aut}(L).$$

Linear forms belonging to the same orbit yield isomorphic algebras. However, this is not the only way to obtain isomorphisms: $u(L, \chi)$ is isomorphic to $u(L)$ precisely when $u(L, \chi)$ is a supplemented algebra. In particular, if L is abelian, then all $u(L, \chi)$ are isomorphic to $u(L)$.

In contrast to modular group algebras information concerning the block structure of reduced enveloping algebras is available only in special cases (cf. [Fa2], [Fa3], [Fe1], [Fe3], [Hu1], [Pe]). In the general setting we will therefore employ the algebro-geometric approach that was introduced by Friedlander and Parshall [FP1].

According to the results of §2 an understanding of the structure of the stable Auslander–Reiten quiver $\Gamma_s(L, \chi)$ of the algebra $u(L, \chi)$ necessitates a detailed knowledge of the periodic modules. Such information also turns out to be useful for the investigation of those blocks of $u(L, \chi)$ that have finite representation type. In this context the structure of the cohomology ring $H^\bullet(u(L), F)$ has proven to be decisive. We shall need the following preparations. In general $u(L, \chi)$ is not a Hopf algebra, yet it is a quotient of the Hopf algebra $\mathcal{U}(L)$ and basic results from that theory can be easily modified to yield the following facts:

(a) Let M_i be a $u(L, \chi_i)$-module $1 \leq i \leq 2$. Then $M_1 \otimes_F M_2$ is a $u(L, \chi_1 + \chi_2)$-module.

(b) If M is a $u(L, \chi)$-module, then its dual M^* is a $u(L, -\chi)$-module.

(c) Given $u(L, \chi)$-modules, M, N there is a graded isomorphism

$$H^\bullet(u(L), \operatorname{Hom}_F(M, N)) \cong \operatorname{Ext}^\bullet_{u(L,\chi)}(M, N)$$

which respects the Yoneda composition.

According to a theorem by Hochschild [Ho1], there is a, with respect to L natural, map $L^* \longrightarrow H^2(u(L), F)^{(-1)}$. Here the superscript indicates the twist of the F-action by the Frobenius homomorphism. Letting $S(L^*)$ denote the symmetric algebra of L^*, we thus obtain an algebra homomorphism $S(L^*) \longrightarrow H^{ev}(u(L), F)^{(-1)}$ of graded algebras (with the elements of L^* having cohomological degree 2). If M is a $u(L)$-module, then $H^\bullet((u(L), M)^{(-1)}$ is, via the Yoneda product, a $H^\bullet(u(L), F)^{(-1)}$-module, hence an $S(L^*)$-module.

Theorem 3.1 (cf. [FP1]). *Let M be a $u(L)$-module. Then $H^\bullet(u(L), M)^{(-1)}$ is a finitely generated $S(L^*)$-module.*

This result has several important consequences:

(a) Let M be a $u(L, \chi)$-module. By applying (3.1) to the $u(L)$-module $\operatorname{Hom}_F(M, M) \cong M^* \otimes_F M$, we see that the Yoneda algebra $\operatorname{Ext}^\bullet_{u(L,\chi)}(M, M)$ is a finitely generated module over a commutative subalgebra that is generated by elements of $\operatorname{Ext}^2_{u(L,\chi)}(M, M)$.

(b) Let M be a $u(L, \chi)$-module. We consider the induced map

$$\Phi_M : S(L^*) \longrightarrow \operatorname{Ext}^\bullet_{u(L,\chi)}(M, M)^{(-1)}.$$

The kernel of this map is an ideal $I_M \subset S(L^*)$. We view the latter algebra as the algebra of polynomial functions on L and associate to I_M its variety

$$\mathcal{V}_L(M) := Z(I_M) := \{x \in L; \ f(x) = 0 \ \ \forall \, f \in I_M\}.$$

Since Φ_M is a homomorphism of degree 0, I_M is a graded ideal, so that the *cohomological support variety* $\mathcal{V}_L(M)$ is conical. By construction the dimension of this variety coincides with the rate of growth of the Yoneda algebra of M, which in turn can be seen to be the complexity $c_{u(L,\chi)}(M)$ of M, i.e. the rate of growth of a minimal projective resolution of M. Consequently, M is projective if and only if $\mathcal{V}_L(M) = \{0\}$. More precisely, if $\dim \mathcal{V}_L(M) = r$, then $\dim_F \operatorname{Ext}^{2n}_{u(L,\chi)}(M, M) \geq \binom{n+r-1}{r-1}$, so that M is projective if one even Ext-group vanishes.

We record the following result, which is anlogous to the situation in the representation theory of finite groups.

Theorem 3.2 ([FP2]). *Let M be an indecomposable $u(L, \chi)$-module. Then the projective variety $\mathrm{Proj}(\mathcal{V}_L(M))$ is connected.*

By way of illustration let us consider a periodic, indecomposable $u(L, \chi)$-module M. Then $c_{u(L,\chi)}(M) = 1$, so that $\mathrm{Proj}(\mathcal{V}_L(M))$ is a point. Consequently, $\mathcal{V}_L(M) = Fx$ is a line.

Support varieties have become an important tool in representation theory. In our context of Auslander–Reiten theory this is attested by the fact that these varieties are invariants of the connected components.

Proposition 3.3. *Let $\Theta \subset \Gamma_s(L, \chi)$ be a connected component of the stable Auslander–Reiten quiver of $u(L, \chi)$. Given $[M], [N] \in \Theta$ we have $\mathcal{V}_L(M) = \mathcal{V}_L(N)$.*

Accordingly, we may define the variety $\mathcal{V}_L(\Theta)$ of the component Θ by means of

$$\mathcal{V}_L(\Theta) := \mathcal{V}_L(M) \quad \forall\, [M] \in \Theta.$$

The relevance of these techniques can be seen by the proof of the succeeding result, which parallels the methods from the representation theory of groups. Note, however, that the result does not obtain for group algebras: for $p = 2$ the group algebra of the quaternion group has periodic modules of period 4.

Theorem 3.4 (cf. [Fa1]). *Let M be an indecomposable $u(L, \chi)$-module with $c_{u(L,\chi)}(M) = 1$. Then $\Omega^2(M) \cong M$.*

Proof. According to the above observations $\mathcal{V}_L(M) = Fx$ is a line. As $x \neq 0$, it does not belong to $Z(X_1, \ldots, X_n)$, the zero set of a basis $\{X_1, \ldots, X_n\}$ of L^*. Hence $X_i(x) \neq 0$ for some $i \in \{1, \ldots, n\}$ and we consider $\zeta := \Phi_F(X_i) \in H^2(u(L), F)^{(-1)}$. As x is contained in $\mathcal{V}_L(F) = Z(\ker \Phi_F)$, ζ does not vanish. By general theory we can view ζ as a map $\zeta : \Omega^2(F) \longrightarrow F$ whose kernel we denote $\mathcal{L}_\zeta$. Owing to [**FP2**, (2.1)] Lemma 4.2 of [**FP1**] holds for arbitrary restricted Lie algebras. Thus we have $\mathcal{V}_L(\mathcal{L}_\zeta) = Z(X_i)$ and $\mathcal{V}_L(\mathcal{L}_\zeta) \cap \mathcal{V}_L(M) = (0)$. We apply [**FP3**, (7.1)] and [**FP3**, (6.2)] successively to see that $\mathcal{L}_\zeta \otimes_F M$ is a projective $u(L, \chi)$-module. Since $u(L, \chi)$ is a Frobenius algebra , the module $\mathcal{L}_\zeta \otimes_F M$ is also injective and the exact sequence

$$(0) \longrightarrow \mathcal{L}_\zeta \otimes_F M \longrightarrow \Omega^2_{u(L)}(F) \otimes_F M \longrightarrow M \longrightarrow (0)$$

induced by ζ splits. Consequently, we have

$$\Omega^2(F) \otimes_F M \cong (\mathcal{L}_\zeta \otimes_F M) \oplus M.$$

By virtue of Schanuel's Lemma there exists a projective $u(L, \chi)$-module P such that the left-hand side is isomorphic to $\Omega^2(M) \oplus P$. Application of Ω therefore yields

$$\Omega^3(M) \cong \Omega(\Omega^2(M) \oplus P) \cong \Omega((\mathcal{L}_\zeta \otimes_F M) \oplus M) \cong \Omega(M).$$

Since, according to [**He**, Prop. 1], Ω defines a bijection on the set of isomorphism classes of nonprojective, indecomposable $u(L, \chi)$-modules, it follows that $\Omega^2(M) \cong M$.   □

By the foregoing result there is an ample supply of modules satisfying $\tau(M) \oplus P \cong M$. In fact, if $\mathcal{V}_L(M)$ has dimension 1, we consider $N := \sum_{i=0}^{p-1} M^{(\mu^i)}$, and (3.4) yields the desired property. It remains to identify varieties. This was done by Jantzen [**Ja1**] for $M = F$ and Friedlander-Parshall [**FP2**] in the general case.

Theorem 3.5 ([FP2], [Ja1]). *Let M be a $u(L, \chi)$-module. Then we have*

$$\mathcal{V}_L(M) := \{x \in L; x^{[p]} = 0 \text{ and } M|_{u(Fx, \chi|_{Fx})} \text{ is not free}\} \cup \{0\}.$$

4. The Auslander–Reiten quiver $\Gamma_s(L, \chi)$

We begin by stydying periodic components, i.e., those that contain a τ periodic module. Recall that the notions of τ-periodicity and Ω-periodicity coincide in our context. According to a general result of Happel–Preiser–Ringel [**HPR**] the infinite periodic components are of the form $\mathbb{Z}[A_\infty]/(\tau^k)$ for some $k \geq 1$. In particular, all modules belonging to a periodic component are periodic (in our context this also follows from our considerations concerning support varieties). Moreover, (3.4) readily yields $k \in \{1, p\}$.

If $\Theta \subset \Gamma_s(L, \chi)$ is a finite component, then, by general theory [**ARS**] we have $\Theta = \Gamma_s(\mathcal{B})$, where $\mathcal{B} \subset u(L, \chi)$ is a block of finite representation type. It now follows from (3.4) that for every simple $\mathcal{B}$-module S, the Auslander–Reiten translate $\tau(S)$ is also simple. General theory [**ARS**] now ensures that $\mathcal{B}$ is a Nakayama algebra (i.e., all principal indecomposable $\mathcal{B}$-modules are uniserial) and $\Gamma_s(\mathcal{B}) \cong \mathbb{Z}[A_\ell]/(\tau^k)$, where $\ell + 1$ is the Loewy length of $\mathcal{B}$ and $k \in \{1, p\}$ is the number of simple $\mathcal{B}$-modules. Thus, if $k = 1$, then $\mathcal{B}$ is primary (one simple module) and $\mathcal{B} \cong \mathrm{Mat}_n(F[X]/(X^{\ell+1}))$. Acordingly, the symmetric reduced enveloping algebras of finite representation type are Morita equivalent to the representation-finite commutative algebras (cf. our introductory examples).

Actually, one can prove a somewhat stronger result, and this turns out to be important in conjunction with the investigation of nonperiodic components (cf. §5). If a simple $u(L, \chi)$-module S is periodic, then the block $\mathcal{B}(S)$ associated to S is a Nakayama algebra and its Gabriel quiver is either a point, a point with a loop, or a diagram of type $\tilde{A}_{p-1}$ (with all arrows oriented clockwise, say) (cf. [**Fa1**]). In the particular case where $\chi = 0$ and $S = F$, this information can be used to obtain the structure of those Lie algebras whose restricted enveloping algebras $u(L) := u(L, 0)$ have finite representation type. The relevant result, which is an analogue of Higman's characterization of representation-finite group algebras [**Hi**], is originally due to Pfautsch and Voigt [**PV**]. Later, Feldvoss and Strade gave a different proof, which does not involve AR-theory. More importantly, in their paper [**FeS**], they employed the information on the structure of L to show that the representation-finite $u(L)$ are Nakayama algebras.

Now let $\Theta \subset \Gamma_s(L, \chi)$ be a nonperiodic component. Then we have

Theorem 4.1 ([Er3], [Fa1]). *The component* Θ *belongs to the following list:*
  (a) $\mathbb{Z}[A_\infty]$, $\mathbb{Z}[D_\infty]$, $\mathbb{Z}[A_\infty^\infty]$,
  (b) $\mathbb{Z}[\tilde{A}_{12}]$, $\mathbb{Z}[\tilde{D}_n]$, $\mathbb{Z}[\tilde{E}_n]$ *or* $\mathbb{Z}[\tilde{A}_n]$ *(Euclidean components).*

Proof. Let $x \neq 0$ be an element of $\mathcal{V}_L(\Theta)$. Given $[X] \in \Theta$, X is not projective for the local algebra $u(Fx, \chi|_{Fx})$, whose simple module we will denote by F_χ. Consequently,

$$\text{Ext}^1_{u(Fx,\chi|_{Fx})}(F_\chi, X) \neq (0).$$

The induced module $M := u(L, \chi) \otimes_{u(Fx,\chi|_{Fx})} F_\chi$ is easily seen to be periodic, so that no constituent of M belongs to the nonperiodic components Θ or $\Omega^{-1}(\Theta)$. Using Frobenius reciprocity, we see that (2.1) applies, so that the tree class T_Θ is determined. The rest follows by computing the automorphism groups of the possible components. $\square$

The above result is the analogue of Webb's theorem **[We]** for group algebras of finite groups. Since the publication of Webb's paper in the early 80s much progress has been made. We only mention two features of the theory that will serve as a motivation for what will follow.

(1) In 1987 T. Okuyama **[Ok]** refined Webb's result by showing that Euclidean components can only occur in characteristic 2. By work of C. Bessenrodt **[Bs]** $\mathbb{Z}[\tilde{A}_{12}]$ and $\mathbb{Z}[\tilde{A}_5]$ are in fact the only Euclidean components.

(2) K. Erdmann recently showed in **[Er2]** that the AR-components of the wild blocks of modular group algebras have tree class A_∞.

We shall see below that Lie algebras admit Euclidean components in any characteristic. The following theorem indicates that K. Erdmann's aforementioned result continues to be valid in principle.

Theorem 4.2 ([Fa5]). *Let* $\Theta \subset \Gamma_s(L, \chi)$ *be a nonperiodic component such that* $\Theta \not\cong \mathbb{Z}[A_\infty]$. *Then* $\dim \mathcal{V}_L(\Theta) = 2$.

In other words, just as in the theory of finite groups, the components $\mathbb{Z}[A_\infty]$ occur "most often". We shall use this information below in connection with Lie algebras of algebraic groups.

5. Simple and solvable Lie algebras

As mentioned in our prefatory remarks, one major shortcoming of the current representation theory of reduced enveloping algebras is the lack of a comprehensive block theory. In fact, information is only available in a few particular cases, and even this often does not suffice to shed light on the Auslander–Reiten theory. The following example, which will also be needed in the sequel, illustrates the techniques introduced above. Moreover, it shows that we cannot hope that descent to subalgebras will provide us with much insight.

Examples.   (a) Let $L = s\ell(2)$ and assume that $p \geq 3$. By early results of Pollack **[Po]** and subsequent work of Fischer **[Fi]**, a block $\mathcal{B} \subset u(L)$ is either simple (with simple module the p-dimensional Steinberg module) or tame. In the latter case $\mathcal{B}$ possesses 2 simple modules S_0, S_1, the Loewy length of the corresponding principal indecomposables P_0, P_1 is 3, and we have $H(P_i) \cong S_{1-i} \oplus S_{1-i}$, $0 \leq i \leq 1$. This implies that the radical $\mathrm{Rad}(P_i)$ of P_i has a predecessor of multiplicity 2, so that the tree class of the corresponding component $\Theta \subset \Gamma_s(L)$ is $\tilde{A}_{12}$. It follows from an application of Ω that the simple modules of $\mathcal{B}$ belong to components of type $\mathbb{Z}[\tilde{A}_{12}]$.

Let Θ be a component containing a simple module. In order to determine the structure of the other components, we consider for an indecomposable $\mathcal{B}$-module M the map

$$\delta_M : \Theta \cup \Omega(\Theta) \longrightarrow \mathbb{N}_0; \quad [X] \mapsto \dim_F \mathrm{Ext}^1_{u(L)}(M, X).$$

According to (2.1) and a result of Webb [We], this function is bounded provided M does not belong to $\Theta \cup \Omega(\Theta)$. Since $u(s\ell(2))$ is symmetric, $\Theta \cup \Omega(\Theta)$ is Ω-invariant. Hence there exists $b \geq 1$ such that

$$\dim_F \mathrm{Ext}^n_{u(L)}(M, X) = \dim_F \mathrm{Ext}^1_{u(L)}(M, \Omega^{1-n}(X)) \leq b \quad \forall\, [X] \in \Theta \cup \Omega(\Theta)$$

for every $n \geq 1$. By applying this to $X = S_i$, $0 \leq i \leq 1$, and using the long exact sequence in cohomology, we obtain the boundedness of the sequence $(\dim_F \mathrm{Ext}^n_{u(L)}(M, M))_{n \geq 0}$. Accordingly, M is a periodic module, so that M belongs to a component of type $\mathbb{Z}[A_\infty]/(\tau)$. One can show in addition that the periodic modules are precisely those of even length and that $\Gamma_s(L) := \Gamma_s(L, 0)$ possesses exactly two components of type $\mathbb{Z}[\tilde{A}_{12}]$.

(b) Now let $K \subset s\ell(2)$ be a proper p-subalgebra. If K has dimension 1, then $u(K)$ is either semisimple (namely precisely when K is a torus), or $u(K) \cong F[X]/(X^p)$ and $\Gamma_s(K) \cong \mathbb{Z}[A_{p-1}]/(\tau)$. Alternatively, K has dimension two and is thus isomorphic to the Borel subalgebra of $s\ell(2)$. In particular, $u(K)$ has finite representation type, and since $u(K)$ has p simple modules, it follows that $\Gamma_s(K) \cong \mathbb{Z}[A_{p-1}]/(\tau^p)$ in this case.

(c) For $p \geq 3$, let H_n denote the $(2n + 1)$-dimensional Heisenberg algebra with basis $\{x_1, \ldots, x_n, y_1, \ldots, y_n, z\}$ and Lie product

$$[x_i, y_j] = \delta_{ij}, \quad [x_i, x_j] = 0 = [y_i, y_j], \quad 1 \leq i, j \leq n, \quad [z, H_n] = (0).$$

The p-map is the zero map. If $\chi(z) \neq 0$, then $u(H_n, \chi)$ has finite representation type. From our introductory remarks of §3 we see that the reduced enveloping algebra $u(K_n, \chi)$ of the p-subalgebra $K_n := \bigoplus_{i=1}^n F x_i$ is isomorphic to

$$u(K_n) \cong F[X_1, \ldots, X_n]/(X_1^p, \ldots, X_n^p).$$

As $p \geq 3$, the latter algebra is wild for $n \geq 2$.

In the foregoing example p-subalgebras of certain types occurred. Let us recall the relevant definitions. A p-subalgebra $K \subset L$ is called a *torus* if $K^{[p]} = K$. We say that K is p-*unipotent* if there is $n \geq 1$ such that $K^{[p]^n} = \{0\}$. The *toral radical* $T(L)$ is the largest p-ideal, which is a torus. It is contained in the center $C(L)$ of L.

There is one general case in which the Auslander–Reiten theory of $u(L)$ is completely understood. If the algebra $u(L)$ is local, that is, if L is p-unipotent, then we have

Theorem 5.1 ([Er3]). *Let L be p-unipotent and $\Theta \subset \Gamma_s(L)$ a nonperiodic component. If $p \geq 3$, then $\Theta \cong \mathbb{Z}[A_\infty]$.*

This result fails for $p = 2$.

(1) The enveloping algebra $u(L)$ of the two-dimensional strongly abelian Lie algebra $L := Fx \oplus Fy$; $L^{[2]} = \{0\}$ is isomorphic to the Kronecker algebra, and the latter admits a component of the form $\mathbb{Z}[\tilde{A}_{12}]$.

(2) The restricted enveloping algebra of the 3-dimensional Heisenberg algebra H_1 is isomorphic to the group algebra of the dihedral group of order 8. The Auslander–Reiten quiver of the latter is known to possess infinitely many components of the form $\mathbb{Z}[A_\infty^\infty]$ as well as infinitely many 1-tubes $\mathbb{Z}[A_\infty]/(\tau)$ (cf. [Er1]).

Throughout the remainder of this section we assume that $p \geq 5$. Let us briefly summarize what is known about the block structure of $u(L, \chi)$.

(a) Suppose $L = \text{Lie}(G)$ is the Lie algebra of a connected algebraic group G. Then G operates on L^* via the coadjoint action

$$g \cdot \chi := \chi \circ \text{Ad}(g^{-1}).$$

Note that this is just the action introduced earlier combined with the adjoint representation $\text{Ad} : G \longrightarrow \text{Aut}(L)$. There is information about two cases:

(a_1) If $\chi = 0$, then the blocks are given by the linkage classes induced by the dot action of the Weyl group (cf. [Hu1]).

(a_2) If χ is *regular nilpotent* (which essentially means that χ corresponds via the Killing form to a nilpotent element of L and that among those elements, the stabilizer in G has minimal possible dimension), then each block of $u(L, \chi)$ is primary (cf. [FP3]).

(b) Suppose that L is *supersolvable*, i.e., its first derived algebra $[L, L]$ is nilpotent. To every block $\mathcal{B} \subset u(L, \chi)$ one can associate a p-subalgebra $K \subset L$ such that every simple $\mathcal{B}$-module S is induced by a one-dimensional $u(K, \chi|_K)$-module. To some extent, the structure of $\mathcal{B}$ is determined by that of its "V-polarization" K. For instance, if K is nilpotent, then $\mathcal{B}$ is primary (cf. [Fa3], [Fe3], [St] for details).

(c) If L is a simple Lie algebra of Cartan type (cf. [SF] for the definition), then $u(L)$ has exactly one block [HN].

Given an arbitrary restricted Lie algebra $(L, [p])$, we let $\mathcal{G}(L)$ be the group of group-like elements of the Hopf algebra $u(L)^*$. For $\lambda \in \mathcal{G}(L)$, we consider the automorphism $\psi_\lambda : u(L, \chi) \longrightarrow u(L, \chi)$ that is given by

$$\psi_\lambda(x) = x + \lambda(x) 1 \quad \forall x \in L.$$

The map $\lambda \mapsto \psi_\lambda$ is an embedding from $\mathcal{G}(L)$ into $\text{Aut}(u(L, \chi))$ and we thus obtain an operation

$$(\lambda, [S]) \mapsto [S^{(\psi_\lambda)}]$$

of $\mathcal{G}(L)$ on the set of isoclasses of simple $u(L, \chi)$-modules. We refer to $\mathcal{B}$ as *transitive* if a subgroup of $\mathcal{G}(L)$ operates transitively on the set of isoclasses of simple $\mathcal{B}$-modules.

(Note that this class of blocks comprises the primary blocks.) In that case all principal indecomposable $\mathcal{B}$-modules have the same length $\ell_{\mathcal{B}}$. For transitive blocks, we have the following result:

Theorem 5.2 (cf. [Fa4]). *Let* $\mathcal{B} \subset u(L, \chi)$ *be a transitive block. Then the following statements hold:*

(1) If $\Theta \subset \Gamma_s(\mathcal{B})$ *is a Euclidean component, then either* $\ell_{\mathcal{B}} = 4$ *and* $\Theta \cong \mathbb{Z}[\tilde{A}_{12}]$ *or* $\ell_{\mathcal{B}} = 8$ *and* $\Theta \cong \mathbb{Z}[\tilde{D}_5]$.

(2) If $\mathcal{B}$ *is primary and symmetric, then* $\Gamma_s(\mathcal{B})$ *possesses a Euclidean component if and only if* $\mathcal{B} \cong \mathrm{Mat}_n(F[X, Y]/(X^2, Y^2))$ *is a full matrix ring over the Kronecker algebra.*

This result applies in two contexts. If L is supersolvable, e.g., the Borel subalgebra of a classical simple Lie algebra, then every block $\mathcal{B} \subset u(L, \chi)$ is transitive and $\ell_{\mathcal{B}}$ is a p-power (cf. [Fa3], [Fe3]). Thus, (5.2) shows that $\Gamma_s(L, \chi)$ does not possess any Euclidean components.

Let L be classical simple and $\chi \in L^*$ a regular nilpotent linear form. Since the blocks of such algebras are primary and symmetric, (2) of (5.2) applies and Euclidean components only occur in exceptional cases. As the length of every principal indecomposable module is given by the Weyl group orbit of a dominant integral weight (cf. [FP3]), one can actually write down the cases in which such blocks occur: the algebra L has to be of type A_3 or B_2 [Fa4]. Note that the resulting blocks are tame algebras whose Auslander–Reiten theory is completely understood.

For the remainder of this section we assume that $\chi = 0$. Let G be a (connected) linear algebraic group with Lie algebra $L = \mathrm{Lie}(G)$. Recall that the group G acts on the Auslander–Reiten quiver $\Gamma_s(L) := \Gamma_s(L, 0)$ via

$$G \times \Gamma_s(L) \longrightarrow \Gamma_s(L); \quad (g, [M]) \mapsto [{}^g M],$$

where ${}^g M := M^{(\mathrm{Ad}(g))}$. Since the functor $M \mapsto {}^g M$ is an equivalence, it commutes with Ω, and an elementary calculation shows that it also commutes with the Nakayama automorphism μ. As a result, G acts on $\Gamma_s(L)$ via automorphisms, so that the components of $\Gamma_s(L)$ are permuted by G. Components that are fixed under this action are of particular interest: their support varieties are invariant under the adjoint representation $\mathrm{Ad} : G \longrightarrow \mathrm{GL}(L)$. Every component Θ containing a vertex $[M]$ with

$$ {}^g M \cong M \quad \forall\, g \in G$$

is of course fixed by this action. In particular, if $\varrho : G \longrightarrow \mathrm{GL}(M)$ is a rational representation, then the $u(L)$-module M has this property. Another way to construct such modules is given by the following observation. We view the adjoint representation as a homomorphism $\mathrm{Ad} : G \longrightarrow \mathrm{Aut}(u(L))$ of algebraic groups. Since the automorphisms of $u(L)$ permute the central primitive idempotents of $u(L)$, and G is connected, it follows that G fixes each block $\mathcal{B}$ of $u(L)$. Thus, if J denotes the Jacobson radical of $u(L)$, then each element of G defines an automorphism of the simple algebra $\mathcal{B}/J\mathcal{B}$.

Hence we have

$$S \cong {}^g S \quad \forall\, g \in G.$$

The same holds for the projective cover $P(S)$ of S and thereby for every Loewy factor of $P(S)$.

According to general theory [We], every Euclidean component Θ is attached (via a standard almost split sequence) to a principal indecomposable module. This means that there is a principal indecomposable module P such that $[\mathrm{Rad}(P)] \in \Theta$. It thus follows that Θ is fixed by the action of G. Moreover, (4.2) tells us that the dimension of the variety of Θ equals 2, and we can check these data for compatibility.

Theorem 5.3 ([Fa6]). *Let G be a reductive group. Then the following statements hold:*
 (1) *If the component Θ contains a simple module, then $\Theta \cong \mathbb{Z}[\tilde{A}_{12}]$ or $\Theta \cong \mathbb{Z}[A_\infty]$.*
 (2) *Every representation-finite block $\mathcal{B} \subset u(L)$ is simple.*
 (3) *If a block $\mathcal{B} \subset u(L)$ is tame, then there exists a block $\mathcal{C} \subset u(s\ell(2))$ such that $\mathcal{B} \cong \mathrm{Mat}_n(\mathcal{C})$.*

This result shows in particular that Euclidean components occur exactly when tame blocks occur and that the representation theory of these blocks follow the pattern given in the example above. If we assume the principal block of $u(L)$ to be tame we obtain $L \cong s\ell(2) \oplus T$, where $T \subset L$ is a toral ideal. This result is related to the following general theorem, due to D. Voigt:

Theorem 5.4 ([Vo]). *Let $(L, [p])$ be a restricted Lie algebra such $u(L)$ is tame. Then $L/C(L) \cong s\ell(2)$.*

Let us elaborate on the connection with tame blocks of reductive Lie algebras. The Lie algebra extensions

$$(0) \longrightarrow C(L) \longrightarrow L \longrightarrow s\ell(2) \longrightarrow (0)$$

are determined by the second Chevalley–Eilenberg cohomology

$$H^2(s\ell(2), C(L)) \cong \mathrm{Ext}^2_{\mathcal{U}(s\ell(2))}(F, C(L))$$

of L with coefficients in the trivial $s\ell(2)$-module $C(L)$. Since $H^2(s\ell(2), F) = (0)$, every such extension splits and we have an isomorphism

$$L \cong s\ell(2) \oplus C(L)$$

of ordinary Lie algebras. However, the extension may not split as an extension of restricted Lie algebras, the difference being accounted for by a p-semilinear map $f : s\ell(2) \longrightarrow C(L)$. In the case discussed above, we have $f = 0$ and $C(L) = T(L)$.

According to the foregoing result restricted enveloping algebras of solvable Lie algebras cannot be tame, yet at the time of this writing it is not known whether they can admit tame blocks. In case L is the Lie algebra of a solvable algebraic group G, there is more information.

Theorem 5.5 ([Fa5]). *Let G be solvable. Then the following statements hold:*

(1) *If a component $\Theta \subset \Gamma_s(L)$ contains a simple module, then either $\Theta \cong \mathbb{Z}[A_\ell]/(\tau^k)$, where $k \in \{1, p\}$, or $\Theta \cong \mathbb{Z}[A_\infty]$.*

(2) *A block $\mathcal{B} \subset u(L)$ is not tame.*

The second part follows from the fact that a block $\mathcal{B}$ of $u(L)$ is isomorphic to $u(L/T(L))$.

We return to (5.1) and assume G to be nilpotent. Then $G \cong T_G \times U_G$, where T_G and U_G are the unique maximal torus and the unipotent radical of G, respectively. Accordingly, we obtain a decomposition $L = \mathrm{Lie}(T_G) \oplus \mathrm{Lie}(U_G)$ of L into a sum of ideals. Since $\mathrm{Lie}(T_G)$ is a torus its (abelian) enveloping algebra is semisimple. Thus, there are isomorphisms

$$u(L) \cong u(\mathrm{Lie}(T_G)) \otimes_F u(\mathrm{Lie}(U_G)) \cong \bigoplus_{i=1}^{r} u(\mathrm{Lie}(U_G)).$$

The algebra $u(\mathrm{Lie}(U_G))$ is local and the structure of the Auslander–Reiten quiver of $u(L)$ is given by (5.1).

6. Related topics and open problems

The second Brauer–Thrall conjecture concerns the question on the number of indecomposable modules of a given dimension. More precisely, it has been shown that an algebra Λ of infinite representation type has, for infinitely many $d > 0$ infinitely many modules of dimension d. In [**Ri**] Ringel asked whether an Auslander–Reiten component contains only finitely many modules in each dimension. There are large classes of algebras, such as tame algebras and hereditary algebras, where this is known to be the case. However, in general the answer is not affirmative (cf. [**LSc**]). The following results, which again illustrate the utility of cohomological methods, provide information concerning the distribution of modules within a given component:

Lemma 6.1. *Let M be a nonperiodic indecomposable $u(L, \chi)$-module with τ-orbit M_τ. Given $d \in \mathbb{N}$ the set M_τ contains only finitely many modules of dimension $\leq d$.*

Proof. Assume for a contradiction that there is $d \in \mathbb{N}$ such that the given set is infinite. Then there are infinitely many numbers $n \geq 0$ such that $\Omega^{2n}(M)$ has dimension $\leq d$. Thus,

$$\dim_F \mathrm{Ext}^{2n}_{u(L,\chi)}(M, M) \leq \dim_F \mathrm{Hom}_{u(L,\chi)}(\Omega^{2n}(M), M) \leq r$$

for infinitely many n. On the other hand, since M is not periodic, we have $\dim \mathcal{V}_L(M) \geq 2$, so that

$$\dim_F \mathrm{Ext}^{2n}_{u(L,\chi)}(M, M) \geq \dim_F F[X, Y]^n \geq n + 1$$

for $n \geq 1$, a contradiction. $\qquad\square$

As a result, any component $\Theta \subset \Gamma_s(L, \chi)$ having a finite tree class (i.e., whose tree class is a finite Dynkin diagram or a Euclidean diagram) possesses only finitely many vertices in each given dimension. One can show that this statement is also valid for components of type $\mathbb{Z}[A_\infty]$ and the methods of [MR] yield the following result, which turns out to be useful in the study of the structure of AR-components of reductive groups.

Theorem 6.2 ([Fa6]). *Let $\Theta \subset \Gamma_s(L, \chi)$ be a component, $d \in \mathbb{N}$. Then Θ contains only finitely many modules of dimension d.*

Example. We return to $L = s\ell(2)$. As we have seen earlier, a nonsimple block has 2 components of type $\mathbb{Z}[\tilde{A}_{12}]$ and infinitely many infinite 1-tubes. By the above, each of these has only finitely many vertices in each given dimension. A more detailed analysis shows that for odd d there are only finitely many modules of length d (since all modules of odd length belong to one of the Euclidean components). The action of SL(2) can be employed to see that there are infinitely many periodic modules of even length d (see [NP] for related work on classical Lie algebras).

Let P be a principal indecomposable $u(L, \chi)$-module. Recall the standard almost split sequence

$$(0) \longrightarrow \mathrm{Rad}(P) \longrightarrow H(P) \oplus P \longrightarrow P/\mathrm{Soc}(P) \longrightarrow (0).$$

This sequence allows us to relate the structure of $H(P)$ to the position of $[\mathrm{Rad}(P)]$ within its Auslander–Reiten component. In most cases $[\mathrm{Rad}(P)]$ is located at an "end" of its component, i.e., $[\mathrm{Rad}(P)]$ has exactly one successor and $H(P)$ is indecomposable. More precisely, one can show:

Theorem 6.3. *Let P be a principal indecomposable $u(L, \chi)$-module. Then the heart $H(P)$ has at most two indecomposable summands. If $H(P)$ has two summands, then $[\mathrm{Rad}(P)]$ belongs to a component of type $\mathbb{Z}[\tilde{A}_{12}]$, $\mathbb{Z}[\tilde{A}_n]$, or $\mathbb{Z}[A_\infty^\infty]$.*

Recall that components of tree class $\tilde{A}_{12}$ occur for $s\ell(2)$, while those of type $\mathbb{Z}[A_\infty^\infty]$ are presently only known to occur only for $p = 2$. If L is the Lie algebra of a reductive or solvable algebraic group, then, by the results of § 5, the heart $H(P)$ is decomposable precisely when it belongs to a block of tame representation type.

In contrast to the modular representation theory groups, the classification of Auslander–Reiten components is far from complete. In particular, one would like to have answers or at least information concerning the following problems:

Problem 1. Let $\Theta \subset \Gamma_s(L, \chi)$ be a Euclidean component. What tree classes can actually occur? Presently, only $\tilde{A}_{12}$ is known.

Problem 2. According to (5.4), $u(L)$ is not tame in case L is solvable and $p \geq 3$. (Recall that this result fails for $p = 2$). What can be said about the representation type of blocks of $u(L, \chi)$ for solvable L?

Problem 3. All known tame blocks of reduced enveloping algebras are *biserial*, that is, every heart is the direct sum of two uniserial modules. (By [CB2] every biserial algebra is tame). It would thus be of interest to see whether a tame block of $u(L, \chi)$ is necessarily biserial.

Problem 4. We have seen that "most" components are of type $\mathbb{Z}[A_\infty]$. It would thus be desirable to have more information concerning these components. A first step in this direction is provided in **[EK]**, where the behavior of the stable Hom-functor is analyzed.

Problem 5. Let $L = \mathrm{Lie}(G)$ be the Lie algebra of an algebraic group, $\chi \in L^*$ a linear form. The general set-up of §5 has the following straightforward generalization. Recall the coadjoint action of G on L^*:

$$g \cdot \chi := \chi \circ \mathrm{Ad}(g^{-1}).$$

Thus, the stabilizer G_χ of χ is a closed subgroup of G and there results an action

$$G_\chi \times \Gamma_s(L, \chi) \longrightarrow \Gamma_s(L, \chi); \quad (g, [M]) \longrightarrow [{}^g M].$$

The information this action provides depends of course on the structure and the size of the stablizer G_χ. If G is almost simple, then the stabilizers are small for the so-called regular elements. The following example illustrates the problems that may occur:

Let $G = \mathrm{SL}(3)$, $p \geq 5$. We let $\chi \in L^*$ be the linear form that corresponds via the Killing form to the matrix $E_{1,2} + E_{2,3}$. An elementary computation shows that $G_\chi = \{a\, I_3 + b\, (E_{1,2} + E_{2,3}) + c E_{1,3};\ a^3 = 1\}$. Thus, the connected component of G_χ is

$$(G_\chi)^0 = \{I_3 + b\, (E_{1,2} + E_{2,3}) + c\, E_{1,3};\ b, c \in F\}.$$

Note that this group is abelian. This, and the fact that the dimension of G_χ coincides with the rank of G is not accidental (cf. **[Hu2, (1.14)]**).

Let S be a simple $u(L, \chi)$-module. Direct computation shows $\mathcal{V}_L(S) = \mathcal{V}_{\mathrm{Lie}(G_\chi)}(S)$, so that the algebraic group G_χ acts trivially on the support variety of S. Hence at the level of these varieties the adjoint action will provide no information. Since the stabilizers of the so-called semisimple linear forms are reductive, the geometric methods appear more promising in this case.

Bibliography

[ARS] M. Auslander, I. Reiten, S. Smalø, Representation theory of Artin algebras, Cambridge Stud. Adv. Math. **36**, Cambridge University Press, Cambridge 1995.

[Be1] D. Benson, Modular representation theory: New trends and methods, Lecture Notes in Math. **1081**, Springer-Verlag, Berlin, Heidelberg, New York 1984.

[Be2] D. Benson, Representations and cohomology I, Cambridge Stud. Adv. Math. **30**, Cambridge University Press, Cambridge 1991.

[Be3] D. Benson, Representations and cohomology II, Cambridge Stud. Adv. Math. **31**, Cambridge University Press, Cambridge 1991.

[Bs] C. Bessenrodt, The Auslander–Reiten quiver of a modular group algebra revisited, Math. Z. **206** (1991), 25–34.

184   R. Farnsteiner

[BG]   K. Bongartz , P. Gabriel, Covering spaces in representation theory, Invent. Math. **65** (1982), 331–378.

[Ca]   J. Carlson, Varieties and the cohomology ring of a module, J. Algebra **85** (1983), 104–143.

[CB1] W. Crawley-Boevey, On tame algebras and bocses, Proc. London Math. Soc. **56** (1988), 451–483.

[CB2] W. Crawley-Boevey, Tameness of biserial algebras, Arch. Math. **65** (1995), 399–407.

[Dr]   Yu. Drozd, Tame and wild matrix problems, Matrix Problems, Amer. Math. Soc. Transl. **128** (1986), 31–55.

[Er1] K. Erdmann, Blocks of tame representation type and related algebras, Lecture Notes in Math. **1428**, Springer-Verlag, Berlin, Heidelberg, New York 1990.

[Er2] K. Erdmann, On Auslander–Reiten components for group algebras, J. Pure Appl. Algebra **104** (1995), 149–160.

[Er3] K. Erdmann, The Auslander–Reiten quiver of restricted enveloping algebras, in: Represetation theory of algebras, R. Bautista et al. (Ed.), CMS Conf. Proc. **18** , Amer. Math. Soc., Providence 1996, 201–214.

[EK]   K. Erdmann, O. Kerner, On the stable module category of a self-injective algebra, to appear in Trans. Amer. Math. Soc.

[ES]   K. Erdmann, A. Skowroński, On Auslander–Reiten components of blocks and self-injective biserial algebras, Trans. Amer. Math. Soc. **330** (1992), 165–189.

[Fa1] R. Farnsteiner, Periodicity and representation type of modular Lie algebras, J. reine angew. Math. **464** (1995), 47–65.

[Fa2] R. Farnsteiner, Representations of Lie algebras with triangular decomposition, in: Symposia Gaussiana, Proc. 2nd Gauss Symposium, Conf. A, M. Behara, R. Fritsch, R. G. Lintz (Ed.), Walter de Gruyter, Berlin 1995, 275–286.

[Fa3] R. Farnsteiner, Representations of blocks associated to induced modules of restricted Lie algebras, Math. Nachr. **179** (1996), 57–88.

[Fa4] R. Farnsteiner, On Auslander–Reiten quivers of enveloping algebras of restricted Lie algebras, to appear in Math. Nachr.

[Fa5] R. Farnsteiner, On support varieties of Auslander–Reiten components, to appear in Indag. Math.

[Fa6] R. Farnsteiner, Auslander–Reiten components for Lie algebras of reductive groups, Manuscript in preparation.

[Fe1] J. Feldvoss, Blocks and projective modules for reduced enveloping algebras of a nilpotent restricted Lie algebra, Arch. Math. **65** (1995), 495–500.

[Fe2] J. Feldvoss, Homological topics in the representation theory of restricted Lie algebras, in: Lie algebras and their representations, S.-J. Kang, M.-H. Kim, I. Lee (Ed.), Contemp. Math. **194**, Amer. Math. Soc., Providence 1996, 69–119.

[Fe3] J. Feldvoss, On the block structure of supersolvable restricted Lie algebras, J. Algebra **183** (1996), 396–419

[FeS] J. Feldvoss, H. Strade, Restricted Lie algebras with bounded cohomology and related classes of algebras, Manuscripta Math. **74** (1992), 47–67

[Fi]   G. Fischer, Darstellungstheorie des ersten Frobeniuskerns der SL_2, Dissertation, Universität Bielefeld, 1982.

[FP1] E. Friedlander, B. Parshall, Geometry of p-unipotent Lie algebras, J. Algebra **109** (1987), 25–45.

[FP2]   E. Friedlander, B. Parshall, Support varieties for restricted Lie algebras, Invent. Math. **86** (1986), 553–562.

[FP3]   E. Friedlander, B. Parshall, Modular representation theory of Lie algebras, Amer. J. Math. **110** (1988), 1055–1094.

[Ga1]   P. Gabriel, Auslander–Reiten sequences and epresentation-finite algebras, in: Representation theory I, V. Dlab, P. Gabriel (Ed.), Lecture Notes in Math. **831**, Springer-Verlag, Berlin, Heidelberg, New York 1980, 1–71.

[Gr]   J. Green, On the indecomposable representations of a finite group, Math. Z. **70** (1959), 430–445.

[HPR]   D. Happel, U. Preiser, C. Ringel, Vinberg's characterization of Dynkin diagrams using subadditive functions with applications to DTr-periodic modules, in: Representation theory II, V. Dlab, P. Gabriel (Ed.), Lecture Notes in Math. **832**, Springer-Verlag, Berlin, Heidelberg, New York 1980, 280–294.

[He]   A. Heller, Indecomposable representations and the loop-space operation, Proc. Amer. Math. Soc. **12** (1961), 640–643.

[Hi]   D. Higman, Indecomposable representations of characteristic p. Duke J. Math. **21** (1954), 377–381.

[Ho1]   G. Hochschild, Cohomology of restricted Lie algebras. Amer. J. Math. **76** (1954), 555–580.

[HN]   R. Holmes, D. Nakano, Block degeneracy and Cartan invariants for graded restricted Lie algebras, J. Algebra **161** (1993), 155–170.

[Hu1]   J. Humphreys, Modular representations of Lie algebras and semisimple groups, J. Algebra **19** (1971), 51–79.

[Hu2]   J. Humphreys, Conjugacy classes in semisimple algebraic groups, Math. Surveys Monogr. **43**. Amer. Math. Soc., Providence 1995.

[Ja1]   J. Jantzen, Kohomologie von p-Lie Algebren und nilpotente Elemente, Abh. Math. Sem. Univ. Hamburg **56** (1986), 191–219.

[LS]   R. Larson, M. Sweedler, An associative orthogonal bilinear form for Hopf algebras, Amer. J. Math. **91** (1969), 75–94.

[LSc]   S. Liu, R. Schulz, The existence of bounded infinite DTr-orbits, Proc. Amer. Math. Soc. **122** (1994), 1003–1005.

[MR]   E. Marmolejo, C. Ringel, Modules of bounded length in Auslander–Reiten components, Arch. Math. **50** (1988), 128–133.

[NP]   D. Nakano, R. Pollack, On the construction of indecomposable modules over restricted enveloping algebras, J. Pure Appl. Algebra **107** (1996), 61–73.

[Ok]   T. Okuyama, On the Auslander–Reiten quiver of a finite group, J. Algebra **110** (1987), 425–430.

[Pa]   B. Pareigis, Kohomologie von p-Lie Algebren, Math. Z. **104** (1968), 281–336.

[Pe]   V. Petrogradsky, On the structure of an enveloping algebra of a nilpotent Lie p-algebra, Moscow Univ. Math. Bull. 45 (1990), 50–52.

[PV]   W. Pfautsch, D. Voigt, The representation-finite algebraic groups of dimension zero, C. R. Acad. Sci. Paris Sér. I Mat. **306** (1988), 685–689.

[Po]   R. Pollack, Restricted Lie algebras of bounded type, Bull. Amer. Math. Soc. **74** (1968), 326–331.

[Ri1]   C. Riedtmann, Algebren, Darstellungsköcher, Ueberlagerungen und zurück, Comment. Math. Helv. **55** (1980), 199–224.

[Ri]   C. Ringel, Representation theory of finite-dimensional algebras, in: Representations of algebras, Proc. Symp., Durham 1985, London Math. Soc. Lecture Note Series **116** (1986), 7–79.

[St]   H. Strade, Darstellungen auflösbarer Lie- p-Algebren, Math. Ann. **232** (1978), 15–32.

[SF]   H. Strade, R. Farnsteiner, Modular Lie algebras and their representations, Monographs Textbooks Pure Appl. Math. **116**, Marcel Dekker, New York 1988.

[VK]   B. Veisfeiler, V. Kac, Irreducible representations of Lie p-algebras, Funct. Anal. Appl. **5** (1971), 111–117.

[Vo]   D. Voigt, The algebraic infinitesimal groups of tame representation type, C. R. Acad. Sci. Paris Sér I Mat. **311** (1990), 757–760.

[We]   P. Webb, The Auslander–Reiten quiver of a finite group, Math. Z. **179** (1982), 97–121.

Department of Mathematics
University of Wisconsin
Milwaukee, WI 53201, U.S.A.
E-mail: rolf@csd.uwm.edu

Chief factors and the principal block of a restricted Lie algebra

Jörg Feldvoss

Let L be a finite-dimensional restricted Lie algebra over a commutative field $\mathbb{F}$ of characteristic p, and let $u(L)$ denote the restricted universal enveloping algebra of L (cf. **[SF**, Theorem 2.5.1]). A complete classification of simple restricted L-modules (over an algebraically closed field) exists for nilpotent restricted Lie algebras and restricted simple Lie algebras of classical and Cartan type, and their distribution into blocks is also known (see **[Voigt]**, **[Fe4]**, **[Hu]**, and **[HN]**). A *block decomposition* of $u(L)$ is a decomposition

$$u(L) = \bigoplus_{j=0}^{r} B_j(L),$$

into indecomposable two-sided ideals $B_j(L)$ of $u(L)$, the so-called *block ideals* of $u(L)$, which are finite-dimensional associative $\mathbb{F}$-algebras with an identity element (see **[HB**, Theorem VII.12.1]). Every indecomposable $u(L)$-module is a (unitary, left) $B_j(L)$-module for some uniquely determined j, and in particular, this induces an equivalence relation "*belonging to the same block*" on the finite set $\mathrm{Irr}_p(L)$ of all isomorphism classes of (irreducible or) simple restricted L-modules

$$\mathrm{Irr}_p(L) = \bigcup_{j=0}^{r} \mathbb{B}_j(L),$$

such that each equivalence class $\mathbb{B}_j(L) = \{[S] \in \mathrm{Irr}_p(L) \mid B_j(L) \cdot S = S\}$, a so-called *block* of $u(L)$, is in one-to-one correspondence with the set of isomorphism classes of simple $B_j(L)$-modules. For more information on block theory we refer the reader to **[CR**, §55] and **[HB**, Chapter VII] or **[Fe4]** and **[Fe5]**.

The block of $u(L)$ containing the one-dimensional trivial L-module is called the *principal block* of L, and the labeling will be chosen such that $\mathbb{B}_0(L)$ is the principal block. The principal block turns out to be the most complicated block of L (see for example **[Fe4**, Corollary 1, Examples 1 and 2]).

The purpose of this note is to apply the methods of **[FK]** in order to relate the projective indecomposable modules and the simple modules of the principal block ideal of a finite-dimensional restricted Lie algebra to tensor products (over the ground field) of (finitely many) composition factors of the adjoint module. The latter can be read off from the structure constants of the Lie algebra and therefore are easily accessible in given examples. Moreover, at the same time we also obtain a different approach to the results of [**Voigt**] and [**Fe4**] for supersolvable (respectively nilpotent) restricted Lie algebras and a Lie theoretic analogue of a theorem of Alperin [**Alp**, Theorem 1] for p-semiprimitive restricted Lie algebras.

It is common from group theory to call a composition factor of the adjoint module of a finite-dimensional Lie algebra L a *chief factor* of L (see [**Ba1**]), and we denote the set of isomorphism classes of chief factors of L by $\mathcal{F}(L)$. If $\mathrm{Ann}_L(M) := \{x \in L \mid x \cdot M = 0\}$ is the *annihilator* of an arbitrary L-module M, then the largest nilpotent ideal $\mathrm{Nil}(L)$ of a finite-dimensional Lie algebra L can be described as follows.

Proposition 1. *Let L be a finite-dimensional Lie algebra. Then*

$$\mathrm{Nil}(L) = \bigcap_{S \in \mathcal{F}(L)} \mathrm{Ann}_L(S).$$

Proof. If $y \in \mathrm{Nil}(L)$, then y is ad-nilpotent with respect to $\mathrm{Nil}(L)$, i.e., $\mathrm{ad}_{\mathrm{Nil}(L)}\, y$ is nilpotent. Hence for any $y \in \mathrm{Nil}(L)$ there exists a positive integer n such that $(\mathrm{ad}_{\mathrm{Nil}(L)}\, y)^n = 0$, and therefore we obtain for every $x \in L$:

$$(\mathrm{ad}_L\, y)^{n+1}(x) = (\mathrm{ad}_L\, y)^n([y, x]) \in (\mathrm{ad}_L\, y)^n(\mathrm{Nil}(L)) = 0.$$

This shows that y is in fact ad-nilpotent on L. In particular, $\mathrm{Nil}(L)$ acts nilpotently on every chief factor S of L. Then the Engel-Jacobson theorem implies

$$S^{\mathrm{Nil}(L)} \neq 0,$$

and the simplicity of S yields $\mathrm{Nil}(L) \subseteq \mathrm{Ann}_L(S)$.

In order to show the other inclusion, let

$$L = L_0 \supset L_1 \supset \cdots \supset L_n = 0$$

be a composition series of L. If $x \in L$ acts trivially on every chief factor of L, then $[x, L_j] \subseteq L_{j+1}$ for every $0 \leq j \leq n - 1$. As a result, $(\mathrm{ad}_L\, x)^n(L) \subseteq L_n = 0$, i.e., by virtue of Engel's theorem, $\bigcap_{S \in \mathcal{F}(L)} \mathrm{Ann}_L(S)$ is a nilpotent ideal of L and thus contained in $\mathrm{Nil}(L)$.   □

Proposition 1 enables us to prove the first main result of this note.

Theorem 1. *Let L be a finite-dimensional restricted Lie algebra. Then every projective indecomposable $u(L/\mathrm{Nil}(L))$-module is a direct summand of a suitable tensor product of (finitely many) chief factors of L.*

Proof. Set

$$V := \bigoplus_{S \in \mathcal{F}(L)} S.$$

According to Proposition 1, we have

$$\mathrm{Ann}_L(V) = \bigcap_{S \in \mathcal{F}(L)} \mathrm{Ann}_L(S) = \mathrm{Nil}(L).$$

Hence V is a faithful $L/\operatorname{Nil}(L)$-module. Then [**FK**, Corollary 3.2] in conjunction with [**FK**, Theorem 2.4] implies that $\mathfrak{T}_n(V)$ is a faithful $u(L/\operatorname{Nil}(L))$-module for some positive integer n ($\leq \dim_{\mathbb{F}} u(L/\operatorname{Nil}(L))$). Since $\mathfrak{T}_n(V)$ is a direct sum of certain tensor products of (at most n) chief factors of L, the assertion follows from [**FK**, Remark 1.4]. $\square$

Remark. Note that the projective indecomposable $u(L)$-modules are *not* necessarily direct summands of a tensor product of chief factors of L. For example, in view of the main result of [**Ho**], the projective cover $P_L(\mathbb{F})$ of the one-dimensional trivial module $\mathbb{F}$ of a finite-dimensional nilpotent restricted Lie algebra L is a direct summand of such a tensor product (i.e., in this case $\mathbb{F}$ itself) if and only if L is a torus.

In particular, we obtain for $\operatorname{Nil}(L) = 0$:

Corollary 1. *Let L be a finite-dimensional semisimple restricted Lie algebra. Then every projective indecomposable $u(L)$-module is a direct summand of a suitable tensor product of (finitely many) chief factors of L.* $\square$

Remark. Note that for a restricted *simple* Lie algebra Corollary 1 is an immediate consequence of [**FK**, Theorem 3.3(a)] because the only chief factor is the adjoint module which itself is faithful.

Using

$$\operatorname{Rad}_p(L) = \bigcap_{S \in \operatorname{Irr}_p(L)} \operatorname{Ann}_L(S)$$

(see [**SF**, Theorem 5.5.3(2)]), it is possible to replace in Theorem 1 the largest nilpotent ideal $\operatorname{Nil}(L)$ by the largest p-nilpotent ideal $\operatorname{Rad}_p(L)$ if we allow in the conclusion that *any* simple restricted L-module can occur in the tensor product. Recall that a restricted Lie algebra is called *p-semiprimitive* if $\operatorname{Rad}_p(L) = 0$ (see [**SF**, p. 220]). Then the just mentioned analogue of Theorem 1 immediately implies that for a finite-dimensional p-semiprimitive restricted Lie algebra L every projective indecomposable $u(L)$-module is a direct summand of a suitable tensor product of (finitely many) simple restricted L-modules (see [**Alp**, Theorem 1] for the analogue in the modular representation theory of finite groups).

In order to derive our second main result from Theorem 1, we need the following necessary condition for a simple restricted L-module to belong to the principal block which slightly generalizes [**Fe4**, Proposition 3]:

Proposition 2. *Let L be a finite-dimensional restricted Lie algebra, and let S be a simple restricted L-module. If S belongs to the principal block of L, then $\operatorname{Nil}(L) \subseteq \operatorname{Ann}_L(S)$.*

Proof. Let x be an arbitrary element in $\operatorname{Nil}(L)$. According to [**SF**, Theorem 2.3.4], there exists a positive integer n such that $x^{[p]^n}$ is semisimple. Hence we obtain from [**Fe4**, Proposition 3] that

$$x^{[p]^n} \in T_p(\operatorname{Nil}(L)) = T_p(L) \subseteq \operatorname{Ann}_L(S).$$

Since S is restricted, $\mathrm{Nil}(L)$ acts nilpotently on S, and by virtue of the Engel-Jacobson theorem, $\mathrm{Nil}(L)$ acts trivially on S. $\square$

Remark. It is an immediate consequence of Proposition 2 that the principal block of a finite-dimensional nilpotent restricted Lie algebra only contains the one-dimensional trivial module (see [**Voigt**, Satz 2.29], [**Pet**], the proof of [**Fe3**, Theorem 1], and [**Fe4**, Theorem 2]).

In particular, if L is solvable, Proposition 2 in conjunction with [**Fe4**, Proposition 2] and Proposition 1 yields the following special case of the analogue of a result for finite-dimensional modular group algebras due to Brauer (see [**Mich**, Theorem 2]).

Corollary 2. *Let L be a finite-dimensional solvable restricted Lie algebra. Then*

$$\mathrm{Nil}(L) = \bigcap_{S \in \mathbb{B}_0(L)} \mathrm{Ann}_L(S).$$ $\square$

Now we are ready to prove the following analogue of [**Will**, Theorem 1.8] (see also [**FG**, Theorem 2.6]) for simple modules in the principal block of a restricted Lie algebra:

Theorem 2. *Let L be a finite-dimensional restricted Lie algebra. Then every simple restricted L-module belonging to the principal block of L is a submodule of a suitable tensor product of (finitely many) chief factors of L.*

Proof. Let S be a simple restricted L-module which belongs to the principal block of L. By Proposition 2, S is a simple restricted $L/\mathrm{Nil}(L)$-module. Since $u(L/\mathrm{Nil}(L))$ is a Frobenius algebra (cf. [**SF**, Corollary 5.4.3]), Ikeda's theorem (cf. [**CR**, Theorem 62.11]) implies that the injective hull of S (with respect to $u(L/\mathrm{Nil}(L))$) is a projective indecomposable $u(L/\mathrm{Nil}(L))$-module. As S can be embedded into its injective hull, the assertion is now an immediate consequence of Theorem 1. $\square$

Remark. Note that Theorem 2 can also be derived directly from *Burnside's theorem* for restricted Lie algebras (see [**FK**, Corollary 3.4]) which is a consequence of [**PQ**, Corollary 10] and the discussion preceding [**PQ**, Theorem 4] but not (explicitly) contained in [**PQ**].

A chief factor I/J of L is called *abelian* if I/J is abelian (as a Lie algebra), i.e., $[I, I] \subseteq J$. The following example shows that for non-solvable restricted Lie algebras it is *not* enough, in Theorem 2, to consider only tensor products of *abelian* chief factors.

Example 1.   Consider $L := \mathfrak{gl}_2(\mathbb{F})$ over an algebraically closed field $\mathbb{F}$ of characteristic $p > 2$. Then the trivial module is the only abelian chief factor of L but the $(p-1)$-dimensional restricted $\mathfrak{sl}_2(\mathbb{F})$-module with trivial central action belongs to the principal block of L.

In order to illustrate Theorem 2, we want to apply it to an important special case. A Lie algebra L is called *supersolvable* if every chief factor of L is one-dimensional. Hence Theorem 2 implies that every simple module in the principal block of a supersolvable restricted Lie algebra is one-dimensional (see also [**Fe2**, Theorem 2] or [**Fe4**, Theorem 1] for a cohomological proof). In this case the principal block $\mathbb{B}_0(L)$ forms a commutative group under

$$[S_1] + [S_2] := [S_1 \otimes_\mathbb{F} S_2] \quad \forall\, [S_1], [S_2] \in \mathbb{B}_0(L),$$

with the isomorphism class $[\mathbb{F}]$ of the one-dimensional trivial module as the identity element and the isomorphism class $[S^*]$ of the dual module as the inverse element of $[S]$. Note that via $\gamma \mapsto [F_\gamma]$ the principal block $\mathbb{B}_0(L)$ is isomorphic to the group G_0^L introduced in [Fe4]. Then Theorem 2 immediately yields the following generalization of [Fe5, Lemma 4.9].

Corollary 3. *Let L be a finite-dimensional supersolvable restricted Lie algebra. Then the principal block of L is generated by the (non-trivial) chief factors of L.*   □

Example 2.   Consider the non-abelian two-dimensional restricted Lie algebra

$$L := \mathbb{F}t \oplus \mathbb{F}e, \quad [t, e] = e, \quad t^{[p]} = t, \quad e^{[p]} = 0.$$

Then the restricted universal enveloping algebra of L has the following Gabriel quiver

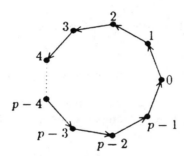

where the vertices corresponding to the one-dimensional restricted L-modules F_τ are labelled by the respective eigenvalues τ of t (see [Fe1, Beispiel II.4.1]). Hence the principal block $\mathbb{B}_0(L)$ of L is generated by the only non-trivial (split) chief factor F_1, i.e., in this case $\mathbb{B}_0(L)$ is a cyclic group of order p.

Recall that a chief factor I/J of L is called *split* if the exact sequence $0 \to I/J \to L/J \to L/I \to 0$ splits (as a sequence of Lie algebras). It is clear that the top chief factor is *always* split. If in addition L is *unimodular*, the proof of [Fe5, Lemma 4.9] shows that then $\mathbb{B}_0(L)$ is even generated by the (non-trivial) *split* chief factors of L. Note that the same proof (using [Sta, Corollary 2] and [Ba2, Theorem 1] instead of [Fe4, Theorem 1] and [Ba1, Theorem 1]) shows that the principal block of a finite p-supersolvable group is *always* generated by its (non-trivial) split chief factors (because the group algebra is *always* symmetric). As a generalization of Example 2 we conclude this note by the following sufficient condition (partly complementary to [Fe5, Lemma 4.9]) pointing towards this stronger result in the case of restricted Lie algebras:

Proposition 3. *Let L be a finite-dimensional supersolvable restricted Lie algebra. If* Nil(L) *is abelian, then every chief factor of L is split.*

Proof. Assume first that the ground field $\mathbb{F}$ is algebraically closed. Since L is supersolvable, there exists a torus T such that $L = T \oplus \text{Nil}(L)$ (see the proof of [**Fe4**, Theorem 4]). Let S be a non-trivial chief factor of L. According to the main result of [**Ho**], the adjoint module L is a semisimple T-module. Since S is not trivial, this in conjunction with the five-term exact sequence for Lie algebra cohomology and Proposition 1 yields

$$\dim_{\mathbb{F}} H^1(L, S) = \dim_{\mathbb{F}} \text{Hom}_T(\text{Nil}(L), S) = \dim_{\mathbb{F}} \text{Hom}_T(L, S) = [L : S] \neq 0,$$

where $[L : S]$ denotes the multiplicity of S as a composition factor of the adjoint module L. If S is trivial, then we obtain

$$H^1(L, S) \cong L/[L, L] \neq 0$$

because $L (\neq 0)$ is solvable. Hence in both cases [**Ba1**, Theorem 1] shows that S is split.

If $\mathbb{F}$ is arbitrary, then let $\overline{\mathbb{F}}$ denote the algebraic closure of $\mathbb{F}$. If S is a chief factor of L, then $S \otimes_{\mathbb{F}} \overline{\mathbb{F}}$ is a direct sum of chief factors of $L \otimes_{\mathbb{F}} \overline{\mathbb{F}}$. Since $L \otimes_{\mathbb{F}} \overline{\mathbb{F}}$ is also supersolvable with Nil($L \otimes_{\mathbb{F}} \overline{\mathbb{F}}$) = Nil($L$) $\otimes_{\mathbb{F}} \overline{\mathbb{F}}$ abelian, we can conclude from the first part of the proof that every direct summand of $S \otimes_{\mathbb{F}} \overline{\mathbb{F}}$ is split. But then

$$H^1(L \otimes_{\mathbb{F}} \overline{\mathbb{F}}, S \otimes_{\mathbb{F}} \overline{\mathbb{F}}) \cong H^1(L, S) \otimes_{\mathbb{F}} \overline{\mathbb{F}}$$

in conjunction with a twofold application of [**Ba1**, Theorem 1] implies that S is split. □

It is easy to see that in general *not* every chief factor of a supersolvable restricted Lie algebra is split, but in the numerous examples (with $\dim_{\mathbb{F}} \text{Nil}(L) \leq 7$) computed by the author it is always true that

$$\text{Nil}(L) = \bigcap_{S \in \mathcal{F}_{\text{split}}(L)} \text{Ann}_L(S),$$

where $\mathcal{F}_{\text{split}}(L)$ denotes the set of isomorphism classes of split chief factors of L. The author was not able to decide whether this slightly stronger version of Proposition 1 does hold (at least) for a finite-dimensional supersolvable restricted Lie algebra, and consequently whether the precise analogue of the group-theoretic result as mentioned before Proposition 3 would or would not be valid.

Acknowledgments. It is a pleasure to thank Joe Ferrar and Tom Gregory for the invitation to the Monster/Lie algebra conference at Ohio State University. Moreover, the author is very grateful to Joe and Sara Ferrar for the hospitality at their house during his stay in Columbus.

Bibliography

[Alp]   J. L. Alperin, Projective modules and tensor products, J. Pure Appl. Algebra **8** (1976), 235–241.

[Ba1]   D. W. Barnes, First cohomology groups of soluble Lie algebras, J. Algebra **46** (1977), 292–297.

[Ba2]   D. W. Barnes, First cohomology groups of p-soluble groups, J. Algebra **46** (1977), 298–302.

[CR]   C. W. Curtis, I. Reiner, Representation theory of finite groups and associative algebras, Wiley Classics Library Edition, John Wiley & Sons, New York 1988.

[Fe1]   J. Feldvoss, Homologische Aspekte der Darstellungstheorie modularer Lie-Algebren, Dissertation, Universität Hamburg, 1989.

[Fe2]   J. Feldvoss, A cohomological characterization of solvable modular Lie algebras, in: Non-associative algebra and its applications, S. González (Ed.), Math. Appl. **303**, Kluwer Academic Publishers, Dordrecht, Boston, London 1994, 133–139.

[Fe3]   J. Feldvoss, Blocks and projective modules for reduced universal enveloping algebras of a nilpotent restricted Lie algebra, Arch. Math. **65** (1995), 495–500.

[Fe4]   J. Feldvoss, On the block structure of supersolvable restricted Lie algebras, J. Algebra **183** (1996), 396–419.

[Fe5]   J. Feldvoss, Homological topics in the representation theory of restricted Lie algebras, in: Lie algebras and their representations, S.-J. Kang, M.-H. Kim, I. Lee (Ed.), Contemp. Math. **194**, Amer. Math. Soc., Providence 1996, 69–119.

[FK]   J. Feldvoss, L. Klingler, Tensor powers and projective modules for Hopf algebras, to appear in Canad. Math. Soc. Conf. Proc.

[FG]   P. Fong, W. Gaschütz, A note on the modular representations of solvable groups, J. reine angew. Math. **208** (1961), 73–78.

[Ho]   G. P. Hochschild, Representations of restricted Lie algebras of characteristic p, Proc. Amer. Math. Soc. **5** (1954), 603–605.

[HN]   R. R. Holmes and D. K. Nakano, Block degeneracy and Cartan invariants for graded Lie algebras of Cartan type, J. Algebra **161** (1993), 155–170.

[Hu]   J. E. Humphreys, Modular representations of classical Lie algebras and semisimple groups, J. Algebra **19** (1971), 51–79.

[HB]   B. Huppert, N. Blackburn, Finite groups II, Grundlehren Math. Wiss. **242**, Springer-Verlag, Berlin, Heidelberg, New York 1982.

[Mich] G. O. Michler, The kernel of a block of a group algebra, Proc. Amer. Math. Soc. **37** (1973), 47–49.

[PQ]   D. S. Passman, D. Quinn, Burnside's theorem for Hopf algebras, Proc. Amer. Math. Soc.**123** (1995), 327–333.

[Pet]   V. M. Petrogradsky, On the structure of an enveloping algebra of a nilpotent Lie p-algebra, Moscow Univ. Math. Bull. **45** (1990), 50–52.

[Sta]   U. Stammbach, Cohomological characterisations of finite solvable and nilpotent groups, J. Pure Appl. Algebra **11** (1977), 293–301.

[SF]   H. Strade, R. Farnsteiner, Modular Lie algebras and their representations, Monographs Textbooks Pure Appl. Math. **116**, Marcel Dekker, New York 1988.

[Voigt] D. Voigt, Induzierte Darstellungen in der Theorie der endlichen, algebraischen Gruppen, Lecture Notes in Math. **592**, Springer-Verlag, Berlin, Heidelberg, New York 1977.

[Will] W. Willems, On p-chief factors of finite groups, Comm. Algebra **13** (1985), 2433–2447.

Mathematisches Seminar der Universität
Bundesstraße 55
D-20146 Hamburg
Federal Republic of Germany
E-mail: feldvoss@math.uni-hamburg.de

On Drinfeld realization of quantum affine algebras

Naihuan Jing

Abstract. We provide a direct proof of the Drinfeld realization for the quantum affine algebras.

1. Introduction

In 1987 Drinfeld [**Dr2**] gave an extremely important realization of quantum affine algebras [**Dr1**], [**Jb**]. This new realization has lead to numerous applications such as the vertex representations [**FJ**], [**J**]. The proof of this realization was not in print until Beck's braid group interpretation for the untwisted types [**B**]. Some of lower rank cases were also studied in [**D**], [**LSS**]. All these work started from the quantum group towards the quantum loop realization, and were based on Lusztig's theory of braid group action on the quantum enveloping algebras [**L**]. However, Drinfeld did give the exact isomorphism between two definitions of quantum affine algebras in [**Dr2**]. In this paper we give another proof directly from the Drinfeld isomorphism. Our proof is self-contained and elementary and works from the opposite direction from the quantum loop algebras towards the quantum groups.

In doing this, we discovered that there are rich structures held by the q-loop algebra realization. We directly deform the argument used by Kac [**K**] to identify the affine Lie algebras and the Kac–Moody algebra defined by generators and relations. Here we must admit that the q-arguments are much more complicated than the classical analog, where we have the root space structure avaliable. It is nontrivial to properly deform usual brackets by q-brackets. As we have shown here in many cases there are strong indications for us to follow. The key are the following identities

$$[a, [b, c]_u]_v = [[a, b]_x, c]_{uv/x} + x\,[b, [a, c]_{v/x}]_{u/x}, \quad x \neq 0 \tag{1.1}$$

$$[[a, b]_u, c]_v = [a, [b, c]_x]_{uv/x} + x\,[[a, c]_{v/x}, b]_{u/x}, \quad x \neq 0 \tag{1.2}$$

where one needs to choose an appropriate x to apply.

We also discuss the Drinfeld realization for the twisted quantum affine algebras using the same approach. The results are used to construct the intertwining operators between level one modules of twisted quantum affine algebras in [**JK**].

I thank Professors Joseph Ferrar and Thomas Gregory for the organization of the conference on Lie algebras, where this paper is reported. I also thank Professor Vyjayanthi Chari for her useful comments on the earlier version. The work is partially supported by NSA grant MDA904-96-1-0087.

2. Quantum affine algebras

Let $A = (a_{ij})$ $(i, j \in I = \{0, 1, \ldots, n\})$ be a generalized Cartan matrix of affine types
[K]. Let $\mathfrak{h}$ be a vector space over $\mathbb{C}(q^{1/2})$ with a basis $\{h_0, h_1, \ldots, h_n, d\}$ and define
the linear functionals $\alpha_i \in \mathfrak{h}^*$ $(i \in I)$ by

$$\alpha_i(h_j) = a_{ji}, \quad \alpha_i(d) = \delta_{i,0} \quad \text{for } j \in I. \tag{2.1}$$

Then the triple $(\mathfrak{h}, \Pi = \{\alpha_i \mid i \in I\}, \Pi^\vee = \{h_i \mid i \in I\})$ is the realization of the matrix
A. The Kac–Moody Lie algebra $\mathfrak{g}$ associated with the matrix A is called the *affine
Kac–Moody algebra of type* A (cf. [K]). The algebra is generated as a Lie algebra by e_i,
f_i, h_i $(i \in I)$ and d subject to the usual relations. The elements of Π (resp. $\Pi^\vee$) are
called the *simple roots* (resp. *simple coroots*) of $\mathfrak{g}$.

The standard nondegenerate symmetric bilinear form $(\ |\ )$ on $\mathfrak{h}^*$ satisfies

$$(\alpha_i|\alpha_i) = d_i a_{ij}, \quad (\delta|\alpha_i) = (\delta|\delta) = 0 \quad \text{for all } i, j \in I. \tag{2.2}$$

where $d_i = (\alpha_i|\alpha_i)/2$ are rational numbers given as follows

$$(ADE)^{(1)} : d_i = 1$$
$$C_n^{(1)} : d_0 = 1, \ d_i = 1/2, \ d_n = 1$$
$$B_n^{(1)} : d_0 = d_i = 1, \ d_n = 1/2$$
$$G_2^{(1)} : d_0 = d_1 = 1, \ d_2 = 1/3$$
$$F_4^{(2)} : d_0 = d_1 = d_2 = 1, \ d_3 = d_4 = 1/2$$

and $q_i = q^{d_i}$ for $i \in I$. The *quantum affine Lie algebra* $U_q(A)$ is the associative algebra
with 1 over $\mathbb{C}(q^{1/2})$ generated by the elements e_i, f_i $(i \in I)$ and q^h $(h \in P^\vee)$ with
the following defining relations:

$$q^0 = 1, \quad q^h q^{h'} = q^{h+h'} \quad \text{for } h, h' \in P^\vee,$$
$$q^h e_i q^{-h} = q^{\alpha_i(h)} e_i, \quad q^h f_i q^{-h} = q^{-\alpha_i(h)} f_i \quad \text{for } h \in P^\vee (i \in I),$$
$$e_i f_j - f_j e_i = \delta_{ij} \frac{t_i - t_i^{-1}}{q_i - q_i^{-1}}, \quad \text{where } t_i = q^{h_i} \text{ and } i, j \in I,$$
$$\sum_{m+k=1-a_{ij}} (-1)^m e_i^{(m)} e_j e_i^{(n)} = 0, \tag{2.3}$$
$$\sum_{m+n=1-a_{ij}} (-1)^m f_i^{(m)} f_j f_i^{(n)} = 0 \quad \text{for } i \neq j,$$

where $e_i^{(k)} = e_i^k/[k]_i!$, $f_i^{(k)} = f_i^k/[k]_i!$, $[m]_i! = \prod_{k=1}^m [k]_i$, and $[k]_i = \frac{q_i^k - q_i^{-k}}{q - q^{-1}}$.

Let Ω be the anti-algebra involution of $U_q(\hat{\mathfrak{g}})$ over $\mathbb{C}$ given by

$$\Omega(e_i) = f_i, \quad \Omega(f_i) = e_i, \quad \Omega(q^h) = q^{-h}, \quad \Omega(q) = q^{-1}.$$

Now we give the Drinfeld realization for the untwisted types.

Let U be the associative algebra with 1 over $\mathbb{C}(q^{1/2})$ generated by the elements $x_i^{\pm}(k)$, $a_i(l)$, $K_i^{\pm 1}$, $\gamma^{\pm 1/2}$, $q^{\pm d}$ $(i = 1, 2, \ldots, n, \ k \in \mathbb{Z}, \ l \in \mathbb{Z} \setminus \{0\})$ with the following defining relations

$$[\gamma^{\pm 1/2}, u] = 0 \quad \text{for all } u \in U,$$

$$K_i K_j = K_j K_i, \quad K_i K_i^{-1} = K_i^{-1} K_i = 1,$$

$$[a_i(k), a_j(l)] = \delta_{k+l,0} \frac{[a_{ij}k]_i}{k} \frac{\gamma^k - \gamma^{-k}}{q_j - q_j^{-1}}, \quad [a_i(k), K_j^{\pm 1}] = [q^{\pm d}, K_j^{\pm 1}] = 0,$$

$$q^d x_i^{\pm}(k) q^{-d} = q^k x_i^{\pm}(k), \quad q^d a_i(l) q^{-d} = q^l a_i(l),$$

$$K_i x_j^{\pm}(k) K_i^{-1} = q^{\pm(\alpha_i|\alpha_j)} x_j^{\pm}(k),$$

$$[a_i(k), x_j^{\pm}(l)] = \pm \frac{[a_{ij}k]_i}{k} \gamma^{\mp |k|/2} x_j^{\pm}(k+l), \tag{2.4}$$

$$x_i^{\pm}(k+1) x_j^{\pm}(l) - q^{\pm(\alpha_i|\alpha_j)} x_j^{\pm}(l) x_i^{\pm}(k+1) = q^{\pm(\alpha_i|\alpha_j)} x_i^{\pm}(k) x_j^{\pm}(l+1) - x_j^{\pm}(l+1) x_i^{\pm}(k),$$

$$[x_i^{+}(k), x_j^{-}(l)] = \frac{\delta_{ij}}{q_i - q_i^{-1}} \left(\gamma^{\frac{k-l}{2}} \psi_i(k+l) - \gamma^{\frac{l-k}{2}} \varphi_i(k+l) \right),$$

where $\psi_i(m)$ and $\varphi_i(-m)$ $(m \in \mathbb{Z}_{\geq 0})$ are defined by

$$\sum_{m=0}^{\infty} \psi_i(m) z^{-m} = K_i \exp\left((q_i - q_i^{-1}) \sum_{k=1}^{\infty} a_i(k) z^{-k} \right),$$

$$\sum_{m=0}^{\infty} \varphi_i(-m) z^{m} = K_i^{-1} \exp\left(-(q_i - q_i^{-1}) \sum_{k=1}^{\infty} a_i(-k) z^{k} \right),$$

and, for $i \neq j$,

$$\text{Sym}_{l_1,\ldots,l_m} \sum_{s=1}^{m=1-a_{ij}} (-1)^s \begin{bmatrix} m \\ s \end{bmatrix}_i x_i^{\pm}(l_1) \ldots x_i^{\pm}(l_s) \ldots x_j^{\pm}(n) x_i^{\pm}(l_{s+1}) \cdots x_i^{\pm}(l_m) = 0,$$

Lemma 2.1. *Let* $I_0 = \{1, 2, \ldots, n\}$ *be the index set for the simple roots of a finite dimensional simple Lie algebra* $\mathfrak{g}_0$. *Then for each* $i \in I_0$, *there exists a sequence of indices* $i = i_1, i_2, \ldots, i_{h-1}$ *such that*

$$(\alpha_{i_1} | \alpha_{i_2}) = \epsilon_{i_1},$$

$$(\alpha_{i_1} + \alpha_{i_2} | \alpha_{i_3}) = \epsilon_{i_2},$$

$$\vdots \tag{2.5}$$

$$(\alpha_{i_1} + \cdots + \alpha_{i_{h-2}} | \alpha_{i_{h-1}}) = \epsilon_{i_{h-2}},$$

where h *is the Coxeter number of the Lie algebra* $\mathfrak{g}_0$, *and* $\epsilon_i \in \mathbb{Q}_-$.

We will call any sequence $i_1, \ldots i_{h-1}$ given by Lemma 2.1 a ϵ sequence associated to the simple *root*.

The twisted commutators $[b_1, \cdots, b_n]_{v_1 \cdots v_{n-1}}$ and $[b_1, \cdots, b_n]'_{v_1 \cdots v_{n-1}}$ is defined inductively by $[b_1, b_2]_v = [b_1, b_2]'_v = b_1 b_2 - v b_2 b_1$ and

$$[b_1, \ldots, b_n]_{v_1 \ldots v_{n-1}} = [b_1, [b_2, \ldots, b_n]_{v_1 \ldots v_{n-2}}]_{v_{n-1}}$$
$$[b_1, \ldots, b_n]'_{v_1 \ldots v_{n-1}} = [[b_1, \cdots, b_{n-1}]'_{v_1 \cdots v_{n-2}}, b_n]_{v_{n-1}}.$$

If A is an antimorphism, then $A([b_1, \ldots, b_n]_{v_1 \ldots v_{n-1}}) = [A(b_n), \ldots, A(b_1)]'_{v_{n-1} \ldots v_1}$. But if B is an antimorphism such that $B(v_i) = v_i^{-1}$, then

$$A([b_1, \ldots, b_n]_{v_1 \cdots v_{n-1}}) = [A(b_n), \ldots, A(b_1)]'_{v_{n-1}^{-1} \ldots v_1^{-1}}$$
$$= v_1^{-1} \ldots v_{n-1}^{-1} [B(b_1), \ldots, B(b_n)]_{v_1 \ldots v_{n-1}}.$$

The following identities follow from the definition.

$$[a, bc]_v = [ab]_x c + x \, b[ac]_{v/x}, \quad x \neq 0$$
$$[ac, b]_v = a[bc]_x + x \, [a, c]_{v/x} b, \quad x \neq 0$$

$$\left[a, [b, c]_u\right]_v = \left[[a, b]_x, c\right]_{uv/x} + x \left[b, [a, c]_{v/x}\right]_{u/x}, \quad x \neq 0 \tag{2.6}$$
$$\left[[a, b]_u, c\right]_v = \left[a, [b, c]_x\right]_{uv/x} + x \left[[a, c]_{v/x}, b\right]_{u/x}, \quad x \neq 0. \tag{2.7}$$

In particular, we have

$$\left[a, [b_1, \cdots, b_n]_{v_1, \cdots, v_{n-1}}\right] = \sum_i [b_1, \cdots, [a, b_i], \ldots, b_n]_{v_1, \cdots, v_{n-1}},$$

$$[a, a, b]_{u\,v} = [a, a, b]_{v\,u} = a^2 b - (u + v)aba + uv \, ba^2. \tag{2.8}$$

The Serre relation for the case of $A_{ij} = -1$ can be written as

$$[x_i^{\pm}(m), x_i^{\pm}(m), x_j^{\pm}(n)]_{q_i, q_i^{-1}} = 0. \tag{2.9}$$

Theorem 2.2 [DR2]. *Fix an ϵ-sequence $i_1, i_2, \ldots, i_{h-1}$, and let $\theta = \sum_{j=1}^{h-1} \alpha_{i_j}$ be the maximal root of the finite dimensional simple Lie algebra $\mathfrak{g}$. Then there is a $\mathbb{C}(q^{1/s})$-algebra isomorphism $\Psi : U_q(\mathfrak{g}) \to U$ defined by*

$$e_i \mapsto x_i^+(0), \quad f_i \mapsto x_i^-(0), \quad t_i \mapsto K_i \quad \text{for } i = 1, \ldots, n,$$
$$e_0 \mapsto [x_{i_{h-1}}^-(0), \ldots, x_{i_2}^-(0), x_{i_1}^-(1)]_{q^{\epsilon_1} \ldots q^{\epsilon_{h-2}}} \gamma K_\theta^{-1},$$
$$f_0 \mapsto a(-q)^{-\epsilon} \gamma^{-1} K_\theta [x_{i_{h-1}}^+(0), \ldots, x_{i_2}^+(0), x_{i_1}^+(-1)]_{q^{\epsilon_1} \ldots q^{\epsilon_{h-2}}} \tag{2.10}$$
$$t_0 \mapsto \gamma K_\theta^{-1}, \quad q^d \mapsto q^d,$$

where $K_\theta = K_{i_1} \ldots K_{i_{h-1}}$, h is the Coxeter number, $\epsilon = \sum_{i=1}^{h-2} \epsilon_i$, and $s = 1, 2, 3$, the quotient of long roots by short roots. The constant a is 1 for simply laced types A_n, D_n, $a = [2]_1$ for $C_n^{(1)}$, and $a = [2]^{1-\delta_{1,i_1}}$ for $B_n^{(1)}$.

Proof. Let E_i, F_i, K_i, D be the images of e_i, f_i, k_i, d in the algebra U. We divide the proof into several steps.

Step 1. The elements E_i, F_i, K_i, D satisfy the defining relations of $U_q(\hat{\mathfrak{g}})$ given in (2.3). Clearly the defining relations of U imply that E_i, F_i, K_i, $i \neq 0$ generate a subalgebra isomorphic to $U_q(\mathfrak{g})$. Thus we are left with relations involving $i = 0$. For $i \neq 0$ we have

$$[E_0, F_i] = \left[[x^-_{i_{h-1}}(0), \ldots, x^-_{i_1}(1)]_{q^{\epsilon_1} \ldots q^{\epsilon_{h-2}}} \gamma K_\theta^{-1}, x^-_i(0)\right]$$

$$= -[x^-_i(0), x^-_{i_{h-1}}(0), \ldots, x^-_{i_1}(1)]_{q^{\epsilon_1} \ldots q^{\epsilon_{h-2} q^{(\theta|\alpha_i)}}} \gamma K_\theta^{-1}.$$

We claim that $[x^-_i(0), x^-_{i_{h-1}}(0), \ldots, x^-_{i_1}(1)]_{q^{\epsilon_1} \ldots q^{\epsilon_{h-2} q^{(\theta|\alpha_i)}}} = 0$ by the Serre relations. In fact this is seen by looking at rank 3 cases. We show the argument by working out all cases of $A_3^{(1)}$. The first two use only one Serre relation, while the third one uses two Serre relations.

$$\left[x^-_1(0), x^-_3(0), x^-_2(0), x^-_1(1)\right]_{q^{-1} q^{-1} q} = \left[x^-_3(0), [x^-_1(0), x^-_2(0), x^-_1(1)]_{q^{-1} q}\right]_{q^{-1}} = 0,$$

$$\left[x^-_3(0), x^-_3(0), x^-_2(0), x^-_1(1)\right]_{q^{-1} q^{-1} q} = \left[x^-_3(0), [[x^-_3(0), x^-_2(0)]_{q^{-1}}, x^-_1(1)]_{q^{-1}}\right]_q$$

$$= \left[[x^-_3(0), x^-_3(0), x^-_2(0)]_{q^{-1} q}, x^-_1(1)\right]_{q^{-1}} = 0,$$

$$\left[x^-_2(0), x^-_3(0), x^-_2(0), x^-_1(1)\right]_{q^{-1} q^{-1} q^{-2}}$$

$$= \left[x^-_2(0), [[x^-_3(0), x^-_2(0)]_{q^{-1}}, x^-_1(1)]_{q^{-1}}\right]_{q^{-2}} \quad \text{by (2.9)}$$

$$= \left[[x^-_2(0), x^-_3(0), x^-_2(0)]_{q^{-1} q^{-1}}, x^-_1(1)\right]_{q^{-2}}$$

$$\quad + q^{-1}\left[[x^-_3(0), x^-_2(0)]_{q^{-1}}, [x^-_2(0), x^-_1(0)]_{q^{-1}}\right] \quad \text{by (2.6)}$$

$$= \left[[x^-_3(0), x^-_2(0), x^-_1(1)]_{q^{-1} q^{-1}}, x^-_2(0)\right]_{q^{-2}} \quad \text{by (2.7)}$$

which implies that $[x^-_2(0), x^-_3(0), x^-_2(0), x^-_1(1)]_{q^{-1} q^{-1} 1} = 0$.

The Serre relations $\sum_{s=0}^{m=1-a_{ij}} (-1)^i \begin{bmatrix} m \\ i \end{bmatrix}_q e^s_i e_j e^{m-s}_i = 0$ (resp. for f_i's) for $i, j \neq 0$ are exactly the Serre relations in the Drinfeld realization. When i or $j = 0$ they boil down to the rank 3 cases. We use $A_n^{(1)}$ to show the idea.

$$e_0 e^2_1 - (q + q^{-1}) e_1 e_0 e_1 + e^2_1 e_0 = q^{-2}([x^-_n(0), \cdots, x^-_1(1)]_{q^{-1} \ldots q^{-1}} x^+_1(0))^2$$

$$- (q^2 + 1) x^+_1(0) [x^-_n(0), \ldots, x^-_1(1)]_{q^{-1} \ldots q^{-1}}$$

$$+ q^2 x^+_1(0)^2 [x^-_n(0), \ldots, x^-_1(1)]_{q^{-1} \ldots q^{-1}}) \gamma K_\theta^{-1}$$

$$= q^{-2} [x^+_1(0), x^+_1(0), x^-_n(0), \ldots, x^-_1(1)]_{q^{-1} \ldots q^{-1} 1 q^2}$$

$$\text{by (2.9)}$$

Using commutation relations it follows that

$$[x_1^+(0), x_1^+(0), x_n^-(0), \ldots, x_1^-(1)]_{q^{-1}\ldots q^{-1}1q^2}$$
$$= \gamma^{-1/2}[x_1^+(0), x_n^-(0), \ldots, x_2^-(0), K_1a_1(1)]_{q^{-1}\ldots q^{-1}q^2}$$
$$= -[x_1^+(0), x_n^-(0), \ldots, x_3^-(0), x_2^-(1)K_1]_{q^{-1}\ldots q^{-1}q^2}$$
$$= [x_1^+(0), x_n^-(0), \ldots, x_3^-(0), x_2^-(1)]_{q^{-1}\ldots q^{-1}1}K_1 = 0$$

Writing $\tilde{e}_0 = e_0\gamma^{-1}K_\theta$, we have

$$e_1e_0^2 - (q + q^{-1})e_0e_1e_0 + e_0^2e_1$$
$$= \left(x_1^+(0)\tilde{e}_0^2q - (1 + q^{-2})\tilde{e}_0x_1^+(0)\tilde{e}_0 + q^{-1}\tilde{e}_0^2x_1^+(0)\right)\gamma^2K_\theta^{-2}$$
$$= q^{-1}[\tilde{e}_0, \tilde{e}_0, x_1^+(0)]_{1,q^2}\gamma^2K_\theta^{-2}$$
$$= q^{-1}\left[\tilde{e}_0, [x_n^-(0), \ldots, x_2^-(0), -\gamma^{-1/2}K_1a_1^+(1)]_{q^{-1}\ldots q^{-1}}\right]_{q^2}\gamma^2K_\theta^{-2}$$
$$= q^{-1}\left[[x_n^-(0), \ldots, x_1^-(1)]_{q^{-1}\ldots q^{-1}}, [x_n^-(0), \cdots, x_2^-(1)]_{q^{-1}\ldots q^{-1}}\right]_q K_1\gamma^2K_\theta^{-2}$$
$$= -\left[[x_n^-(0), \ldots, x_2^-(1)]_{q^{-1}\ldots q^{-1}}, e_0\right]\gamma K_1K_\theta^{-1}$$
$$= 0$$

where we used our earlier result: $[e_0, f_i] = 0$ for $i \neq 0$ and another identity $[e_0, x_2^-(1)] = 0$ by a similar argument as given in (2.7).

Finally we check the relations $[e_i, f_i] = \frac{t_i - t_i^{-1}}{q_i - q_i^{-1}}$. Again it suffices to see the case of $i = 0$. We want to give two cases to show the argument.

First we consider the case of $A_n^{(1)}$.

$$\left[[x_n^-(0), \ldots, x_1^-(1)]_{q^{-1}\ldots q^{-1}}, [x_n^+(0), \ldots, x_1^+(-1)]_{q^{-1}\ldots q^{-1}}\right]$$
$$= \left[[\tilde{e}_0, x_1^+(0)], \ldots, x_n^+(-1)\right]_{q^{-1}\ldots q^{-1}} + \left[x_n^+(0), \ldots, [\tilde{e}_0, x_1^+(-1)]\right]_{q^{-1}\ldots q^{-1}}$$
$$= \left[[x_{n-1}^-(0), \ldots, x_1^-(1)]_{q^{-1}\ldots q^{-1}}K_n, x_{n-1}^+(0), \ldots, x_1^+(1)\right]_{q^{-1}\ldots q^{-1}}$$
$$\quad + \left[x_n^+(0), \ldots, x_2^+(0), K_1^{-1}[x_{n-1}^-(0), \ldots, x_2^-(0)]_{q^{-1}\ldots q^{-1}}\right]\gamma$$
$$= (-q^{-1})[e_0(n-1), f_0(n-1)] + q^{-1}K_1^{-1}\gamma\left[x_{n-1}^-(0), \ldots, x_3^+(0),\right.$$
$$\left. [x_2^+(0), x_{n-1}^-(0), \ldots, x_2^-(0)]_{q^{-1}\ldots q^{-1}1}\right]_{q^{-1}\ldots q^{-1}}$$
$$= (-q)^{-n}\frac{\gamma K_\theta^{-1}K_n - \gamma^{-1}K_\theta}{q - q^{-1}} + (-q)^{-n+1}K_1^{-1}\ldots K_{n-1}^{-1}\gamma[x_n^+(0), x_n^-(0)]$$
$$= (-q)^{-n}\frac{\gamma K_\theta^{-1} - \gamma^{-1}K_\theta}{q - q^{-1}}$$

where we have reasoned as follows: the simplest Serre relations $[e_0, e_i] = 0$ for $i \geq 2$; an induction on rank n as well as another induction on the second bracket in line two. The elements $e_0(n-1)$, $f_0(n-1)$ refer to the corresponding ones for $A_{n-1}^{(1)}$.

The computations in other cases are similar. I just give $C_2^{(1)}$ to show some flavor. We write

$$\bar{e}_0 = [x_2^-(0), x_1^-(-1)]_{q^{-1}}, \qquad \bar{f}_0 = [x_2^+(0), x_1^+(1)]_{q^{-1}},$$

$$[[x_1^-(0), x_2^-(0), x_1^-(1)]_{q^{-1}\,1\,1}, [x_1^+(0), x_2^+(0), x_1^+(-1)]_{q^{-1}\,1\,1}]$$
$$= [[x_1^-(0), x_2^-(1)]_{q^{-1}}K_1, \bar{f}_0] + [x_1^+(0), [x_1^+(0), [x_1^-(0), \bar{e}_0], \bar{f}_0]]$$
$$= -[\bar{e}_0, \bar{f}_0] + [x_1^+(0), [-x_2^+(-1)K_1^{-1}, \bar{e}_0]]$$
$$= q^{-1}\gamma \frac{K_1^{-1}K_2^{-1} - \gamma^{-1}K_1K_2}{q_1 - q_1^{-1}} + q^{-1}[x_1^-(0), x_1^+(0)]\gamma K_1^{-1}K_2^{-1}$$
$$= q^{-1}[2]_1 \frac{\gamma K_\theta^{-1} - \gamma^{-1}K_\theta}{q - q^{-1}}.$$

Step 2. We show that the algebra U is generated by E_i, F_i, $K_i^{\pm 1}$, $D^{\pm 1}$. Write $U' = \langle E_i, F_i, K_i^{\pm 1}, D^{\pm 1} \rangle$. The Cartan subalgebra is clearly generated by $K_i^{\pm 1}$ and $D^{\pm 1}$. Rewriting (2.13) we have

$$x_{i_1}^-(1) = a[E_{i_2}, E_{i_3}, \ldots, E_{i_{h-1}}, E_0]_{q^{\epsilon_1}, \ldots, q^{\epsilon_{h-2}}}$$
$$x_{i_1}^+(-1) = b[F_{i_2}, F_{i_3}, \ldots, F_{i_{h-1}}, F_0]_{q^{\epsilon_1}, \ldots, q^{\epsilon_{h-2}}}$$

where a, b are constants. It then follows from Drinfeld relations that

$$a_{i_1}(1) = K_{i_1}^{-1}\gamma^{1/2}[x_{i_1}^+(0), x_{i_1}^-(1)]$$
$$a_{i_1}(-1) = K_{i_1}\gamma^{-1/2}[x_{i_1}^+(-1), x_{i_1}^-(0)]$$

which implies that $a_{i_1}(n) \in U'$, subsequently $x_{i_1}^\pm(n) \in U'$. Then we follow the ϵ-sequence and get that $x_j^\pm(n) \in U'$. Thus $U' = U$.

Step 3. We now have an algebra epimorphism $\Phi \colon U_q(\mathfrak{g}) \to U$. It is clear that $\mathrm{Ker}\,\Phi = 1$ when $q \to 1$ using Gabber–Kac **[GK]**. Since quantization does not change the multiplicity we obtain that Φ is an automorphism.

Remark. If $\mathfrak{g}$ is simply-laced, then $\epsilon = -h + 2$. The sequences are by no means unique, though ϵ is independent from the choice of the sequences. For example, we also have for E_6:

$$\alpha_1 \xrightarrow{-1} \cdots \xrightarrow{-1} \alpha_6 \xrightarrow{-1} \alpha_3 \xrightarrow{-1} \alpha_2 \xrightarrow{-1} \alpha_4 \xrightarrow{0} \alpha_6 \xrightarrow{-2} \alpha_3$$

Table 2.1. ϵ-sequences for simple Lie algebras

$\mathfrak{g}$	ϵ-sequence $\quad \epsilon = \sum \epsilon_i$	ϵ
A_n	$\alpha_1 \xrightarrow{-1} \cdots \xrightarrow{-1} \alpha_n$	$-n+1$
B_n	$\alpha_1 \xrightarrow{-1} \cdots \xrightarrow{-1} \alpha_{n-1} \xrightarrow{0} \alpha_n \xrightarrow{-1} \cdots \xrightarrow{-1} \alpha_2$	$-2n+4$
C_n	$\alpha_1 \xrightarrow{-1/2} \cdots \xrightarrow{-1} \alpha_n \xrightarrow{-1/2} \cdots \xrightarrow{-1/2} \alpha_2 \xrightarrow{0} \alpha_1$	$-n+1$
D_n	$\alpha_1 \xrightarrow{-1} \cdots \xrightarrow{-1} \alpha_n \xrightarrow{-1} \alpha_{n-2} \xrightarrow{-1} \cdots \xrightarrow{-1} \alpha_2$	$-2n+4$
E_6	$\alpha_1 \xrightarrow{-1} \cdots \xrightarrow{-1} \alpha_6 \xrightarrow{-1} \alpha_3 \xrightarrow{-1} \alpha_2 \xrightarrow{-1} \alpha_4 \xrightarrow{-1} \alpha_3 \xrightarrow{-1} \alpha_6$	-10
E_7	$1\,2\,3\,4\,5\,6\,7\,3\,2\,4\,5\,3\,4\,7\,3\,2\,1, \quad \epsilon_i = -1$	-16
E_8	$1\,2\,3\,4\,5\,6\,7\,8\,5\,4\,3\,2\,6\,5\,8\,4\,3\,5\,6\,7\,4\,5\,8\,6\,5\,4\,3\,2\,1, \quad \epsilon_i = -1$	-16
F_4	$1 \xrightarrow{-1} 2 \xrightarrow{-1} 3 \xrightarrow{-1/2} 4 \xrightarrow{-1/2} 3 \xrightarrow{-1} 2 \xrightarrow{-1} 3 \xrightarrow{-1} 4 \xrightarrow{0} 3 \xrightarrow{-1} 2 \xrightarrow{-1} 1$	-7
G_2	$\alpha_1 \xrightarrow{-1} \alpha_2 \xrightarrow{-1/3} \alpha_2 \xrightarrow{0} \alpha_1 \xrightarrow{-2/3} \alpha_2$	-2

3. Twisted quantum affine algebras

We now derive the Drinfeld realizations for the twisted types.

Let $X_N^{(1)}$ be a simply laced affine Cartan matrix. Let α_i' be the basis of the simple roots. The standard invariant bilinear form is normalized as

$$(\alpha_i' | \alpha_i') = 2r, \quad i = 0, 1, \ldots, N.$$

Let $L_q(X_N^{(1)})$ be the quantum affine algebra associated with $X_N^{(1)}$ realized in the Drinfeld quantum loop form. We denote the corresponding generators be putting an extra prime to distinguish. Clearly the diagram automorphism σ acts on the quantum affine algebra. We will use a different indexing for the type $A_{2n}^{(2)}$ from [**K**]. The action of σ is given as follows:

$$A_N : \sigma(i) = N - i$$
$$D_N : \sigma(i) = i, \quad 1 \leq i \leq N - 2; \ \sigma(N-1) = N$$
$$E_6 : \sigma(i) = 6 - i, \quad 1 \leq i \leq 5; \ \sigma(6) = 6$$
$$D_4 : \sigma(1, 2, 3, 4) = (3, 2, 4, 1)$$

We construct a special invariant subalgebra U_σ generated by

$$a_i(l) = \frac{1}{[d_i]\sqrt{r}} \sum_{s=0}^{r-1} a_{\sigma^s(i)}(l)' \omega^{-ls}, \quad K_i = \prod_{s=0}^{r-1} K'_{\sigma^s(i)} \tag{3.1}$$

$$x_i^\pm(k) = \frac{1}{[d_i]\sqrt{r}} \sum_{s=0}^{r-1} x_{\sigma^s(i)}(k)' \omega^{-ks} \tag{3.2}$$

where ω be a primitive rth root, and where $(d_0, \ldots, d_n) = (1, \ldots, 1, 2)$, $(1, 2, \ldots, 2, 1)$, $(2, 1, \ldots, 1, 1/2)$, $(2, 1/2)$, $(1, 1, 1, 2, 2)$ and $(1, 1, 3)$, for $A_{2n-1}^{(2)}$, $D_{n+1}^{(2)}$, $A_{2n}^{(2)}$, $A_2^{(2)}$, $E_6^{(2)}$ and $D_4^{(3)}$, respectively. We denote $[k]_j = \frac{q_i^k - q_i^{-k}}{q_i - q_i^{-1}}$ and $\alpha_j = \alpha_i$ if j belongs to the σ-orbit of i, then $[k]_j$ or α_j is defined for all $j = 1, \ldots, N$ though we use only $[k]_i$ or α_i for $i \in \{0\} \cup \Gamma_\sigma = \{0, 1, \ldots, n\}$. Note that (3.1) implies that $(\alpha_i | \alpha_j)$ for $i, j = 1, \ldots, n$ (in fact N) is the standard inner product on the corresponding twisted Kac–Moody algebra $X_N^{(r)}$. Easy and long calculation will lead to the following relations presented in the Drinfeld realization of twisted quantum affine algebras.

Theorem 3.1. *The algebra U_σ is the associative algebra with 1 over $\mathbb{C}(q^{1/2})$ generated by the elements $x_i^\pm(k)$, $a_i(l)$, $K_i^{\pm 1}$, $\gamma^{\pm 1/2}$, $q^{\pm d}$ $(i = 1, 2, \ldots, N, \ k \in \mathbb{Z}, \ l \in \mathbb{Z} \setminus \{0\})$ with the following defining relations:*

$$x_{\sigma(i)}^\pm(k) = \omega^k x_i^\pm(k), \quad a_{\sigma(i)}(l) = \omega^l a_i(l), \quad K_{\sigma(i)} = K_i$$

$$[\gamma^{\pm 1/2}, u] = 0 \quad \text{for all } u \in U,$$

$$[a_i(k), a_j(l)] = \delta_{k+l,0} \sum_{s=0}^{r-1} \frac{[k(\alpha_i' | \sigma^s(\alpha_j'))/rd_i]_i}{k} \omega^{ks} \frac{\gamma^k - \gamma^{-k}}{q_j - q_j^{-1}},$$

$$[a_i(k), K_j^{\pm 1}] = [q^{\pm d}, K_j^{\pm 1}] = 0,$$

$$q^d x_i^\pm(k) q^{-d} = q^k x_i^\pm(k), \quad q^d a_i(l) q^{-d} = q^l a_i(l),$$

$$K_i x_j^\pm(k) K_i^{-1} = q^{\pm(\alpha_i | \alpha_j)} x_j^\pm(k),$$

$$[a_i(k), x_j^\pm(l)] = \pm \sum_{s=0}^{r-1} \frac{[k(\alpha_i' | \sigma^s(\alpha_j'))/rd_i]_i}{k} \omega^{ks} \gamma^{\mp |k|/2} x_j^\pm(k+l),$$

$$\prod_s (z - \omega^s q^{\pm(\alpha_i' | \sigma^s(\alpha_j'))/r} w) x_i^\pm(z) x_j^\pm(w) = \prod_s (zq^{\pm(\alpha_i' | \sigma^s(\alpha_j'))/r} - \omega^s w) x_j^\pm(w) x_i^\pm(z)$$

$$[x_i^+(k), x_j^-(l)] = \sum_{s=0}^{r-1} \frac{\delta_{\sigma^s(i),j} \omega^{sl}}{q_i - q_i^{-1}} (\gamma^{\frac{k-l}{2}} \psi_i(k+l) - \gamma^{\frac{l-k}{2}} \varphi_i(k+l)),$$

where $\psi_i(m)$ and $\varphi_i(-m)$ $(m \in \mathbb{Z}_{\geq 0})$ are defined by

$$\sum_{m=0}^{\infty} \psi_i(m) z^{-m} = K_i \exp\left((q_i - q_i^{-1}) \sum_{k=1}^{\infty} a_i(k) z^{-k}\right),$$

$$\sum_{m=0}^{\infty} \varphi_i(-m) z^m = K_i^{-1} \exp\left(-(q_i - q_i^{-1}) \sum_{k=1}^{\infty} a_i(-k) z^k\right),$$

$$\mathrm{Sym}_{z_1, z_2} P_{ij}^{\pm}(z_1, z_2) \sum_{s=0}^{2} (-1)^s \begin{bmatrix} 2 \\ s \end{bmatrix}_{q^{d_{ij}}} x_i^{\pm}(z_1) \ldots x_i^{\pm}(z_s) x_j^{\pm}(w) x_i^{\pm}(z_{s+1}) \ldots x_i^{\pm}(z_2) = 0$$

for $A_{ij} = -1$, $\sigma(i) \neq j$, and

$$\mathrm{Sym}_{z_1, z_2, z_3} \left[(q^{\mp 3r/4} z_1 - q^{r/4} + q^{-r/4}) z_2 + q^{\pm 3r/4} z_3) x_i^{\pm}(z_1) x_i^{\pm}(z_2) x_i^{\pm}(z_3) \right] = 0$$

for $A_{i,\sigma(i)} = -1$, where Sym means the symmetrization over z_i, $P_{ij}^{\pm}(z, w)$ and d_{ij} are defined as follows:

If $\sigma(i) = i$, then $P_{ij}^{\pm}(z, w) = 1$ and $d_{ij} = r$.

If $A_{i,\sigma(i)} = 0$ and $\sigma(j) = j$, then $P_{ij}^{\pm}(z, w) = \dfrac{z^r q^{\pm 2r} - w^r}{z q^{\pm 2} - w}$ and $d_{ij} = r$.

If $A_{i,\sigma(i)} = 0$ and $\sigma(j) \neq j$, then $P_{ij}^{\pm}(z, w) = 1$ and $d_{ij} = 1/2$.

If $A_{i,\sigma(i)} = -1$, then $P_{ij}^{\pm}(z, w) = z q^{\pm r/2} + w$ and $d_{ij} = r/4$. □

To state the isomorphism we generalize the notion of the ϵ-sequences to the twisted cases. Under the map σ the Lie algebra $\mathfrak{g} = X_N$ is decomposed into a graded Lie algebra

$$\mathfrak{g} = \mathfrak{g}_0 \oplus \cdots \oplus \mathfrak{g}_{r-1},$$

where $\mathfrak{g}_i = \{x \in \mathfrak{g} \mid \sigma(x) = \omega^i(x)\}$. The subalgebra $\mathfrak{g}_0$ is a simple Lie algebra, and $\mathfrak{g}_{r-1}$ is a highest weight $\mathfrak{g}_0$-module with the highest weight θ (cf. [**K**]).

Lemma 3.2. *Let $I_0 = \{1, 2, \ldots, n\}$ be the index set for the simple roots of the finite dimensional simple Lie algebra $\mathfrak{g}_0$. Then for each $i \in I_0$, there exists a sequence of indices $i = i_1, i_2, \ldots, i_{h-1}$ such that*

$$\begin{aligned}
(\alpha_{i_1} | \alpha_{i_2}) &= \epsilon_{i_1}, \\
(\alpha_{i_1} + \alpha_{i_2} | \alpha_{i_3}) &= \epsilon_{i_2}, \\
&\vdots \\
(\alpha_{i_1} + \cdots + \alpha_{i_{h-2}} | \alpha_{i_{h-1}}) &= \epsilon_{i_{h-2}},
\end{aligned} \tag{3.3}$$

so that the highest weight $\theta = \alpha_{i1} + \cdots + \alpha_{i_{h-1}}$, and $\epsilon_i \in \mathbb{Q}_-$. □

In Table 3.1 we list the ϵ-sequences for the twisted cases.

Table 3.1. ϵ-sequences for twisted cases

$\mathfrak{g}^{(r)}$	$\mathfrak{g}_{(0)}$	ϵ-sequence	ϵ	a
$A_{2n-1}^{(2)}$	C_n	$\alpha_1 \xrightarrow{-1} \cdots \xrightarrow{-1} \alpha_{n-1} \xrightarrow{-2} \alpha_n \xrightarrow{-1} \cdots \xrightarrow{-1} \alpha_2$	$-2n+2$	-2
$D_{n+1}^{(2)}$	B_n	$\alpha_n \xrightarrow{-2} \cdots \xrightarrow{-2} \alpha_1$	$-2n+2$	$(-2)^{n+1}$
$A_{2n}^{(2)}$	B_n	$\alpha_1 \xrightarrow{-1} \cdots \xrightarrow{-1} \alpha_n \xrightarrow{0} \alpha_{n-1} \xrightarrow{-1} \cdots \xrightarrow{-1} \alpha_2 \xrightarrow{0} \alpha_1$	$-2n+3$	$-[2]^{2n-2}$
$A_2^{(2)}$	A_1	$\alpha_1 \xrightarrow{1} \alpha_1$	1	-1
$D_4^{(3)}$	G_2	$\alpha_1 \xrightarrow{-3} \alpha_2 \xrightarrow{-1} \alpha_1$	-4	3
$E_6^{(2)}$	F_4	$\alpha_1 \xrightarrow{-1} \cdots \xrightarrow{-1} \alpha_6 \xrightarrow{-1} \alpha_3 \xrightarrow{-1} \alpha_2 \xrightarrow{-1} \alpha_4 \xrightarrow{-1} \alpha_3 \xrightarrow{-1} \alpha_6$	-10	

Theorem 3.3 ([Dr2]). *Fix an ϵ-sequence $i_1, i_2, \ldots, i_h$, and let $\theta = \sum_{j=1}^{h-1} \alpha_{i_j}$ be the highest weight of the $\mathfrak{g}_0$-module $\mathfrak{g}_{r-1}$. Then there is a $\mathbb{C}(q^{1/r})$ algebra isomorphism $\Psi : U_q(\hat{\mathfrak{g}}^{(r)}) \to U$ defined by*

$$e_i \mapsto x_i^+(0), \quad f_i \mapsto \frac{1}{p_i} x_i^-(0), \quad t_i \mapsto K_i \quad \text{for } i = 1, \ldots, n,$$

$$e_0 \mapsto [x_{i_{h-1}}^-(0), \ldots, x_{i_2}^-(0), x_{i_1}^-(1)]_{q^{\epsilon_1} \ldots q^{\epsilon_{h-2}}} \gamma K_\theta^{-1},$$

$$f_0 \mapsto a(-q)^{-\epsilon} \gamma^{-1} K_\theta [x_{i_{h-1}}^+(0), \ldots, x_{i_2}^+(0), x_{i_1}^+(-1)]_{q^{\epsilon_1} \ldots q^{\epsilon_{h-2}}} \qquad (3.4)$$

$$t_0 \mapsto \gamma K_\theta^{-1}, \quad q^d \mapsto q^d,$$

where $p_i = 1$ for $\sigma(i) \neq i$, $p_i = i$ otherwise, $K_\theta = K_{i_1} \ldots K_{i_{h-1}}$, $\epsilon = \sum_{i=1}^{h-2} \epsilon_i$, and a is a constant given by Table 3.1.

Proof. The proof is similar to that of the untwisted case. $\qquad \square$

Bibliography

[B]   J. Beck, Braid group action and quantum affine algebras, Comm. Math. Phys. **165** (1994), 555–568.

[Da]   I. Damiani, A basis of type Poincare–Birkhoff–Witt for the quantum affine algebra of $\hat{sl}(2)$, J. Algebra **161** (1993), 291–310.

[Dr1]   V. G. Drinfeld, Quantum groups, Proc. Internat. Cong. Math., Berkeley, Amer. Math. Soc., Providence 1987, 798–820.

[Dr2]   V. G. Drinfeld, A new realization of Yangians and quantized affine algebras, Soviet Math. Dokl. **36** (1988), 212–216.

[FJ]   I. Frenkel, N. Jing, Vertex representations of quantum affine algebras, Proc. Nat. Acad. Sci. U.S.A. **85** (1988), 9373–9377.

[GK]   O. Gabber, V. G. Kac, On defining relations of certain inifinite dimensional Lie algebras, Bull. Amer. Math. Soc. **5** (1981), 185–189.

[Jb]   M. Jimbo, A q-difference analog of $U(\mathfrak{g})$ and the Yang–Baxter equation, Lett. Math. Phys. **10** (1985), 63–69.

[J]   N. Jing, Twisted vertex representations of quantum affine algebras, Invent. Math. **102** (1990), 663–690.

[JM]   N. Jing, K. Misra, Vertex operators for twisted quantum affine algebras, Trans. Amer. Math. Soc. (1997), to appear; q-alg/9701034.

[K]   V. Kac, Infinite dimensional Lie algebras, 3rd ed., Cambridge Univ. Press, 1990.

[L]   G. Lusztig, Introduction to quantum groups, Birkhäuser, Boston 1993.

[LSS]   S. Lewendorskii, Y. Soilbelman and V. Stukopin, Quantum Weyl group and universal quantum R-matrix for affine Lie algebra $A_1^{(1)}$, Lett. Math. Phys. **27** (1993) 253–264.

Department of Mathematics
North Carolina State University
Raleigh, NC 27695-8205, U.S.A.
E-mail: jing@eos.ncsu.edu

Free Lie superalgebras and the generalized Witt formula

Seok-Jin Kang[1]

Abstract. In this paper, we interpret the product identities for normalized formal power series as the denominator identities for free Lie superalgebras, and derive a superdimension formula for the homogeneous subspaces of free Lie superalgebras.

1. Product identities

The purpose of this paper is to investigate the relation of graded Lie superalgebras and product identities for normalized formal power series. Let us begin with the binomial expansion

$$(1-q)^r = \sum_{k=0}^{r} (-1)^k \binom{r}{k} q^k. \tag{1.1}$$

Let $L = \mathbb{C}x_1 \oplus \cdots \oplus \mathbb{C}x_r$ be the r-dimensional abelian Lie algebra with basis $\{x_1, \ldots, x_r\}$. Then we have

$$H_k(L) = \Lambda^k(L) = \mathrm{Span}\{x_{i_1} \wedge \cdots \wedge x_{i_k} \mid 1 \le i_1 < \cdots < i_k \le r\},$$

and hence the identity (1.1) can be interpreted as the Euler–Poincaré principle for the abelian Lie algebra L.

In [**K2**], Jacobi's triple product identity

$$\prod_{n=1}^{\infty}(1 - p^n q^n)(1 - p^{n-1}q^n)(1 - p^n q^{n-1}) = \sum_{k \in \mathbb{Z}}(-1)^k p^{\frac{k(k-1)}{2}} q^{\frac{k(k+1)}{2}} \tag{1.2}$$

was interpreted as the denominator identity for the affine Kac–Moody algebra of type $A_1^{(1)}$ (cf. [**K4**]). In fact, Kac showed that all the Macdonald identities ([**M**]) are equivalent to the denominator identities for affine Kac–Moody algebras ([**K2**]). Since then, the theory of infinite dimensional Lie algebras and their representations has been the focus of extensive research activities due to its rich and significant connections to many areas of mathematics and mathematical physics.

Recently, Borcherds completed the proof of the *Moonshine Conjecture* by constructing an infinite dimensional Lie algebra, called the *Monster Lie algebra* ([**B2**]). The main

[1] This research was supported in part by Basic Science Research Institute Program, Ministry of Education, BSRI-96-1414, and GARC-KOSEF at Seoul National University.

ingredient of his proof is the following product identity

$$p^{-1} \prod_{\substack{m>0 \\ n \in \mathbb{Z}}} (1 - p^m q^n)^{c(mn)} = j(p) - j(q), \tag{1.3}$$

where $c(n)$ are the coefficients of the elliptic modular function

$$j(q) - 744 = \sum_{n=-1}^{\infty} c(n)q^n = q^{-1} + 196884q + 21493760q^2 + \cdots. \tag{1.4}$$

It was shown in [**B2**] that the Monster Lie algebra is a generalized Kac–Moody algebra (cf. [**B1**]) and the identity (1.3) was interpreted as the denominator identity for the Monster Lie algebra.

More generally, let Γ be a countable abelian semigroup such that every element $\alpha \in \Gamma$ can be written as a sum of elements in Γ in only finitely many ways, and consider a normalized formal power series

$$1 - \sum_{\alpha \in \Gamma} d(\alpha)e^\alpha \quad \text{with} \quad d(\alpha) \in \mathbb{Z} \quad \text{for all} \quad \alpha \in \Gamma. \tag{1.5}$$

Suppose we have a product identity for the above formal power series:

$$\prod_{\alpha \in \Gamma} (1 - e^\alpha)^{A(\alpha)} = 1 - \sum_{\alpha \in \Gamma} d(\alpha)e^\alpha \tag{1.6}$$

with $A(\alpha) \in \mathbb{Z}_{\geq 0}$ for all $\alpha \in \Gamma$.

If we could construct a Γ-graded Lie algebra $L = \bigoplus_{\alpha \in \Gamma} L_\alpha$ such that the character of the homology space $H = \sum_{k=1}^{\infty} (-1)^{k+1} H_k(L)$ is given by

$$\operatorname{ch} H = \sum_{\alpha \in \Gamma} (\dim H_\alpha)e^\alpha = \sum_{\alpha \in \Gamma} d(\alpha)e^\alpha, \tag{1.7}$$

then the identity (1.6) can be interpreted as the Euler–Poincaré principle for the Lie algebra L. In particular, we would have

$$\dim L_\alpha = A(\alpha) \quad \text{for all} \quad \alpha \in \Gamma. \tag{1.8}$$

In this case, we call (1.6) the *denominator identity* for the Lie algebra L.

If $d(\alpha) \in \mathbb{Z}_{\geq 0}$ for all $\alpha \in \Gamma$, then any product identity of the form (1.6) can always be interpreted as the denominator identity for some suitably defined free Lie algebra. More precisely, let $V = \bigoplus_{\alpha \in \Gamma} V_\alpha$ be a Γ-graded vector space over $\mathbb{C}$ with $\dim V_\alpha = d(\alpha)$ for all $\alpha \in \Gamma$, and let $L = \bigoplus_{\alpha \in \Gamma} L_\alpha$ be the free Lie algebra generated by V. Then, since $H_1(L) = V$ and $H_k(L) = 0$ for all $k \geq 2$, the right-hand side of (1.6) can be regarded as $1 - \operatorname{ch} H$, and hence the identity (1.6) can be interpreted as the denominator identity for the free Lie algebra L ([**KK1**], [**KK2**]).

For example, the product identity (1.3) can be rephrased as

$$\prod_{m,n=1}^{\infty} (1 - p^m q^n)^{c(mn)} = 1 - \sum_{i,j=1}^{\infty} c(i+j-1)p^i q^j. \tag{1.9}$$

Let $V = \bigoplus_{i,j=1}^{\infty} V_{(i,j)}$ be a $(\mathbb{Z}_{>0} \times \mathbb{Z}_{>0})$-graded vector space over $\mathbb{C}$ with $\dim V_{(i,j)} = c(i+j-1)$, and let $L = \bigoplus_{m,n=1}^{\infty} L_{(m,n)}$ be the free Lie algebra generated by V. Then, the product identity (1.8) for the elliptic modular function j is the denominator identity for the free Lie algebra L, and hence we have

$$\dim L_{(m,n)} = c(mn) \quad \text{for all} \quad m, n \geq 1.$$

However, in general, it is quite complicated and difficult to construct a graded Lie algebra corresponding to the product identity (1.6). In this paper, we will by-pass the difficulty by considering Lie superalgebras instead of Lie algebras, and show that any product identity of the form (1.6) with $A(\alpha)$, $d(\alpha) \in \mathbb{Z}$ ($\alpha \in \Gamma$) can be interpreted as the denominator identity for some suitably defined free Lie superalgebra. In [KW], Kac and Wakimoto investigated the relation of affine Lie superalgebras and many interesting product identities arising from number theory. These product identities can also be interpreted as the denominator identities for some free Lie superalgebras. A more extensive study of graded Lie superalgebras will appear in [Ka].

2. Generalized Witt formula

We first recall some of the basic facts about Lie superalgebras (cf. [Ba], [K3], [S]). A $\mathbb{Z}_2$-graded vector space $V = V_{(0)} \oplus V_{(1)}$ is called a *superspace*, and a $\mathbb{Z}_2$-graded associative algebra $A = A_{(0)} \oplus A_{(1)}$ satisfying $A_{(i)} A_{(j)} \subset A_{(i+j)}$ for $i, j \in \mathbb{Z}_2$ is called a *superalgebra*.

Definition 2.1. A *Lie superalgebra* is a $\mathbb{Z}_2$-graded superspace $L = L_{(0)} \oplus L_{(1)}$ together with a bilinear operation $[\ ,\ ] : L \times L \longrightarrow L$ satisfying

$$[L_{(i)}, L_{(j)}] \subset L_{(i+j)},$$
$$[x, y] = -(-1)^{ij}[y, x],$$
$$[x, [y, z]] = [[x, y], z] + (-1)^{ij}[y, [x, z]] \qquad (2.1)$$

for all $x \in L_{(i)}$, $y \in L_{(j)}$, $i, j \in \mathbb{Z}_2$.

The *universal enveloping algebra* of a Lie superalgebra $L = L_{(0)} \oplus L_{(1)}$ is the pair $(U(L), \iota)$, where $U(L)$ is a superalgebra and $\iota : L \to U(L)$ is a linear mapping satisfying

$$\iota([x, y]) = \iota(x)\iota(y) - (-1)^{ij}\iota(y)\iota(x) \quad \text{for} \quad x \in L_{(i)}, \ y \in L_{(j)}$$

such that for any superalgebra $A = A_{(0)} \oplus A_{(1)}$ and a linear mapping $j : L \to A$ satisfying

$$j([x, y]) = j(x)j(y) - (-1)^{ij} j(y)j(x) \quad \text{for} \quad x \in L_{(i)}, \ y \in L_{(j)}$$

there exists a unique superalgebra homomorphism $\psi : U(L) \to A$ satisfying $\psi \circ \iota = j$.

The uniqueness of the universal enveloping algebra $U(L)$ can be proved in the usual way. For the existence, consider the tensor algebra $\mathcal{T}(L) = \bigoplus_{k=0}^{\infty} L^{\otimes k}$ of L and let $\mathcal{R}$

be the ideal of $\mathcal{T}(L)$ generated by the elements of the form

$$[x, y] - x \otimes y + (-1)^{ij} y \otimes x \quad (x \in L_{(i)}, y \in L_{(j)}).$$

Then the quotient algebra $U(L) = \mathcal{T}(L)/\mathcal{R}$ together with the natural mapping $\iota :$ $L \to U(L)$ is the universal enveloping algebra of L. For the structure of the universal enveloping algebra of a Lie superalgebra, we have the following version of Poincaré–Birkhoff–Witt Theorem.

Theorem 2.2 (cf. [Ba], [K3], [S]). *Let $L = L_{(0)} \oplus L_{(1)}$ be a Lie superalgebra, and let $X = \{x_\alpha \mid \alpha \in \Lambda\}$ (resp. $Y = \{y_\beta \mid \beta \in \Omega\}$) be a homogeneous basis of $L_{(0)}$ (resp. $L_{(1)}$). Then the elements of the form*

$$x_{\alpha_1} x_{\alpha_2} \dots x_{\alpha_k} y_{\beta_1} y_{\beta_2} \dots y_{\beta_l} \quad \text{with} \quad \alpha_1 \le \dots \le \alpha_k, \ \beta_1 < \dots < \beta_l$$

together with 1 form a basis of the universal enveloping algebra $U(L)$ of L.   □

From now on, we will concentrate on free Lie superalgebras. Let Γ be a countable abelian semigroup satisfying the finiteness condition given in Section 1, and let $V = \bigoplus_{\alpha \in \Gamma} V_\alpha$ be a Γ-graded vector space over $\mathbb{C}$ with $\dim V_\alpha < \infty$ for all $\alpha \in \Gamma$. A partition $\Gamma = \Gamma_0 \cup \Gamma_1$ of Γ into a disjoint union of two subsets Γ_0 and Γ_1 yields a $(\Gamma \times \mathbb{Z}_2)$-gradation $V = \bigoplus_{(\alpha,i) \in \Gamma \times \mathbb{Z}_2} V_{(\alpha,i)}$ of the vector space V, which makes V a superspace. Recall that the $(\Gamma \times \mathbb{Z}_2)$-graded character of V is given by

$$\text{ch}_{\Gamma \times \mathbb{Z}_2}(V) = \sum_{(\alpha,i) \in \Gamma \times \mathbb{Z}_2} (\dim V_{(\alpha,i)}) e^{(\alpha,i)}, \tag{2.3}$$

where $e^{(\alpha,i)}$ are the basis elements of the semigroup algebra $\mathbb{C}[\Gamma \times \mathbb{Z}_2]$ with the multiplication $e^{(\alpha,i)} e^{(\beta,j)} = e^{(\alpha+\beta,i+j)}$ for all $\alpha, \beta \in \Gamma$, $i, j \in \mathbb{Z}_2$.

We define the *superdimension* of the homogeneous subspace $V_{(\alpha,i)}$ by

$$\text{Dim } V_{(\alpha,i)} = (-1)^i \dim V_{(\alpha,i)}, \tag{2.4}$$

and introduce another basis elements of $\mathbb{C}[\Gamma \times \mathbb{Z}_2]$ by setting

$$E^{(\alpha,i)} = (-1)^i e^{(\alpha,i)} \quad \text{for} \quad i \in \mathbb{Z}_2. \tag{2.5}$$

Then it is easy to verify that $E^{(\alpha,i)} E^{(\beta,j)} = E^{(\alpha+\beta,i+j)}$ for all $\alpha, \beta \in \Gamma$, $i, j \in \mathbb{Z}_2$. We define the $(\Gamma \times \mathbb{Z}_2)$-graded *supercharacter* of V by

$$\text{Ch}_{\Gamma \times \mathbb{Z}_2}(V) = \sum_{(\alpha,i) \in \Gamma \times \mathbb{Z}_2} (\text{Dim } V_{(\alpha,i)}) E^{(\alpha,i)}. \tag{2.6}$$

Note that $\text{Ch}_{\Gamma \times \mathbb{Z}_2}(V) = \text{ch}_{\Gamma \times \mathbb{Z}_2}(V)$. The only (but important) difference is that, in the supercharacter, we are allowed to have negative coefficients.

For a $(\Gamma \times \mathbb{Z}_2)$-graded superspace $V = \bigoplus_{(\alpha,i) \in \Gamma \times \mathbb{Z}_2} V_{(\alpha,i)}$, let $V_\alpha = V_{(\alpha,0)} \oplus V_{(\alpha,1)}$, and define

$$\text{Dim } V_\alpha = \text{Dim } V_{(\alpha,0)} + \text{Dim } V_{(\alpha,1)} = \dim V_{(\alpha,0)} - \dim V_{(\alpha,1)}.$$

Then, by specializing $E^{(\alpha,i)} = E^\alpha$ for all $i \in \mathbb{Z}_2$, we are reduced to the Γ-graded supercharacter of V:

$$\mathrm{Ch}_\Gamma(V) = \sum_{\alpha \in \Gamma} (\mathrm{Dim}\, V_\alpha) E^\alpha. \tag{2.7}$$

Let $L = \mathcal{F}(V)$ be the free Lie superalgebra generated by V. Then L has a $\Gamma \times \mathbb{Z}_2$-gradation $L = \bigoplus_{(\alpha,i) \in \Gamma \times \mathbb{Z}_2} L_{(\alpha,i)}$ and a Γ-gradation $L = \bigoplus_{\alpha \in \Gamma} L_\alpha$ induced by those of V. By the Poincaré–Birkhoff–Witt Theorem, we have

$$\mathrm{ch}_{\Gamma \times \mathbb{Z}_2} U(L) = \frac{\displaystyle\prod_{(\beta,1) \in \Gamma \times \{1\}} (1 + e^{(\beta,1)})^{\dim L_{(\beta,1)}}}{\displaystyle\prod_{(\alpha,0) \in \Gamma \times \{0\}} (1 - e^{(\alpha,0)})^{\dim L_{(\alpha,0)}}}. \tag{2.8}$$

By taking the superdimensions, (2.8) can be written as

$$\mathrm{Ch}_{\Gamma \times \mathbb{Z}_2} U(L) = \prod_{(\alpha,i) \in \Gamma \times \mathbb{Z}_2} (1 - E^{(\alpha,i)})^{-\mathrm{Dim}\, L_{(\alpha,i)}}. \tag{2.9}$$

On the other hand, since L is free on V, the universal enveloping algebra $U(L)$ of L is equal to the tensor algebra $\mathcal{T}(V) = \mathbb{C} \oplus V \oplus (V \otimes V) \oplus \cdots$ of V. Consequently, we have

$$\mathrm{Ch}_{\Gamma \times \mathbb{Z}_2} U(L) = 1 + \mathrm{Ch}_{\Gamma \times \mathbb{Z}_2}(V) + (\mathrm{Ch}_{\Gamma \times \mathbb{Z}_2}(V))^2 + \cdots$$
$$= \frac{1}{1 - \mathrm{Ch}_{\Gamma \times \mathbb{Z}_2}(V)}. \tag{2.10}$$

Therefore, we obtain the following product identity

$$\prod_{(\alpha,i) \in \Gamma \times \mathbb{Z}_2} (1 - E^{(\alpha,i)})^{\mathrm{Dim}\, L_{(\alpha,i)}} = 1 - \mathrm{Ch}_{\Gamma \times \mathbb{Z}_2}(V) = 1 - \sum_{(\alpha,i) \in \Gamma \times \mathbb{Z}_2} d(\alpha,i) E^{(\alpha,i)}, \tag{2.11}$$

where $d(\alpha,i) = \mathrm{Dim}\, V_{(\alpha,i)}$ ($\alpha \in \Gamma, i \in \mathbb{Z}_2$). The identity (2.11) will be called the *denominator identity* for the $(\Gamma \times \mathbb{Z}_2)$-graded free Lie superalgebra L.

Let $P(V, \Gamma \times \mathbb{Z}_2) = \{(\alpha,i) \in \Gamma \times \mathbb{Z}_2 \mid d(\alpha,i) \neq 0\}$ and let $\{(\tau_i, j) \in \Gamma \times \mathbb{Z}_2 \mid i = 1,2,3,\ldots, j = 0,1\}$ be an enumeration of $P(V, \Gamma \times \mathbb{Z}_2)$. For $(\tau, k) \in \Gamma \times \mathbb{Z}_2$, we define

$$T(\tau,k) = \left\{ s = (s_{ij})_{i \geq 1, j=0,1} \mid s_{ij} \in \mathbb{Z}_{\geq 0}, \sum s_{ij}(\tau_i, j) = (\tau,k) \right\}, \tag{2.12}$$

which is the set of all partitions of (τ, k) into a sum of (τ_i, j)'s, and define the *Witt partition function* $W(\tau, k)$ by

$$W(\tau,k) = \sum_{s \in T(\tau,k)} \frac{(|s| - 1)!}{s!} \prod d(\tau_i, j)^{s_{ij}}, \tag{2.13}$$

where $|s| = \sum s_{ij}$ and $s! = \prod s_{ij}!$.

By taking the logarithm of the left-hand side of (2.11), we have

$$\log \left(\prod_{(\alpha,i)\in\Gamma\times\mathbb{Z}_2} (1 - E^{(\alpha,i)})^{-\operatorname{Dim} L_{(\alpha,i)}} \right) = \sum_{(\alpha,i)\in\Gamma\times\mathbb{Z}_2} \sum_{r=1}^{\infty} \frac{1}{r} \operatorname{Dim} L_{(\alpha,i)} E^{r(\alpha,i)}.$$

On the other hand, the right-hand side yields

$$\log \left(\frac{1}{1 - \operatorname{Ch}_{\Gamma\times\mathbb{Z}_2} V} \right) = \log \left(\frac{1}{1 - \sum_{\substack{i\geq 1 \\ j=0,1}} d(\tau_i, j) E^{(\tau_i, j)}} \right)$$

$$= \sum_{(\tau,k)\in\Gamma\times\mathbb{Z}_2} \left(\sum_{s\in T(\tau,k)} \frac{(|s| - 1)!}{s!} \prod d(\tau_i, j)^{s_{ij}} \right) E^{(\tau,k)}$$

$$= \sum_{(\tau,k)\in\Gamma\times\mathbb{Z}_2} W(\tau, k) E^{(\tau,k)}.$$

Therefore, we have

$$W(\tau, k) = \sum_{(\alpha,i)|(\tau,k)} \frac{(\alpha, i)}{(\tau, k)} \operatorname{Dim} L_{(\alpha,i)}.$$

Hence, by Möbius inversion, we obtain the following superdimension formula for the homogeneous subspaces $L_{(\alpha,i)}$ of L:

Theorem 2.3. *Let* $V = \bigoplus_{(\alpha,i)\in\Gamma\times\mathbb{Z}_2} V_{(\alpha,i)}$ *be a* Γ-*graded vector space over* $\mathbb{C}$ *with* $\dim V_{(\alpha,i)} < \infty$ *for all* $\alpha \in \Gamma$, $i \in \mathbb{Z}_2$, *and let* $L = \bigoplus_{(\alpha,i)\in\Gamma\times\mathbb{Z}_2} L_{(\alpha,i)}$ *be the free Lie superalgebra generated by* V. *Then we have*

$$\operatorname{Dim} L_{(\alpha,i)} = \sum_{(\tau,k)|(\alpha,i)} \frac{(\tau, k)}{(\alpha, i)} \mu \left(\frac{(\alpha, i)}{(\tau, k)} \right) W(\tau, k), \tag{2.14}$$

where μ *is the classical Möbius function, and* $(\tau, k)|(\alpha, i)$ *means* $(\alpha, i) = d(\tau, k)$ *for some* $d \in \mathbb{Z}_{>0}$, *in which case* $\dfrac{(\alpha, i)}{(\tau, k)} = d$, $\dfrac{(\tau, k)}{(\alpha, i)} = \dfrac{1}{d}$. $\qquad\square$

If $(\tau, k)|(\alpha, 0)$, then $(\alpha, 0) = d(\tau, k)$ for some $d \in \mathbb{Z}_{>0}$, which implies $\alpha = d\tau$ and $dk \equiv 0 \pmod{2}$. Hence, if $k = 0$, then d can be any positive integer dividing α, and if $k = 1$, then d must be even, dividing α. Therefore, we have

$$\operatorname{Dim} L_{(\alpha,0)} = \sum_{d|\alpha} \frac{1}{d} \mu(d) W \left(\frac{\alpha}{d}, 0 \right) + \sum_{\substack{d|\alpha \\ d:\text{ even}}} \frac{1}{d} \mu(d) W \left(\frac{\alpha}{d}, 1 \right). \tag{2.15}$$

Similarly, if $(\tau, k)|(\alpha, 1)$, then k must be 1 and d must be odd, dividing α. Therefore, we have

$$\operatorname{Dim} L_{(\alpha,1)} = \sum_{\substack{d|\alpha \\ d:\text{ odd}}} \frac{1}{d} \mu(d) W \left(\frac{\alpha}{d}, 1 \right). \tag{2.16}$$

Let $\text{Dim } L_\alpha = \text{Dim } L_{(\alpha,0)} + \text{Dim } L_{(\alpha,1)}$, $\text{Dim } V_\alpha = \text{Dim } V_{(\alpha,0)} + \text{Dim } V_{(\alpha,1)}$, $d(\alpha) = d(\alpha, 0) + d(\alpha, 1)$, and $E^{(\alpha,0)} = E^{(\alpha,1)} = E^{(\alpha)}$. Then, the denominator identity (2.11) yields

$$\prod_{\alpha \in \Gamma}(1 - E^\alpha)^{\text{Dim } L_\alpha} = 1 - \sum_{\alpha \in \Gamma} d(\alpha)E^\alpha, \tag{2.17}$$

which will also be called the *denominator identity* for the Γ-graded free Lie superalgebra $L = \bigoplus_{\alpha \in \Gamma} L_\alpha$.

Let $P(V, \Gamma) = \{\alpha \in \Gamma \mid d(\alpha) \neq 0\}$ and let $\{\tau_1, \tau_2, \tau_3, \ldots\}$ be an enumeration of $P(V, \Gamma)$. For $\tau \in \Gamma$, we define

$$T(\tau) = \Big\{s = (s_i)_{i \geq 1} \mid s_i \in \mathbb{Z}_{\geq 0}, \sum s_i \tau_i = \tau\Big\}, \tag{2.18}$$

which is the set of all partitions of τ into a sum of τ_i's, and define the *Witt partition function* $W(\tau)$ by

$$W(\tau) = \sum_{s \in T(\tau)} \frac{(|s| - 1)!}{s!} \prod d(\tau_i)^{s_i}, \tag{2.19}$$

where $|s| = \sum s_i$ and $s! = \prod s_i!$. Then, as a corollary of Theorem 2.3, we obtain the following superdimension formula for the homogeneous subspaces L_α of L:

Corollary 2.4. *Let* $V = \bigoplus_{\alpha \in \Gamma} V_\alpha$ *be a* Γ*-graded superspace over* $\mathbb{C}$ *with* $\dim V_\alpha = d(\alpha) < \infty$ *for all* $\alpha \in \Gamma$*, and let* $L = \bigoplus_{\alpha \in \Gamma} L_\alpha$ *be the free Lie superalgebra generated by* V*. Then we have*

$$\text{Dim } L_\alpha = \sum_{d \mid \alpha} \frac{1}{d} \mu(d) W\left(\frac{\alpha}{d}\right) = \sum_{d \mid \alpha} \frac{1}{d} \mu(d) \sum_{s \in T(\frac{\alpha}{d})} \frac{(|s| - 1)!}{s!} \prod d(\tau_i)^{s_i}. \tag{2.20}$$

Remark. The superdimension formulas (2.14) and (2.20) will be called the *generalized Witt formulas* for free Lie superalgebras.

Example 2.5. Let $V = V_{(0)} \oplus V_{(1)}$ be a superspace with $\dim V_{(0)} = r$ and $\dim V_{(1)} = s$, and let $\deg v = 1$ for all $v \in V$. Thus, $V = V_{(1,0)} \oplus V_{(1,1)}$ with $\text{Dim } V_{(1,0)} = r$, $\text{Dim } V_{(1,1)} = -s$, and $\text{Dim } V = r - s$. Let $L = \mathcal{F}(V)$ be the free Lie superalgebra generated by V. Then, the Lie superalgebra L has a $(\mathbb{Z}_{>0} \times \mathbb{Z}_2)$-gradation $L = \bigoplus_{(n,i) \in \mathbb{Z}_{>0} \times \mathbb{Z}_2} L_{(n,i)}$ and a $\mathbb{Z}_{>0}$-gradation $L = \bigoplus_{n=1}^\infty L_n$ induced by that of V. Since $P(V, \mathbb{Z}_{>0}) = \{1\}$ with $\text{Dim } V = r - s$, the denominator identity for the $\mathbb{Z}_{>0}$-graded free Lie superalgebra $L = \bigoplus_{n=1}^\infty L_n$ is the same as

$$\prod_{n=1}^\infty (1 - q^n)^{\text{Dim } L_n} = 1 - (r - s)q,$$

and the Witt partition function $W(k)$ is given by

$$W(k) = \frac{(k - 1)!}{k!}(r - s)^k = \frac{1}{k}(r - s)^k.$$

Hence, by the generalized Witt formula (2.20), we have

$$\operatorname{Dim} L_n = \sum_{d|n} \frac{1}{d}\mu(d) W\left(\frac{n}{d}\right) = \frac{1}{n}\sum_{d|n}\mu(d)(r-s)^{\frac{n}{d}}.$$

Next, consider the $(\mathbb{Z}_{>0} \times \mathbb{Z}_2)$-gradation on the free Lie superalgebra $L = \bigoplus_{(n,i)\in\mathbb{Z}_{>0}\times\mathbb{Z}_2} L_{(n,i)}$. Since $P(V, \mathbb{Z}_{>0}\times\mathbb{Z}_2) = \{(1,0), (1,1)\}$ with $\operatorname{Dim} V_{(1,0)} = r$ and $\operatorname{Dim} V_{(1,1)} = -s$, by specializing $p = e^{(1,0)}$ and $q = e^{(0,1)}$, the denominator identity for the $(\mathbb{Z}_{>0} \times \mathbb{Z}_2)$-graded free Lie superalgebra L is equal to

$$\prod_{\substack{n\in\mathbb{Z}_{>0}\\i=0,1}} (1 - p^n q^i)^{\operatorname{Dim} L_{(n,i)}} = 1 - rp + spq = 1 - p(r - sq).$$

For $m \in \mathbb{Z}_{>0}$, we have

$$\begin{aligned}
T(m, 0) &= \{s = (a, b) \mid a, b \in \mathbb{Z}_{\geq 0},\ a(1,0) + b(1,1) = (m,0)\} \\
&= \{s = (a, b) \mid a, b \in \mathbb{Z}_{\geq 0},\ a + b = m,\ b \equiv 0 \ (\mathrm{mod}\ 2)\} \\
&= \left\{s = (m - 2j, 2j) \mid 0 \leq j \leq \frac{m}{2}\right\},
\end{aligned}$$

which yields

$$W(m, 0) = \sum_{0\leq j\leq \frac{m}{2}} \frac{(m-1)!}{(m-2j)!(2j)!} r^{m-2j}s^{2j}.$$

Similarly, we obtain

$$T(m, 1) = \left\{s = (m - 2j - 1, 2j + 1) \mid 0 \leq j \leq \frac{m-1}{2}\right\}$$

and

$$W(m, 1) = -\sum_{0\leq j\leq \frac{m-1}{2}} \frac{(m-1)!}{(m-2j-1)!(2j+1)!} r^{m-2j-1}s^{2j+1}.$$

Therefore, by (2.15) and (2.16), we obtain

$$\begin{aligned}
\operatorname{Dim} L_{(n,0)} = &\sum_{d|n} \frac{1}{d}\mu(d) \sum_{0\leq j\leq \frac{n}{2d}} \left(\frac{n}{d} - 1\right)!\left(\frac{n}{d} - 2j\right)!(2j)!r^{\frac{n}{d}-2j}s^{2j} \\
&- \sum_{\substack{d|n\\d:\ \mathrm{even}}} \frac{1}{d}\mu(d) \sum_{0\leq j\leq \frac{n-d}{2d}} \frac{(\frac{n}{d} - 1)!}{\frac{n}{d} - 2j - 1)!(2j + 1)!} r^{\frac{n}{d}-2j-1}s^{2j+1},
\end{aligned}$$

$$\operatorname{Dim} L_{(n,1)} = -\sum_{\substack{d|n\\d:\ \mathrm{odd}}} \frac{1}{d}\mu(d) \sum_{0\leq j\leq \frac{n-d}{2d}} \frac{(\frac{n}{d} - 1)!}{(\frac{n}{d} - 2j - 1)!(2j + 1)!} r^{\frac{n}{d}-2j-1}s^{2j+1}.$$

Note that

$$\text{Dim } L_{(n,0)} + \text{Dim } L_{(n,1)} = \frac{1}{n} \sum_{d|n} \mu(d)(r-s)^{\frac{n}{d}} = \text{Dim } L_n,$$

as expected.   □

3. Free Lie superalgebras and product identities

We now return to the product identity of the form

$$\prod_{\alpha \in \Gamma} (1 - E^\alpha)^{A(\alpha)} = 1 - \sum_{\alpha \in \Gamma} d(\alpha) E^\alpha \tag{3.1}$$

with $A(\alpha), d(\alpha) \in \mathbb{Z}$ for all $\alpha \in \Gamma$.

Consider a Γ-graded vector space $V = \bigoplus_{\alpha \in \Gamma} V_\alpha$ over $\mathbb{C}$ with $\dim V_\alpha = |d(\alpha)|$ for all $\alpha \in \Gamma$. Let $V_{(0)} = \bigoplus_{\alpha:d(\alpha)>0} V_\alpha$ and $V_{(1)} = \bigoplus_{\alpha:d(\alpha)<0} V_\alpha$. Then $V = V_{(0)} \oplus V_{(1)} = \bigoplus_{(\alpha,i)\in\Gamma\times\mathbb{Z}_2} V_{(\alpha,i)}$ becomes a $(\Gamma \times \mathbb{Z}_2)$-graded superspace with $\text{Dim } V_{(\alpha,i)} = d(\alpha)$ for all $\alpha \in \Gamma$, $i \in \mathbb{Z}_2$. Thus $\text{Dim } V_\alpha = d(\alpha)$ for all $\alpha \in \Gamma$ and the right-hand side of (3.1) can be interpreted as $1 - \text{Ch } V$. Let $L = \mathcal{F}(V)$ be the free Lie superalgebra generated by V. Then the free Lie superalgebra L has a Γ-gradation induced by that of V, and the denominator identity for $L = \bigoplus_{\alpha \in \Gamma} L_\alpha$ is the same as the product identity (3.1). In particular, we have

$$\text{Dim } L_\alpha = A(\alpha) \quad \text{for all} \quad \alpha \in \Gamma. \tag{3.2}$$

On the other hand, let $P(V, \Gamma) = \{\alpha \in \Gamma \mid d(\alpha) \neq 0\}$ and let $\{\tau_1, \tau_2, \tau_3, \ldots\}$ be an enumeration of $P(V, \Gamma)$. Then, by the generalized Witt formula (2.20), we have

$$\text{Dim } L_\alpha = \sum_{d|\alpha} \frac{1}{d} \mu(d) \sum_{s \in T(\frac{\alpha}{d})} \frac{(|s|-1)!}{s!} \prod d(\tau_i)^{s_i}, \tag{3.3}$$

which yields a combinatorial identity

$$A(\alpha) = \sum_{d|\alpha} \frac{1}{d} \mu(d) \sum_{s \in T(\frac{\alpha}{d})} \frac{(|s|-1)!}{s!} \prod d(\tau_i)^{s_i}. \tag{3.4}$$

Example 3.1. (a) Consider the generating function for the partition function $p(n)$:

$$\prod_{n=1}^{\infty} (1 - q^n)^{-1} = \sum_{i=0}^{\infty} p(i) q^i = 1 - \sum_{i=1}^{\infty} (-p(i)) q^i. \tag{3.5}$$

Let $V = \bigoplus_{i=1}^{\infty} V_i$ be a $\mathbb{Z}_{>0}$-graded superspace with $\text{Dim } V_i = -p(i)$ for all $i \geq 1$ (thus $V_{(0)} = 0$, $V_{(1)} = V$), and let $L = \bigoplus_{n=1}^{\infty} L_n$ be the free Lie superalgebra generated by V. Then (3.5) can be interpreted as the denominator identity for the free Lie superalgebra L, and hence we have

$$\text{Dim } L_n = -1 \quad \text{for all} \quad n \geq 1.$$

Therefore, we obtain

$$\sum_{d|n} \frac{1}{d} \mu(d) \sum_{s \in T(\frac{n}{d})} \frac{(|s|-1)!}{s!} (-1)^{|s|} \prod p(i)^{s_i} = -1. \tag{3.6}$$

(b) Recall the definition of the Ramanujan tau function

$$\Delta(q) = q \prod_{n=1}^{\infty} (1-q^n)^{24} = \sum_{i=1}^{\infty} \tau(i) q^i$$

$$= q - 24q^2 + 252q^3 - 1472q^4 - \cdots. \tag{3.7}$$

We can rewrite it as

$$\prod_{n=1}^{\infty} (1-q^n)^{24} = 1 - \sum_{i=1}^{\infty} (-\tau(i+1)) q^i. \tag{3.8}$$

Let $V = \bigoplus_{i=1}^{\infty} V_i$ be a $\mathbb{Z}_{>0}$-graded superspace with $\mathrm{Dim}\, V_i = -\tau(i+1)$ for all $i \geq 1$ (thus $V_{(0)} = \bigoplus_{i: \tau(i+1)<0} V_i$, $V_{(1)} = \bigoplus_{i: \tau(i+1)>0} V_i$), and let $L = \bigoplus_{n=1}^{\infty} L_n$ be the free Lie superalgebra generated by V. Then (3.8) can be interpreted as the denominator identity for the free Lie superalgebra L, and hence we have

$$\mathrm{Dim}\, L_n = 24 \quad \text{for all} \quad n \geq 1.$$

Therefore, we obtain

$$\sum_{d|n} \frac{1}{d} \mu(d) \sum_{s \in T(\frac{n}{d})} \frac{(|s|-1)!}{s!} (-1)^{|s|} \prod \tau(i+1)^{s_i} = 24. \tag{3.9}$$

Actually, the relation (3.9) allows us to determine the values of the Ramanujan tau function $\tau(n)$ recursively. $\qquad\square$

In [B3], Borcherds gives a very important method for constructing automorphic forms on $O_{s+2,2}(\mathbb{R})^+$ with infinite product expansions. In particular, he proved:

Theorem 3.2 ([B3]). *Consider*

$$H(\tau) = \sum_{n \in \mathbb{Z}} H(n) q^n = -\frac{1}{12} + \frac{1}{3} q^3 + \frac{1}{2} q^4 + q^7 + q^8 \cdots,$$

where $H(n)$ is the Hurwitz class number for the discriminant $-n$ if $n > 0$ and $H(0) = -\frac{1}{12}$. Suppose that $f_0(\tau) = \sum_{n \in \mathbb{Z}} c_0(n) q^n$ is a meromorphic modular form of weight $\frac{1}{2}$ for $\Gamma_0(4)$ with integral coefficients, with poles only at cusps, and with $c_0(n) = 0$ if $n \equiv 2, 3 \pmod 4$. Put

$$\Psi(\tau) = q^{-h} \prod_{n>0} (1-q^n)^{c_0(n^2)}, \tag{3.10}$$

where h is the constant term of $f_0(\tau)H(\tau)$. Then Ψ is a meromorphic modular form for some characters of $\mathrm{SL}(2,\mathbb{Z})$ of integral weight with leading coefficient 1, all of whose zeros and poles are either at cusps or imaginary quadratic irrationals. $\qquad\square$

For example, let

$$F(\tau) = \sum_{n \in \mathbb{Z}_{>0}, \text{ odd}} \sigma_1(n)q^n = q + 4q^3 + 6q^5 + \cdots, \tag{3.11}$$

$$\theta(\tau) = \sum_{n \in \mathbb{Z}} q^{n^2} = 1 + 2q + 2q^4 + \cdots, \tag{3.12}$$

and define

$$f_0(\tau) = F(\tau)\theta(\tau)\frac{(\theta(\tau)^4 - 2F(\tau))(\theta(\tau)^4 - 16F(\tau))E_6(4\tau)}{\Delta(4\tau)} + 56\theta(\tau)$$

$$= \sum_{n \in \mathbb{Z}} c_0(n)q^n = q^{-3} - 248q + 26752q^4 - \cdots, \tag{3.13}$$

$$g_0(\tau) = (j(4\tau) - 876)\theta(\tau) - 2F(\tau)\theta(\tau)\frac{(\theta(\tau)^4 - 2F(\tau))(\theta(\tau)^4 - 16F(\tau))E_6(4\tau)}{\Delta(4\tau)}$$

$$= \sum_{n \in \mathbb{Z}} b_0(n)q^n = q^{-4} + 6 + 504q + 143388q^4 - \cdots. \tag{3.14}$$

Then Theorem 3.2 yields product identities for the elliptic modular function j and the Eisenstein series E_4, E_6, E_8, E_{10}, and E_{14} (**[B3]**):

$$j(\tau) = q^{-1} + 744 + 196884q + \cdots = q^{-1} \prod_{n=1}^{\infty} (1 - q^n)^{3c_0(n^2)}, \tag{3.15}$$

$$E_4(\tau) = 1 + 240 \sum_{n=1}^{\infty} \sigma_3(n)q^n = \prod_{n=1}^{\infty} (1 - q^n)^{c_0(n^2)+8}, \tag{3.16}$$

$$E_6(\tau) = 1 - 504 \sum_{n=1}^{\infty} \sigma_5(n)q^n = \prod_{n=1}^{\infty} (1 - q^n)^{b_0(n^2)}, \tag{3.17}$$

$$E_8(\tau) = 1 + 480 \sum_{n=1}^{\infty} \sigma_7(n)q^n = \prod_{n=1}^{\infty} (1 - q^n)^{2c_0(n^2)+16}, \tag{3.18}$$

$$E_{10}(\tau) = 1 - 264 \sum_{n=1}^{\infty} \sigma_9(n)q^n = \prod_{n=1}^{\infty} (1 - q^n)^{b_0(n^2)+c_0(n^2)+8}, \tag{3.19}$$

$$E_{14}(\tau) = 1 - 24 \sum_{n=1}^{\infty} \sigma_{13}(n)q^n = \prod_{n=1}^{\infty} (1 - q^n)^{b_0(n^2)+2c_0(n^2)+16}. \tag{3.20}$$

The product identity for the modular function j can be written as

$$\prod_{n=1}^{\infty} (1 - q^n)^{3c_0(n^2)} = 1 + 744q + 196884q^2 + \cdots$$

$$= 1 - \sum_{i=1}^{\infty} (-c(i-1)) q^i, \qquad (3.21)$$

where $c(i)$ are the coefficients of j. (Note that $c(0) = 744$.) Let $V = \bigoplus_{i=1}^{\infty} V_i$ be the $\mathbb{Z}_{>0}$-graded superspace with Dim $V_i = -c(i-1)$ for all $i \geq 1$ (thus $V_{(0)} = 0$, $V_{(1)} = V$), and let $L = \bigoplus_{n=1}^{\infty} L_n$ be the free Lie superalgebra generated by V. Then (3.21) can be interpreted as the denominator identity for the free Lie superalgebra L, and hence

$$\text{Dim } L_n = 3c_0(n^2) \quad \text{for all} \quad n \geq 1.$$

Therefore, we obtain a combinatorial identity:

$$3c_0(n^2) = \sum_{d|n} \frac{1}{d} \mu(d) \sum_{s \in T(\frac{n}{d})} \frac{(|s|-1)!}{s!} (-1)^{|s|} \prod c(i-1)^{s_i}. \qquad (3.22)$$

Similarly, the product identities for the Eisenstein series can be interpreted as the denominator identities for the free Lie superalgebras $L = \bigoplus_{n=1}^{\infty} L_n$ generated by the $\mathbb{Z}_{>0}$-graded superspaces $V = \bigoplus_{i=1}^{\infty} V_i$ with Dim $V_i = -240\sigma_3(i)$, $504\sigma_5(i)$, $-480\sigma_7(i)$, $264\sigma_9(i)$, and $24\sigma_{13}(i)$, respectively. Therefore, the superdimensions of the homogeneous subspaces L_n are given by

$$\text{Dim } L_n = c_0(n^2) + 8 \qquad \text{(corresponding to } E_4(\tau)), \qquad (3.23)$$
$$\text{Dim } L_n = b_0(n^2) \qquad \text{(corresponding to } E_6(\tau)), \qquad (3.24)$$
$$\text{Dim } L_n = 2c_0(n^2) + 16 \qquad \text{(corresponding to } E_8(\tau)), \qquad (3.25)$$
$$\text{Dim } L_n = b_0(n^2) + c_0(n^2) + 8 \qquad \text{(corresponding to } E_{10}(\tau)), \qquad (3.26)$$
$$\text{Dim } L_n = b_0(n^2) + 2c_0(n^2) + 16 \qquad \text{(corresponding to) } E_{14}(\tau)), \qquad (3.27)$$

and (3.4) yields the following combinatorial identities:

$$c_0(n^2) + 8 = \sum_{d|n} \frac{1}{d} \mu(d) \sum_{s \in T(\frac{n}{d})} \frac{(|s|-1)!}{s!} (-240)^{|s|} \prod \sigma_3(i)^{s_i}, \qquad (3.28)$$

$$b_0(n^2) = \sum_{d|n} \frac{1}{d} \mu(d) \sum_{s \in T(\frac{n}{d})} \frac{(|s|-1)!}{s!} 504^{|s|} \prod \sigma_5(i)^{s_i}, \qquad (3.29)$$

$$2c_0(n^2) + 16 = \sum_{d|n} \frac{1}{d} \mu(d) \sum_{s \in T(\frac{n}{d})} \frac{(|s|-1)!}{s!} (-480)^{|s|} \prod \sigma_7(i)^{s_i}, \qquad (3.30)$$

$$b_0(n^2) + c_0(n^2) + 8 = \sum_{d|n} \frac{1}{d} \mu(d) \sum_{s \in T(\frac{n}{d})} \frac{(|s|-1)!}{s!} 264^{|s|} \prod \sigma_9(i)^{s_i}, \qquad (3.31)$$

$$b_0(n^2) + 2c_0(n^2) + 16 = \sum_{d|n} \frac{1}{d} \mu(d) \sum_{s \in T(\frac{n}{d})} \frac{(|s|-1)!}{s!} 24^{|s|} \prod \sigma_{13}(i)^{s_i}. \qquad (3.32)$$

Acknowledgements. I would like to express my sincere gratitude to Professors Joe Ferrar and Koichiro Harada for their hospitality during the conference on the Monster simple group and Lie algebras at Ohio State University in May, 1996.

References

[Ba]   Yu. A. Bahturin, A. A. Mikhalev, V. M. Petrogradsky, M. V. Zaicev, Infinite dimensional Lie superalgebras, de Gruyter Exp. Math. **7**, Walter de Gruyter, Berlin 1992.

[B1]   R. E. Borcherds, Generalized Kac–Moody algebras, J. Algebra **115** (1988), 501–512.

[B2]   R. E. Borcherds, Monstrous moonshine and monstrous Lie superalgebras, Invent. Math. **109** (1992), 405–444.

[B3]   R. E. Borcherds, Automorphic forms on $O_{s+2,2}(\mathbb{R})$ and infinite products, Invent. Math. **120** (1995), 161–213.

[Bo]   N. Bourbaki, Lie groups and Lie algebras, Part 1, Herman, Paris 1975.

[CE]   H. Cartan, S. Eilenberg, Homological algebra, Princeton University Press, Princeton 1956.

[CN]   J. H. Conway, S. Norton, Monstrous moonshine, Bull. London Math. Soc. **11** (1979), 308–339.

[K1]   V. G. Kac, Simple irreducible graded Lie algebras of finite growth, Math. USSR-Izv. **2** (1968), 1271–1311.

[K2]   V. G. Kac, Infinite-dimensional Lie algebras and Dedekind's η-function, Funct. Anal. Appl. **8** (1974), 68–70.

[K3]   V. G. Kac, Lie superalgebras, Adv. Math. **26** (1977), 8–96.

[K4]   V. G. Kac, Infinite Dimensional Lie Algebras (3rd ed.), Cambridge University Press, Cambridge 1990.

[KW]   V. G. Kac, M. Wakimoto, Integrable highest weight modules over affine superalgebras and number theory, in: Lie theory and geometry, Progr. Math. **123**, Birkhäuser, Boston 1994, 415–456.

[Ka]   S.-J. Kang, Graded Lie superalgebras and the superdimension formula, to appear in J. Algebra.

[KK1]   S.-J. Kang, M.-H. Kim, Free Lie algebras, generalized Witt formula, and the denominator identity, J. Algebra **183** (1996), 560–594.

[KK2]   S.-J. Kang, M.-H. Kim, Dimension formula for graded Lie algebras and its applications, to appear in Trans. Amer. Math. Soc.

[M]   I. G. Macdonald, Affine root systems and Dedekind's η-function, Invent. Math. **15** (1972), 91–143.

[S]   M. Scheunert, The theory of Lie superalgebras, Lecture Notes in Math. **716**, Springer-Verlag, Berlin, Heidelberg, New York 1979.

[Se]   J. P. Serre, Lie algebras and Lie groups, Benjamin, New York 1965.

Department of Mathematics
Seoul National University
Seoul 151-742, Korea
E-mail: sjkang@math.snu.ac.kr

A generalization of the Jordan approach to symmetric Riemannian spaces

I. L. Kantor

0. Introduction

The Jordan approach to symmetric Riemannian spaces was started approximately 30 years ago. This method is not applicable to all symmetric spaces but only to symmetric R-spaces. However this class of symmetric Riemannian spaces is very important. Particularly, a realization of a noncompact symmetric space as a bounded domain in the dual compact space is constructed only for R-spaces. Moreover it can be given in the terms of Jordan triple systems.

Roughly speaking, the Jordan approach is based on the following two mappings. The first one $\mathcal{L}$ associates with every Jordan algebra A, or in a more general situation with every Jordan triple system ϕ, a graded Lie algebra of the form

$$\mathcal{L}(\phi) = U_{-1} \oplus U_0 \oplus U_1. \tag{0.1}$$

The other one S associates with A or ϕ two subalgebras of the Lie algebra (0.1). The subalgebras $S_+(\phi)$ and $S_-(\phi)$ have the structure of the Lie algebra of a symmetric space:

$$\begin{aligned} S_+(\phi) &= H \oplus E_+, \\ S_-(\phi) &= H \oplus E_-. \end{aligned} \tag{0.2}$$

Consider a homogeneous R-space $\mathcal{M}$ defined by the Lie algebra $\mathcal{L}(\phi)$ as a Lie algebra of transformations with the subalgebra $U_+ = U_0 \oplus U_1$ as a stationary subalgebra. The space $\mathcal{M}$, up to some points at "infinity", can be identified with the space of the Jordan triple system ϕ. The mapping S gives an opportunity to realize symmetric spaces corresponding to symmetric pairs $S_+(\phi) \supset H$ and $S_-(\phi) \supset H$ as subsets (which are in fact domains) of $\mathcal{M}$. Thus, the symmetric spaces can be realized as domains in the space of the Jordan triple system ϕ.

The goal of this paper is to develop a theory applying some generalizations of Jordan algebras to the investigation of a larger class of Riemannian spaces in the same way as symmetric spaces are treated by the Jordan approach. So we have to generalize the notions of a Jordan triple system and a symmetric space.

In this paper we show that most of the good properties of the Jordan approach can be preserved if the notions of a Jordan triple system and a symmetric space simultaneously transfer to a generalized Jordan triple system of second order and a bisymmetric space.

If now ϕ is such a triple system then formulas (0.1) and (0.2) look as

$$\mathcal{L}(\phi) = U_{-2} \oplus U_{-1} \oplus U_0 \oplus U_1 \oplus U_2. \tag{0.3}$$

and

$$S_+(\phi) = H_0 \oplus E_1 \oplus E_2,$$
$$S_-(\phi) = H_0 \oplus E_1^- \oplus E_2. \tag{0.4}$$

where the right hand sides in (0.4) have the structure of a Lie algebra of a homogeneous bisymmetric space.

Some historical remarks. The mapping $\mathcal{L}$ for the case of a Jordan algebra was introduced consecutively by J. Tits, I. L. Kantor and M. Koecher in [1], [2], [3], [4] and for a Jordan triple system by K. Meyberg in [5]. The mapping S for a Jordan algebra was introduced by I.L.Kantor, A. I. Sirota, A. S. Solodovnikov in [7], [8] and M. Koecher in [9]. In the subsequent active period the most contribution was done by O. Loos (see for example [10]). The approach was summarized in the encyclopedic books [11] and [12].

The notion of a Jordan triple system of second order and the mapping $\mathcal{L}$ in this case was introduced in [13], [14] and then developed in [15], [16], [17], [18], [19], [20]. The notion of a bisymmetric space was introduced in [21].

The paper is organized as follows. In the first chapter a sketch of the Jordan approach is given. Then in the second chapter, in a parallel way under the same numbers, definitions and theorems are presented for the more general situation.

1. A sketch of the Jordan approach to symmetric Riemannian spaces

Let $\mathcal{U}$ be a linear space and ϕ a trilinear operation on $\mathcal{U}$, i.e., a linear mapping $\mathcal{U} \times \mathcal{U} \times \mathcal{U} \mapsto \mathcal{U}$. We will also write (abc) in place of $\phi(a, b, c)$.

Definition 1.1. A linear space $\mathcal{U}$ with a trilinear operation $\phi(a, b, c) \equiv (abc)$ is said to be a *Jordan triple system* ϕ if the following identities are satisfied:

$$(ab(cdx) = ((abc)dx) - (c(bad)x) + (cd(abx)), \tag{1.1}$$
$$abc) = (cba). \tag{1.2}$$

Let ϕ be a Jordan triple system. To define the Lie algebra $\mathcal{L}(\phi)$ we introduce the following notations: S_{uv} is the linear operator defined by the formula

$$S_{uv}(x) = (uvx); \tag{1.3}$$

$\mathcal{S}$ is the linear space spanned by all linear operators S_{uv}.

Now consider the following direct sum of linear spaces

$$\mathcal{U} \oplus \mathcal{S} \oplus \bar{\mathcal{U}}, \tag{1.4}$$

where $\bar{\mathcal{U}}$ is a "second" copy of $\mathcal{U}$. In this space we will define the structure of a graded Lie algebra of the form

$$\mathcal{L}(\phi) = U_{-1} \oplus U_0 \oplus U_1, \tag{1.5}$$

where the spaces U_{-1}, U_0, U_1 are the spaces $\mathcal{U}$, $\mathcal{S}$, $\bar{\mathcal{U}}$ respectively.

This algebra $\mathcal{L}(\phi)$ will possess an involutory graded automorphism τ which is defined by formulas

$$\tau(a) = \bar{a}, \quad \tau(\bar{a}) = a, \quad \tau(S_{uv}) = -S_{vu} \tag{1.6}$$

where $a \in \mathcal{U}$.

We define commutators on the space (1.4) by the following formulas

$$[S_1, S_2] = S_1 S_2 - S_2 S_1, \quad [S, a] = S(a), \quad [a, \bar{b}] = S_{ab},$$
$$[S, \bar{a}] = \overline{S(a)}, \quad [a, b] = 0, \quad [\bar{a}, \bar{b}] = 0, \tag{1.7}$$

where $a, b \in \mathcal{U}$, $S, S_1, S_2 \in \mathcal{S}$; the bar over an element denotes the application of τ.

Theorem 1.1. *Let* $U = U_{-1} \oplus U_0 \oplus U_1$ *be a graded Lie algebra with a graded involutory automorphism* τ *such that* $\tau(U_\alpha) = U_{-\alpha}$. *Then the trilinear operation*

$$\phi(a, b, c) = [[a, \tau(b)], c], \tag{1.8}$$

where $a, b, c \in U_{-1}$, *defines on the space* U_{-1} *the structure of a Jordan triple system.*

Conversely, for any Jordan triple system ϕ *there exists a graded Lie algebra* $\mathcal{L}(\phi) = U_{-1} \oplus U_0 \oplus U_1$ *with graded involutory automorphism* τ *such that the operation* ϕ *can be expressed by formula* (1.8). *This graded Lie algebra* $\mathcal{L}(\phi)$ *and involutory graded automorphism* τ *are defined on the space* (1.4) *by formulas* (1.7), (1.6).

Definition 1.2. The construction of a graded Lie algebra $\mathcal{L}(\phi)$ is called a *Tits–Kantor–Koecher construction.*

The construction of the Lie algebra $\mathcal{L}(\phi)$ in terms of Jordan algebras was given in **[1]**, **[2]**, **[3]**, **[4]** and in terms of Jordan triple systems in **[5]**.

Definition 1.3. A *symmetric pair of Lie algebras* is a pair of Lie algebras $G \supset H$ over $\mathbb{R}$ such that there is a decomposition of G in a direct sum

$$G = H \oplus E \tag{1.9}$$

with the following relations of commutators:

$$[H, H] \subset H, \quad [H, E] \subset E, \quad [E, E] \subset H. \tag{1.10}$$

If in addition the subalgebra H is compact, then the symmetric pair $G \supset H$ is called *Riemannian.*

If the decomposition (1.9) gives a symmetric pair of Lie algebras $G \supset H$ then the space

$$G_d = H \oplus i E, \tag{1.11}$$

where i is the imaginary unit, with commutators defined in a natural way, gives also a symmetric pair of Lie algebras $G_d \supset H$.

Definition 1.4. The symmetric pair (1.11) is called the *dual symmetric pair* to the symmetric pair (1.9). The symmetric pairs (1.9) and (1.11) are also called *mutually dual.*

Definition 1.5. A homogeneous space $\mathcal{M} = \mathcal{G}/\mathcal{H}$ is called *symmetric* if
  1. the Lie algebras G and H of the Lie groups $\mathcal{G}$ and $\mathcal{H}$ form a symmetric pair;
  2. there is an involutory element $\sigma \in \mathcal{G}$ such that the eigensubspaces with eigenvalues 1 and -1 of the induced action of σ on G are the subspaces the H and E respectively.
  If in addition the symmetric pair is Riemannian, then the symmetric space $\mathcal{G}$ is called *Riemannian*.
  Two symmetric spaces with mutually dual symmetric pairs are also called *mutually dual*.

Proposition 1.1. *Let* $G \supset H$ *be a Riemannian symmetric pair of Lie algebras. Then there is a Riemannian symmetric space* $\mathcal{M} = \mathcal{G} / \mathcal{H}$ *such that G and H are the Lie algebras of the Lie groups* $\mathcal{G}$ *and* $\mathcal{H}$.

  See proof in [6].
  Let ϕ be a Jordan triple system. Consider the following two subspaces $S_+(\phi)$ and $S_-(\phi)$ of the Lie algebra $\mathcal{L}(\phi)$:

$$S_+(\phi) = H \oplus E_+,$$
$$S_-(\phi) = H \oplus E_-,$$
(1.12)

where

$$H = \{A \in S \mid \tau(A) = A\},$$
$$E_+ = \{v + \bar{v} \mid \forall v \in V\},$$
$$E_- = \{v - \bar{v} \mid \forall v \in V\}.$$
(1.13)

Theorem 1.2. *The pairs* $S_+(\phi) \supset H$ *and* $S_-(\phi) \supset H$ *are two mutually dual symmetric pairs of Lie algebras.*

Proof. Using formulas (1.6), (1.7), (1.13) one can easily check that relations (1.10) are fulfilled and Definition 1.2 is satisfied.

Definition 1.6. The symmetric pairs $S_+(\phi) \supset H$ and $S_-(\phi) \supset H$ are called *mutually dual symmetric pairs defined by a Jordan triple system* ϕ.

Definition 1.7. Two Jordan triple systems ϕ and ϕ_1 *are equivalent* if the corresponding Lie algebras $\mathcal{L}(\phi)$ and $\mathcal{L}(\phi_1)$ are graded isomorphic.

  There are equivalent Jordan triple systems ϕ_i such that the corresponding Lie algebras $S_+(\phi_i)$ (and also $S_-(\phi_i)$) are not isomorphic.
  From now on we suppose that the Lie algebra $U = U_{-1} \oplus U_0 \oplus U_1$ is semisimple. Consider the Lie group $\mathcal{U} = \text{Aut}\, U$ and the homogeneous space $\mathcal{M} = \mathcal{U}/\mathcal{U}_+$, where the parabolic subgroup $\mathcal{U}_+$ corresponds to the parabolic subalgebra $U_+ = U_0 \oplus U_1$. The space $\mathcal{M}$ is compact and is called an *R-space*. One can identify almost the whole of $\mathcal{M}$ with the linear space U_{-1} in the following way. The orbit $\mathcal{M}_-$ of the subgroup corresponding to the subalgebra U_{-1} is a "big" cell of $\mathcal{M}$ which is in one-to-one correspondence with U_{-1}. So when U is presented as $U = \mathcal{L}(\phi)$ one can identify the

compact R-space $\mathcal{M}$ up to some points lying at "infinity" with the space of the triple system ϕ.

Consider now the subgroup corresponding to the subalgebra $S_+(\phi)$. The orbit of this subgroup in the space $\mathcal{M}$ has the structure of a symmetric space. So we have a symmetric subset (which is in fact a domain) in the space $\mathcal{M}$. We denote this symmetric domain $D_+(\phi)$. In the same way, one can define the dual symmetric domain $D_-(\phi)$.

To formulate Theorem 1.3 we need also the following definition.

Definition 1.8. Let $\phi(x, y, z)$ be a Jordan triple system. The *square of the norm* $\|x\|_\phi^2$ of an element x is the absolute value of the largest eigenvalue of the linear operator

$$L_{xx}(z) = \frac{1}{2}\phi(x, z, x).$$ (1.14)

Theorem 1.3. *Let U be a semisimple graded Lie algebra of the form*

$$U = U_{-1} \oplus U_0 \oplus U_1.$$ (1.15)

Each isomorphic class among the equivalent Jordan triple systems ϕ such that $\mathcal{L}(\phi) = U$ corresponds to the mutually dual pairs of symmetric domains $D_+(\phi)$ and $D_-(\phi)$ in the space $\mathcal{M}$.

Among these Jordan triple systems ϕ there is one (and only one) ϕ_0 such that the subalgebra $S_+(\phi_0)$ is compact and the domain $D_+(\phi_0)$ is the whole space $\mathcal{M}$. (Hence $\mathcal{M}$ has the structure of a Riemannian symmetric R-space.) The dual domain $D_-(\phi_0)$ is the bounded Riemannian symmetric domain dual to Riemannian symmetric space $\mathcal{M}$.

The domain $D_-(\phi_0)$ can be described as a bounded domain of the space of ϕ_0 by the formula

$$D_-(\phi_0) = \{x \in \phi_0 \mid \|x\|_{\phi_0} < 1\}.$$ (1.16)

2. Extension to bisymmetric Riemannian spaces

Definition 2.1. A linear space $\mathcal{U}$ with a trilinear operation $\phi(a, b, c) \equiv (abc)$ is said to be a *generalized Jordan triple system of second order* if the following identities are satisfied:

$$(ab(cdx)) = ((abc)dx) - (c(bad)x) + (cd(abx)),$$ (2.1)

$$[((avb)uc) - (cu(avb)) - (cv(aub)) - (a(vcu)b)]_{a,b} = 0,$$ (2.2)

where the symbol $[\ldots]_{a,b}$ denotes the alternation with respect to a and b.

Let ϕ be a generalized Jordan triple system of second order. To define the Lie algebra $\mathcal{L}(\phi)$ we introduce the following notation: S_{uv} and K_{uv} are the linear operators defined by the formulas

$$S_{uv}(x) = (uvx), \quad K_{uv}(x) = (uxv) - (vxu);$$ (2.3)

S and K are the linear spaces spanned by all linear operators S_{uv} and K_{uv} (respectively).

Now consider the following direct sum of linear spaces:

$$\mathcal{K} \oplus \mathcal{U} \oplus \mathcal{S} \oplus \bar{\mathcal{U}} \oplus \bar{\mathcal{K}}, \tag{2.4}$$

where $\bar{\mathcal{K}}$ and $\bar{\mathcal{U}}$ are "second" copies of $\mathcal{K}$ and $\mathcal{U}$. In this space we will define the structure of a graded Lie algebra of the form

$$\mathcal{L}(\phi) = U_{-2} \oplus U_{-1} \oplus U_0 \oplus U_1 \oplus U_2, \tag{2.5}$$

where the spaces U_{-2}, U_{-1}, U_0, U_1, U_2 are the spaces $\mathcal{K}$, $\mathcal{U}$, $\mathcal{S}$, $\bar{\mathcal{U}}$, $\bar{\mathcal{K}}$ respectively.

This algebra $\mathcal{L}(\phi)$ will possess an involutory graded automorphism τ which is defined by formulas

$$\tau(K) = \bar{K}, \quad \tau(\bar{K}) = K, \quad \tau(a) = \bar{a}, \quad \tau(\bar{a}) = a, \quad \tau(S_{uv}) = -S_{vu}, \tag{2.6}$$

where $a \in \mathcal{U}$, $K \in \mathcal{K}$.

Taking into account that the algebra to be constructed from (24) must be graded and the mapping τ is an automorphism, it suffices to define the commutators $[U_0, U_0]$, $[U_0, U_{-1}]$, $[U_{-1}, U_1]$, $[U_{-1}, U_{-1}]$, $[U_0, U_{-2}]$, $[U_{-2}, U_2]$, $[U_{-2}, U_1]$.

We define these by means of the following formulas

$$[S_1, S_2] = S_1 S_2 - S_2 S_1, \quad [S, a] = S(a), \quad [a, \bar{b}] = S_{ab}, \quad [a, b] = K_{ab},$$
$$[S, K] = SK - K\bar{S}, \quad [K_1, \bar{K}_2] = K_1 K_2, \quad [K, \bar{a}] = K(a), \tag{2.7}$$

where $a, b \in \mathcal{U}$, $S, S_1, S_2 \in \mathcal{S}$, $K, K_1, K_2 \in \mathcal{K}$; the bar over an element denotes the application of τ.

Theorem 2.1. *Let* $U = U_{-2} \oplus U_{-1} \oplus U_0 \oplus U_1 \oplus U_2$ *be a graded Lie algebra with a graded involutory automorphism* τ *such that* $\tau(U_\alpha) = U_{-\alpha}$. *Then the trilinear operation*

$$\phi(a, b, c) = [[a, \tau(b)], c], \tag{2.8}$$

where $a, b, c \in U_{-1}$, *defines on the space* U_{-1} *the structure of a generalized Jordan triple system of second order.*

Conversely, for any generalized Jordan triple system of second order ϕ *there exists a graded Lie algebra* $\mathcal{L}(\phi) = U = U_{-2} \oplus U_{-1} \oplus U_0 \oplus U_1 \oplus U_2$ *with graded involutory automorphism* τ *such that the operation* ϕ *can be expressed by formula (2.8). This graded Lie algebra* $\mathcal{L}(\phi)$ *and involutory graded automorphism* τ *are defined on the space (2.4) by formulas (2.7), (2.6).*

See proof in **[13]**, **[14]**.

Definition 2.2. We shall call the graded Lie algebra $\mathcal{L}(\phi)$ the *Lie algebra defined by generalized Jordan triple system of second order.*

Definition 2.3. A Lie algebra G over $\mathbb{R}$ is called a *Lie algebra of a bisymmetric fiber bundle* or the triple $G \supset H \supset H_0$ over $\mathbb{R}$ is called a *bisymmetric triple* if G can be presented in the form

$$G = H_0 \oplus E_1 \oplus E_2 \tag{2.9}$$

such that $H = H_0 \oplus E_2$ and $G = H \oplus E_1$ are Lie algebras of symmetric spaces or in other words

$$G \supset H \quad \text{and} \quad H \supset H_0 \tag{2.10}$$

are symmetric pairs.

If in addition the subalgebra H is compact then the symmetric triple $G \supset H \supset H_0$ is called *Riemannian*.

If the decomposition (2.9) gives a bisymmetric triple $G \supset H \supset H_0$, then the space

$$G_d = H_0 \oplus i E_1 \oplus E_2, \tag{2.11}$$

where i is the imaginary unit, with commutators defined in a natural way gives also a bisymmetric triple $G_d \supset H \supset H_0$.

Definition 2.4. The bisymmetric triple (2.11) is called the *dual bisymmetric triple* to the bisymmetric triple (2.9). The bisymmetric triples (2.9) and (2.11) are also called *mutually dual*.

Definition 2.5. We shall call a homogeneous space $\mathcal{M} = \mathcal{G}/\mathcal{H}_0$ *bisymmetric* if there is a subgroup $\mathcal{H}$ such that $\mathcal{G} \supset \mathcal{H} \supset \mathcal{H}_0$ and the spaces $\mathcal{G}/\mathcal{H}$ and $\mathcal{H}/\mathcal{H}_0$ are symmetric.

If in addition, the symmetric triple is Riemannian, then the bisymmetric space $\mathcal{M}$ is called *Riemannian*.

Two bisymmetric spaces with mutually dual symmetric triples are also called *mutually dual*.

Remark. By definition, a bisymmetric space is a fiber-space such that both the base and the fiber are symmetric spaces. The symmetric pair $G \supset H$ corresponds to the base of $\mathcal{M}$ and the symmetric pair $H \supset H_0$ corresponds to the fiber of $\mathcal{M}$. There are some properties for the commutators of subspaces H_0, E_1, E_2, which follow from the (2.10) (see [21]).

Proposition 2.1. *Let $G \supset H \supset H_0$ be a Riemannian bisymmetric triple of Lie algebras. Then there is a Riemannian homogeneous fiber space $\mathcal{M}$ with symmetric base and locally symmetric fiber such that the corresponding triple of Lie algebras coincides with the given triple $G \supset H \supset H_0$.*

See proof in [21].

Let ϕ be a generalized Jordan triple system of second order. Consider the following subspaces $S_+(\phi)$ and $S_-(\phi)$ of Lie algebra $\mathcal{L}(\phi)$:

$$\begin{aligned}
S_+(\phi) &= H_0 \oplus E_1 \oplus E_2, \\
S_-(\phi) &= H_0 \oplus E_1^- \oplus E_2,
\end{aligned} \tag{2.12}$$

where

$$H_0 = \{A \in S \mid \tau(A) = A\},$$
$$E_1^+ = \{v + \bar{v} \mid \forall v \in V\},$$
$$E_1^- = \{v - \bar{v} \mid \forall v \in V\}, \tag{2.13}$$
$$E_2 = \{k + \bar{k} \mid \forall k \in K\}.$$

Theorem 2.2. *The triples $S_+(\phi) \supset H \supset H_0$ and $S_-(\phi) \supset H \supset H_0$, where $H = H_0 \oplus E_2$, are mutually dual bisymmetric triples of Lie algebras.*

Proof. Using formulas (2.6), (2.7), (2.13) one can easily check that the pairs $S_+(\phi) \supset H$, $S_-(\phi) \supset H$ and $H \supset H_0$ are symmetric and Definition 2.2 is satisfied.

Definition 2.6. The bisymmetric triples $S_+(\phi) \supset H \supset H_0$ and $S_-(\phi) \supset H \supset H_0$ are called *mutually dual bisymmetric triples defined by a generalized Jordan triple system of second order* ϕ.

Definition 2.7. Two generalized Jordan triple systems of second order ϕ and ϕ_1 *are equivalent* if the corresponding Lie algebras $\mathcal{L}(\phi)$ and $\mathcal{L}(\phi_1)$ are graded isomorphic.

As in previous case, there are equivalent generalized Jordan triple systems of second order ϕ_i such that the corresponding Lie algebras $S_+(\phi_i)$ (and also $S_-(\phi_i)$) are not isomorphic. We will discuss below a geometric sense of this phenomena.

From now on we suppose that the Lie algebra $U = U_{-2} \oplus U_{-1} \oplus U_0 \oplus U_1 \oplus U_2$ is semisimple. Consider the Lie group $\mathcal{U} = \operatorname{Aut} U$ and the homogeneous space $\mathcal{M} = \mathcal{U}/\mathcal{U}_+$ where the parabolic subgroup $\mathcal{U}_+$ corresponds to the parabolic subalgebra $U_+ = U_0 \oplus U_1 \oplus U_2$. The space $\mathcal{M}$ is compact and is called an *R-space*. One can identify almost the whole of $\mathcal{M}$ with the linear space $U_- = U_{-1} \oplus U_{-2}$ in the following way. The orbit $\mathcal{M}_-$ of the subgroup corresponding to subalgebra U_- is a "big" cell of $\mathcal{M}$ which is in one-to-one correspondence with U_-. So when U is presented as $U = \mathcal{L}(\phi)$ one can identify the compact R-space $\mathcal{M}$ up to some points lying at "infinity" with the linear fiber space with base ϕ and fiber U_{-2}.

Consider now the subgroup corresponding to the subalgebra $S_+(\phi)$. The orbit of this subgroup in the space $\mathcal{M}$ has the structure of a bisymmetric space. So we have a bisymmetric subset (which is in fact a domain) in the space $\mathcal{M}$. We denote this bisymmetric domain $D_+(\phi)$. In the same way, one can define the dual bisymmetric domain $D_-(\phi)$.

To formulate an analogy of Theorem 1.3 we need the following definitions.

Definition 2.8. Let $\phi(x, y, z)$ be a generalized Jordan triple system of second order.

The *square of the norm* $\|x\|_\phi^2$ of an element x is the absolute value of the largest eigenvalue of the linear operator

$$L_{xx}(z) = \frac{1}{2}\phi(x, z, x) \tag{2.14}$$

Definition 2.9. A *projectively bounded domain* is a subset of a fiber space such that under canonical projection to the base it becomes a bounded domain of the base.

In this paper the analogy of the general part of Theorem 1.3 and the description of the domain will be given in the following case.

Definition 2.10. A generalized Jordan triple system of second order (abc) is called *weakly commutative* if

$$K_{u,(uuu)} = 0 \quad \forall u.$$

Theorem 2.3. *Let U be a semisimple graded Lie algebra of the form*

$$U = U_{-2} \oplus U_{-1} \oplus U_0 \oplus U_1 \oplus U_2. \tag{2.15}$$

Each isomorphic class among the equivalent Jordan triple systems of second order ϕ such that $\mathcal{L}(\phi) = U$ corresponds to the mutually dual pairs of bisymmetric domains $D_+(\phi)$ and $D_-(\phi)$ in the space $\mathcal{M}$.

Among these generalized Jordan triple systems of second order ϕ there is one (and only one) ϕ_0 such that the subalgebra $S_+(\phi_0)$ is compact and the domain $D_+(\phi_0)$ is the whole space $\mathcal{M}$. (Hence $\mathcal{M}$ has the structure of a Riemannian bisymmetric R-space.) The dual domain $D_-(\phi_0)$ is the Riemannian bisymmetric domain dual to the Riemannian bisymmetric space $\mathcal{M}$. The projection of the dual domain $D_-(\phi_0)$ to the base can be described in the case ϕ is weakly commutative as a bounded domain of the space of ϕ_0 by the formula

$$D_-(\phi_0) = \{x \in \phi_0 \mid \|x\|_{\phi_0} < 1\}. \tag{2.16}$$

Proof. Let U be a semisimple graded Lie algebra of the form (2.15). Consider a triple system ϕ such that $\mathcal{L}(\phi) = U$. According to Theorem 2.1, it is defined by formula (2.8), where τ is some graded involutory automorphism. Then the Lie subalgebras $S_+(\phi)$ and $S_-(\phi)$ are defined by formulas (2.13), where bar means the acting of this involutory automorphism τ.

It follows from formulas (2.13) that the subgroups of the group Aut U corresponding to $S_+(\phi)$ and $S_-(\phi)$ acting on the space $\mathcal{M}$ are locally transitive in a neighborhood of the point x_0 whose stationary subalgebra is $U_+ = U_0 \oplus U_1 \oplus U_2$. Indeed, according to (2.13), the point x_0 can be moved in each direction $\lambda v + \mu k$ by a one-parameter group corresponding to an element of $E_1 \oplus E_2$ or $E_1^- \oplus E_2$. So the orbits of the groups $D_+(\phi)$ and $D_-(\phi)$ are domains in the space $\mathcal{M}$. Theorem 2.2 implies that $D_+(\phi)$ and $D_-(\phi)$ are mutually dual bisymmetric spaces.

It was proved in **[12]** that such a grading as (2.15) for the semisimple Lie algebra in the complex case is defined by a subset $\delta \subset \Delta$, where $\Delta = \{\alpha_1, \alpha_2, \ldots, \alpha_n, \}$ is the system of simple roots in the Cartan subalgebra H of U. The subset δ has to have the following property: if $\tilde{\alpha} = \nu_1\alpha_1 + \nu_2\alpha_2 + \cdots + \nu_n\alpha_n$ is the highest root then $\sum_{\alpha_t \in \delta} \nu_t = 2$. (So in fact δ consist of one or two roots.) Now let β be an arbitrary root

$$\beta = \mu_1\alpha_1 + \mu_2\alpha_2 + \cdots + \mu_n\alpha_n$$

and e_β be a rootvector corresponding to β. Then

U_i is a subspace of U spanned by $\{e_\beta \mid \sum_{\alpha_t \in \delta} \mu_t = i\}$ if $i \neq 0$,

U_0 is a subspace of U spanned by $\{H \cup e_\beta \mid \sum_{\alpha_t \in \delta} \mu_t = 0\}$.

It is easy to see that in the real case the grading is obtained in the same way (in fact it comes from the complex grading). The only difference is that the subset δ has to consist of such simple roots which correspond to white circles in the Satake diagram (see [6] and [22] of U and also if two simple roots have the same restrictions to a maximal abelian noncompact subalgebra (in this case they are joined by curved arrow in Satake diagram), then both roots can not be separately in the subset δ.

Consider the Cartan involution τ of the algebra U which is associated to the given Satake diagram. For all simple roots α_t corresponding to the white circles one has $\tau(e_{\alpha_t}) = e_{-\alpha_t}$. So τ has the property $\tau(U_\alpha) = U_{-\alpha}$ and according to Theorem 2.1 defines on the space U_{-1} the structure of a generalized Jordan triple system of second order ϕ_0 by formula (2.8).

It is evident that $\mathcal{L}(\phi_0) = U$ and that subalgebra $\mathcal{S}_+(\phi_0)$ is the maximal compact subalgebra of U because $\mathcal{S}_+(\phi_0) = \{a \in U \mid \tau(a) = a\}$ (the eigen subspace with the eigenvalue 1 of a Cartan involution τ of the simple Lie algebra U is a maximal compact subalgebra of U).

The subgroup corresponding to the maximal compact subalgebra $\mathcal{S}_+(\phi_0)$ in the group Aut U is a maximal compact subgroup of Aut U. According to the Montgomery theorem on groups acting transitively on compact manifolds (see [24]) this subgroup acts transitively on the space $\mathcal{M}$. So the domain $D_+(\phi_0)$ is the whole space $\mathcal{M}$. Hence $\mathcal{M}$ has a structure of a Riemannian bisymmetric R-space.

As it was previously shown, the dual domain $D_+(\phi_0)$ is a Riemannian bisymmetric domain dual to the Riemannian bisymmetric space $\mathcal{M}$. To prove the last assertion we identify the space $\mathcal{M}_-$ with the linear space $U_- = U_{-1} \oplus U_{-2}$. Let (x, K) where $x \in U_{-1}$ and $K \in U_{-2}$ be a generic point of U_-. The subalgebra $H_0 \oplus E_2$ is a maximal compact subalgebra of $\mathcal{S}_-(\phi_0)$. So according to the Iwasawa decomposition theorem, the action of one parameter subgroups generated by elements $a \in E_1$ on the point $(0, 0)$ gives representatives of the whole base.

To find this variety one have to solve the system of differential equations

$$dx/dt = a - K(a) - 1/2(xax),$$
$$dK/dt = 1/2K_{a,x} - 1/2K_{K(a),x} + 1/4K_{(xax),x} \tag{2.17}$$

with initial conditions $x(0) = 0$, $K(0) = 0$, $\frac{dx}{dt}(0) = a$, $\frac{dK}{dt}(0) = 0$ for all $a \in E_1$. The right hand sides in (2.17) are components of the vector field generated by $a \in E_1$ in the system of coordinates (x, K). (See [25], where formulas for vector fields in such coordinates are considered in a more general situation and the case of the graded Lie algebra of depth two is specially given in terms of commutators of the Lie algebra.)

It is not difficult to check directly that in the weakly commutative case the solution of (2.17) is given by

$$x(t) = 2B^{-1}\tanh(Bt/2)(a),$$
$$K(t) = 0, \tag{2.18}$$

where B is a linear operator defined by $B^2(x) = 2(axa)$.

Indeed, in the weakly commutative case, the triple system restricted on a space generated by one element a becomes a Jordan triple system. So it is an associative triple system consisting of all linear combinations of odd degrees of element a. Then substituting the formulas (2.18) in (2.17) shows that it is a solution. Hence the formula (2.16) is true because $\tanh(t) < 1$. Thus the theorem is proved.

Example.   Consider the generalized Jordan triple system of second order D_{nk} over the field $\mathbb{R}$. It is the space of all $n \times k$ matrices over $\mathbb{R}$ with the operation

$$(X, Y, Z) = XY^\top Z + ZY^\top X - YX^\top Z.$$

The Lie algebra $\mathcal{L}(D_{nk})$ is $SO(n+k, k)$. This linear Lie algebra consists of the following matrices of order $n + 2k$:

$$\begin{pmatrix} A & \vdots & -V^T & \vdots & K_1 \\ \cdots & \cdots & \cdots & \cdots & \cdots \\ W & \vdots & K & \vdots & V \\ \cdots & \cdots & \cdots & \cdots & \cdots \\ K_2 & \vdots & -W^T & \vdots & -A^T \end{pmatrix},$$

where V and W are arbitrary $n \times k$ matrices, A is an arbitrary matrix of order k, K_1, K_2 are skew-symmetric square matrices of order k, K is a skew-symmetric square matrix of order n. The grading of this Lie algebra is given by $K_1 \in U_{-2}$, $V \in U_{-1}$, $A, K \in U_0$, $W \in U_1$, $K_2 \in U_2$.

The Lie algebra $S_+(D_{nk})$ is $SO(n+k) \oplus SO(k)$. This subalgebra of $\mathcal{L}(D_{nk})$ consists of the following matrices

$$\begin{pmatrix} K_0 & \vdots & -V^T & \vdots & K_1 \\ \cdots & \cdots & \cdots & \cdots & \cdots \\ V & \vdots & K & \vdots & V \\ \cdots & \cdots & \cdots & \cdots & \cdots \\ K_1 & \vdots & -V^T & \vdots & K_0 \end{pmatrix}$$

where K_0 is a skew-symmetric matrix of order k. The ideals $SO(n + k)$ and $SO(k)$ of the Lie algebra $S_+(D_{nk})$ consist respectively of matrices of the form

$$\begin{pmatrix} K_0 & \vdots & -V^T & \vdots & K_0 \\ \cdots & \cdots & \cdots & \cdots & \cdots \\ V & \vdots & K & \vdots & V \\ \cdots & \cdots & \cdots & \cdots & \cdots \\ K_0 & \vdots & -V^T & \vdots & K_0 \end{pmatrix}, \quad \begin{pmatrix} K_0 & \vdots & 0 & \vdots & -K_0 \\ \cdots & \cdots & \cdots & \cdots & \cdots \\ 0 & \vdots & 0 & \vdots & 0 \\ \cdots & \cdots & \cdots & \cdots & \cdots \\ -K_0 & \vdots & 0 & \vdots & K_0 \end{pmatrix}$$

The structure of a Lie algebra of a bisymmetric space (2.9) is given by $K_0, K \in H_0$, $V \in E_1$, $K_1 \in E_2$.

The corresponding bisymmetric triple of Lie algebras is

$$SO(n + k) \oplus SO(k) \supset SO(n) \oplus SO(k) \oplus SO(k) \supset SO(n) \oplus SO(k),$$

where the last $SO(k)$ is embedded in the tensor sum $SO(k) \oplus SO(k)$ as the diagonal.

Hence, the base of the bisymmetric space is the symmetric space $(SO(n + k) \oplus SO(k))/(SO(n) \oplus SO(k) \oplus SO(k)) = SO(n + k)/(SO(n) \oplus SO(k))$, which is the space of all k-dimensional planes in $(n + k)$-dimensional Euclidean space.

The fiber is the symmetric space $(SO(n) \oplus SO(k) \oplus SO(k))/(SO(n) \oplus SO(k)) = (SO(k) \oplus SO(k))/SO(k)$, where the last $SO(k)$ is embedded in the tensor sum as the diagonal. This symmetric space is the orthogonal group $SO(k)$ on which the tensor sum $SO(k) \oplus SO(k)$ acts by left and right translations. The orthogonal group $SO(k)$ can be identified with the set of all orthogonal bases in the k-dimensional Euclidean space.

An geometric interpretation in this case is the following. The bisymmetric space $D_+(D_{nk})$ is the variety of all pairs $(\tilde{V}, \tilde{K})$ where $\tilde{V}$ is a k-dimensional plane in $(n + k)$-dimensional Euclidean space and $\tilde{K}$ is an orthogonal basis in this plane with orthogonal group $SO(n + k) \oplus SO(k)$ acting on this variety in natural way.

The canonical projection of the point of bisymmetric space $(\tilde{V}, \tilde{K})$ is the forgetting of the orthogonal basis $\tilde{K}$ and taking only the plane $\tilde{V}$.

The dual noncompact bisymmetric space $D_-(D_{nk})$ is constructed in the same way. But now the Lie algebra $S_-(D_{nk})$ is $SO(n, k) \oplus SO(k)$. This subalgebra of $\mathcal{L}(D_{nk})$ consists of the following matrices

$$\begin{pmatrix} K_0 & \vdots & V^T & \vdots & K_1 \\ \cdots & \cdots & \cdots & \cdots & \cdots \\ V & \vdots & K & \vdots & V \\ \cdots & \cdots & \cdots & \cdots & \cdots \\ K_1 & \vdots & V^T & \vdots & K_0 \end{pmatrix},$$

The structure of a Lie algebra of a bisymmetric space (2.9) is also given by $K_0, K \in H_0$, $V \in E_1$, $K_1 \in E_2$.

The corresponding bisymmetric triple of Lie algebras is

$$SO(n, k) \oplus SO(k) \supset SO(n) \oplus SO(k) \oplus SO(k) \supset SO(n) \oplus SO(k),$$

where the last $SO(k)$ is embedded in the tensor sum $SO(k) \oplus SO(k)$ as the diagonal.

An geometric interpretation is the following. The bisymmetric space $D_-(D_{nk})$ is the variety of all pairs $(\tilde{V}, \tilde{K})$ where $\tilde{V}$ is a k-dimensional plane in $(n + k)$-dimensional pseudo-Euclidean space of type (n, k) such that the induced metric on $\tilde{V}$ has to be positively defined, and $\tilde{K}$ is an orthogonal basis in this plane. The group acting on this variety in natural way is $SO(n, k) \oplus SO(k)$.

The bounded domain (2.16) on the space of D_{nk}, which is the space of $(n \times k)$-matrices, is given by

$$V^T V < E. \tag{2.19}$$

This means that eigenvalues of the square matrix $V^T V$ of order k less then 1.

References

[1]   J. Tits, Une classe d'algèbres de Lie en relation avec les algèbres de Jordan, Indag. Math. **24** (1962), 530–535.

[2]   I. L. Kantor, Classification of irreducible transitive differential groups, Dokl. Akad. Nauk SSSR **158** (1964), 1271–1274.

[3]   I. L. Kantor, Transitive-differential groups and invariant connections on homogeneous spaces, Trudy Sem. Vektor. Tenzor. Anal. **13** (1966), 310–398.

[4]   M. Koecher, Imbedding of Jordan algebras on Lie algebras I, Amer. J. Math. **89** (1967), 787–875.

[5]   K. Meyberg, Lectures on algebras and triple systems, Lecture Notes, University of Virginia, Charlottesville 1972.

[6]   S. Helgason, Differential geometry and symmetric spaces, Academic Press, New York 1962.

[7]   I. L. Kantor, A. I. Sirota, A. S. Solodovnikov, Homogeneous open domains on homogeneous spaces, Abstracts of Moscow International Mathematical Congress, Sect. 9, p. 36, Moscow 1966.

[8]   I. L. Kantor, A. I. Sirota, A. S. Solodovnikov, A class of symmetric spaces with extendable group of motions and a generalization of the Poincaré model, Dokl. Akad. Nauk SSSR **173** (1967), 511–514; Soviet Math. Dokl. **8** (1967), 423–426.

[9]   M. Koecher, Imbedding of Jordan algebras into Lie algebras II, Amer. J. Math. **90** (1968), 476–510.

[10]   O. Loos, Jordan triple systems, R-spaces and bounded symmetric domains, Bull. Amer. Math. Soc. **77** (1971), 558–561.

[11]   O. Loos, Jordan pairs and bounded symmetric domains, University of California, 1977.

[12]   I. Satake, Algebraic structures of symmetric domains, Princeton Univ. Press, Princeton 1980.

[13]   I. L. Kantor, Some generalizations of Jordan algebras, Trudy Sem. Vektor Tenzor Anal. **16** (1972), 407–499.

[14]   I. L. Kantor, Models of exceptional Lie algebras, Dokl. Akad. Nauk SSSR **108** (1973), 1276–1279; Soviet Math. Dokl. **14** (1973), 254–258.

[15]   H. Asano, K. Yamaguti, A construction of Lie by generalized Jordan triple systems of second order, Proc. Kon. Nederl. Akad. Wetensch. **83** (1980), 249–253.

[16]   K. Yamaguti, Constructions of Lie (super)algebras from Freudenthal–Kantor triple system $U(\epsilon, \delta)$, in: Group theoretical methods in physics, Proc. 14th Internat. Colloq., Seoul/Korea, World Scientific, 1986, 222–225.

[17]   K. Yamaguti, A. Ono, On representations of Freudenthal–Kantor triple system $U(\epsilon, \delta)$, Bull. Fac. Sch. Educ. Hiroshima Univ. **7** (2) (1986), 43–51.

[18]   Y. Kakiichi, Another construction of Lie (super)algebras by associative triple systems and Freudenthal–Kantor (super) triple systems, Internat. symp. on nonassociative algebras and related topics, World Scientific, Singapore 1991, 59–64.

[19]   N. Kamiya, A structure theory of Freudenthal–Kantor triple systems, J. Algebra **110** (1987), 108–123.

[20]   N. Kamiya, On (ϵ, δ)-Freudenthal–Kantor systems, Internat. symposium on nonassociative algebras and related topics, World Scientific, Singapore 1991, 65–76.

[21]   I. L. Kantor, A. I. Sirota, A. S. Solodovnikov, Bisymmetric Riemannian spaces, Izv. Ross. Akad. Nauk Ser. Mat. **59** (1995), 85–92; Russian Acad. Sci. Izv. Math. **59** (1995), 963–970.

[22]   S. I. Araki, On root systems and infinitesimal classification of irreducible symmetric spaces. J. Math. Osaka City Univ. **13** (1962), 1–34.

[23]   D. Montgomery, Simply connected homogeneous spaces, Proc. Amer. Math. Soc. **1** (1950), 467–469.

[24]   I. L. Kantor, Formulas for infinitesimal operators of Lie algebra of homogeneous space, Trudy Sem. Vektor Tenzor Anal. **17** (1974), 243–249.

Department of Mathematics
University of Lund
Box 118
S-22100 Lund
Sweden

Representation theory of Lie algebras of Cartan type

*Daniel K. Nakano**

1. Introduction

1.1. Let k be an algebraically closed field of characteristic $p > 0$. Block and Wilson [**BW**] have shown that all restricted simple Lie algebras over k are either classical or of Cartan type provided that the characteristic of k is larger than 7. The classical Lie algebras arise naturally as the Lie algebra of a reductive connected algebraic group. These algebras can also be realized over fields of modular characteristic by taking the complex semisimple Lie algebras with generators and relations over $\mathbb{Z}$ and tensoring by k. The Lie correspondence between reductive groups and classical Lie algebras allows one to pass between the representation theory of algebraic groups and the representation theory of restricted enveloping algebras of classical Lie algebras. In recent times, a general scheme-theoretic setting, which encapsulates this correspondence when $r = 1$, has been used to relate the representation theory of the algebraic group G with the representation theories of the infinitesimal Frobenius kernels G_r. The central results of this theory have been well-documented by Jantzen [**Jan2**].

On the other hand, the Lie algebras of Cartan type do not arise as the Lie algebra of any algebraic group. These simple algebras can be realized by considering the derivation algebra of a truncated polynomial ring and looking at certain subalgebras which stabilize a given differential form. There are four infinite classes of Lie algebras of Cartan type denoted by W, S, H and K. A detailed construction of these algebras can be found in [**BW**], [**SF**]. We will provide a brief description of the algebras and the automorphism groups in the following section. The representation theory for Lie algebras of Cartan type is not as well understood as their classical counterparts. Early results in this area focused primarily on the structure of the center of the universal enveloping algebra [**Er1–6**], [**Pan1–4**], and the description of the irreducible representations of the algebras [**Ch**], [**Gr1–2**], [**Kor1–4**], [**Kry1–2**], [**Mil1–3**], [**St1–3**]. Although some of the methods introduced in these early works were successful for low rank algebras, there seemed to be no systematic way to treat the general representation theory of these algebras like the representation theory of reductive groups. The purpose of this paper is provide an exposition of work done on the representation theory of Lie algebras of Cartan type in the last ten years. It will be shown that there is a natural setting where the representations of these algebras can be viewed systematically as in the classical theory. Moreover,

* Research of the author partially supported by NSF grant 9500715 and Utah St. University research grant 84968

we show that in this framework there are connections to the significant problems in the representation theory of reductive algebraic groups.

1.2.   This paper will deal primarily with restricted representations for the restricted Lie algebras of Cartan type. This is equivalent to studying the module theory for the finite-dimensional restricted enveloping $u(\mathfrak{g})$ where $\mathfrak{g}$ is the Lie algebra of Cartan type. In the following section we will construct "hyperalgebras" $D(G_r)u(\mathfrak{g})$ for Lie algebras of Cartan type. The case when $r = 1$ basically coincides with the restricted enveloping algebra $u(\mathfrak{g})$. Most of the results will be stated in this generality. In several sections some results about non-restricted representations for the reduced enveloping algebra $u(\mathfrak{g}, \chi)$ where χ is a linear functional in the dual of $\mathfrak{g}$ will be noted.

The paper is organized as follows. In the next section we review the construction of the algebras and their automorphism groups. With this information we will show how to construct Hopf algebras whose representation theory encompasses the representation theory of $u(\mathfrak{g})$. The representation theory of reductive groups is used extensively in the next two sections. In Section 3, the irreducible representations for these Hopf algebras are described. The dimensions and characters of these representations can be determined up to knowing the validity of the Lusztig conjecture. The fourth section is devoted to examining the block theory and structure of projectives modules for these Hopf algebras. Once again it will be shown that this information is completely determined up to the validity of the Lusztig conjecture. Finally we survey some recent results on the cohomology and support varieties for these algebras.

The author would like to thank Joseph Ferrar and Thomas Gregory for the invitation to speak at the Monster/Lie algebra conference and for their hospitality while visiting Ohio State University.

2. Algebras and automorphisms

2.1.   Let $A(m, n)$ where $n = (n_1, n_2, \ldots, n_m)$ be the commutative associative algebra with basis $\{X^\alpha : 0 \le \alpha_i \le p^{n_i} - 1, i = 1, 2, \ldots, m\}$. The multiplication is given by $X^\alpha X^\beta = \binom{\alpha+\beta}{\alpha} X^{\alpha+\beta}$. Let ϵ_j be the m-tuple with 1 in the jth position and 0 elsewhere, and let D_i be the derivation which takes X^α to $X^{\alpha-\epsilon_i}$ for $i = 1, 2, \ldots, m$. The Lie algebra $W(m, n)$ of type W consists of all continuous divided power derivations of $A(m, n)$. A basis for $W(m, n)$ is given by $\{X^\alpha D_i : 0 \le \alpha_i \le p^{n_i} - 1, i = 1, 2, \ldots, m\}$, with Lie multiplication

$$[X^\alpha D_i, X^\beta D_j] = X^\alpha D_i(X^\beta)D_j - X^\beta D_j(X^\alpha)D_i.$$

Set $R = \mathrm{Hom}_{A(m,n)}(W(m, n), A(m, n))$ and let $d : A(m, n) \to R$ be the map given by evaluation. Since $da = \sum_{i=1}^{m}(D_i a)dx_i$, it follows that R is a free $A(m, n)$-module with base $\{dx_1, dx_2, \ldots, dx_m\}$ dual to $\{D_1, \ldots, D_m\}$.

Let $\Omega(m)$ be the exterior algebra of R over $A(m, n)$. The action of $W(m, n)$ on R is given by $D.f = D \circ f - f \circ (\mathrm{ad}\, D)$ for $D \in W(m, n)$ and $f \in R$. Furthermore, the action of $W(m, n)$ can be extended to $\Omega(m)$ by letting $D.(\alpha \wedge \beta) = D.\alpha \wedge \beta + \alpha \wedge D.\beta$.

The other infinite families of Lie algebras of Cartan type are defined by considering the action of $W(m, \boldsymbol{n})$ on the differential forms:

$$\omega_S = dx_1 \wedge dx_2 \wedge \cdots \wedge dx_m,$$

$$\omega_H = \sum_{i=1}^{r} dx_i \wedge dx_{i+r} \quad \text{for} \quad m = 2r, \quad \text{and}$$

$$\omega_K = dx_{2r+1} + \sum_{i=1}^{r} (x_i dx_{i+r} - x_{i+r} dx_i) \quad \text{for} \quad m = 2r + 1.$$

The Lie algebras of type S, CS, H, CH and K are subalgebras of a Lie algebra of type W defined by

$$S(m, \boldsymbol{n}) = \{ D \in W(m, \boldsymbol{n}) : D.\omega_S = 0 \},$$

$$CS(m, \boldsymbol{n}) = \{ D \in W(m, \boldsymbol{n}) : D.\omega_S \in k.\omega_S \},$$

$$H(2r, \boldsymbol{n}) = \{ D \in W(m, \boldsymbol{n}) : D.\omega_H = 0 \}, \quad (m = 2r)$$

$$CH(2r, \boldsymbol{n}) = \{ D \in W(m, \boldsymbol{n}) : D.\omega_H \in k.\omega_H \}, \quad (m = 2r)$$

$$K(2r + 1, \boldsymbol{n}) = \{ D \in W(m, \boldsymbol{n}) : D.\omega_K \in A(m, \boldsymbol{n}) \cdot \omega_K \}, \quad (m = 2r + 1).$$

There is a natural grading on each of the Lie algebras induced by a grading on $A(m, \boldsymbol{n})$. For $\mathfrak{g}$ of types W, S, CS, H and CH, $\mathfrak{g} = \mathfrak{g}_{-1} \oplus \mathfrak{g}_0 \oplus \mathfrak{g}_1 \oplus \cdots \oplus \mathfrak{g}_s$. The $\mathfrak{g}_0$ component is a classical Lie algebra and described below in each case:

$$\mathfrak{g}_0 \cong \begin{cases} \mathfrak{gl}_m(k) & \mathfrak{g} \cong W(m, \boldsymbol{n}), CS(m, \boldsymbol{n}) \\ \mathfrak{sl}_m(k) & \mathfrak{g} \cong S(m, \boldsymbol{n}) \\ \mathfrak{csp}_{2r}(k) & \mathfrak{g} \cong CH(2r, \boldsymbol{n}) \\ \mathfrak{sp}_{2r}(k) & \mathfrak{g} \cong H(2r, \boldsymbol{n}). \end{cases}$$

We remark that the $\mathfrak{g}_{-1}$ is the natural representation of $\mathfrak{g}_0$. For type K, $\mathfrak{g} = \mathfrak{g}_{-2} \oplus \mathfrak{g}_{-1} \oplus \mathfrak{g}_0 \oplus \mathfrak{g}_1 \oplus \cdots \oplus \mathfrak{g}_s$ where $\mathfrak{g}_0 = \mathfrak{csp}_{2r}(k)$ where $m = 2r + 1$. For the remainder of this paper, set $\mathfrak{n}^- = \oplus_{i<0} \mathfrak{g}_i$, $\mathfrak{n}^+ = \oplus_{i>0} \mathfrak{g}_i$, and $\mathfrak{b}^\pm = \mathfrak{g}_0 \oplus \mathfrak{n}^\pm$.

2.2. Let $\mathfrak{g} = X(m, \boldsymbol{n})$ generically be a Lie algebra of Cartan type where X is W, S, H or K, and $CX(m, \boldsymbol{n})$ be the corresponding algebras in types S and H. If $p > 5$, then $X(m, \mathbf{1})^{[2]}$ yields a restricted simple Lie algebra. For the sake of simplicity, we will assume that $\mathfrak{g} = X(m, \mathbf{1})^{[2]}$. Many of the results to follow also hold in the general context when $\boldsymbol{n} \neq \mathbf{1}$. Results along these lines will be indicated to the reader.

Let $G = \mathrm{Aut}(X(m, \boldsymbol{n})^{[2]})$ be the automorphism group of $\mathfrak{g}$. Jacobson calculated G when $\mathfrak{g} \cong W(m, \mathbf{1})$, and Wilson [W1] subsequently extended this result to the case when $\mathfrak{g} = W(m, \boldsymbol{n})$. This was accomplished by using natural identification given by Ree

$$\mathrm{Aut}(W(m, \boldsymbol{n})) \cong \{ g \in \mathrm{Aut}(A(m, \boldsymbol{n})) : gDg^{-1} \in W(m, \boldsymbol{n}) \text{ for all } D \in W(m, \boldsymbol{n}) \}$$

and knowledge about $\mathrm{Aut}(A(m, \boldsymbol{n}))$ to calculate the structure of the group. In later work, Wilson [W2] calculated G for X of type S, CS, H, CH, and K. In general

$G \cong G_0 \ltimes U$ where U is a unipotent algebraic group and

$$G_0 \cong \begin{cases} \mathrm{GL}_m(k) & X \text{ is of type } W, S, CS \\ CSP_{2r}(k) & X \text{ is of type } CH, H, K. \end{cases}$$

In all these cases Lie $G \subseteq \mathfrak{b}^+$ and Lie $G \cap \mathfrak{b}^+ = \mathfrak{b}^+$. For X of type W, CS, CH, and K, we have Lie $G = \mathfrak{b}^+$. Even though the Lie algebra of the automorphism group is not all of $\mathfrak{g}$, we can still construct interesting Hopf algebras associated with the representation theory of Lie algebras of Cartan type. These constructions will be shown in the following subsection.

2.3.   We now consider $G = \mathrm{Aut}(\mathfrak{g})$, $\mathfrak{g} = X(m, \mathbf{1})^{[2]}$ as an affine algebraic group scheme over k. For a closed subgroup scheme H of G, let $D(H)$ be the distribution algebra of H [**Jan2**, I 7.7]. If $\mathfrak{l}$ is a Lie subalgebra of $\mathfrak{g}$ which is invariant under the action of H, then by [**LN1**, Thm 2.10] there exists a cocommutative Hopf algebra $D(H)u(\mathfrak{l})$ which contains $D(H)$ as a Hopf subalgebra and $u(\mathfrak{l})$ as normal Hopf subalgebra. The algebra $D(H)u(\mathfrak{l})$ is also generated by $D(H)$ and $u(\mathfrak{l})$. We first apply this construction by setting $G = H$ and $\mathfrak{l} = \mathfrak{g}$. This yields an infinite-dimensional cocommutative Hopf algebra $A = D(G)u(\mathfrak{g})$ containing $u(\mathfrak{g})$ as a normal Hopf subalgebra. This is reminiscent of the familiar setting for reductive groups. The restricted enveloping algebra of a classical Lie algebra is a normal Hopf subalgebra of the distribution algebra of the corresponding reductive algebraic group.

Now let $F : G \to G$ be the Frobenius morphism of group schemes and F^r be the r th iteration of the Frobenius morphism. Set $H = G_r = \ker F^r$ and $\mathfrak{l} = \mathfrak{g}$. Since $D(G_r)$ is finite-dimensional, it follows that $A_r = D(G_r)u(\mathfrak{g})$ is a finite-dimensional cocommutative Hopf algebra. Moreover, $u(\mathfrak{g}) \subseteq A_1$ with equality holding for X of type W, CS, CH and K. Furthermore,

$$A_1 \subset A_2 \subset A_3 \subset \cdots \subset A.$$

The algebra A_r can be regarded as "hyperalgebra" for Lie algebras of Cartan type analogous to the ones for associated to reductive groups by taking the distribution algebra of the r th Frobenius kernel. A graded version of this construction also exists. Let T be a maximal torus for G_0 and $H = G_r T = (F^r)^{-1}(T)$ and $\mathfrak{l} = \mathfrak{g}$. Then the algebra $D(G_r T)u(\mathfrak{g})$ is a graded version of the algebra $D(G_r)u(\mathfrak{g})$. The advantage to having this construction is the ability to distinguish weights and has its origins from work of Jantzen [**Jan1**].

3. Irreducible representations

3.1.   One of the earliest results on the representation theory of Cartan type Lie algebras was due to Chang in 1941 [**Ch**]. He was able to completely determine the finite-dimensional irreducible representations for the toral rank one algebra $W(1, \mathbf{1})$. Subsequent work has been done by Koreshkov [**Kor1–4**] on determining the irreducible representations for other classes of Cartan type Lie algebras. Strade [**St1**] later provided another approach to determining irreducible representations for $W(1, \mathbf{1})$. This section will be devoted to describing all the restricted irreducible representations of $X(m, 1)$ where X is of type W, S, CS, CH, H and K. The restricted irreducible representations for types W, S and H were first determined by Shen [Sh1–3] by constructing a "mixed product". Later, Holmes and the author [N1] were able to recover Shen's results for type W, by concretely realizing the action on certain induced modules. Holmes [Ho1–3] later used some of these methods to determine all the restricted irreducible representations for $K(m, 1)$. Independently, Hu [Hu1–2] also has results along these lines for type K. We should remark that the determination of the restricted irreducible representations is given up to knowing the irreducible representations for $u(\mathfrak{g}_0)$ where $\mathfrak{g}_0$ is the classical zero component of $\mathfrak{g}$. Determining the irreducible representations for $u(\mathfrak{g}_0)$ is still an outstanding problem in the representation theory of reductive groups. There is a famous conjecture due to Lusztig which provides the character of these representations [Jan2, II 7.20].

3.2. We will now shift to a module theoretic viewpoint and use the term simple module in place of irreducible representation. Let T be a maximal torus for G_0 and $X_1(T)$ be the set of restricted weights. The simple $u(\mathfrak{g})$ and $u(\mathfrak{g}_0)$-modules are indexed by weights in $X_1(T)$. For each $\lambda \in X_1(T)$, let $\mathcal{L}_1(\lambda)$ (resp. $L_1(\lambda)$) be the corresponding simple $u(\mathfrak{g})$-module (resp. $u(\mathfrak{g}_0)$-module). The reader is referred to [Jan2, II 3] for this correspondence. Set

$$\mathcal{V}^+_{\mathrm{irr},1}(\lambda) = \mathrm{coind}^{u(\mathfrak{g})}_{u(\mathfrak{b}^+)} L_1(\lambda) = u(\mathfrak{g}) \otimes_{u(\mathfrak{b}^+)} L_1(\lambda).$$

For type W, S, CS, CH and H, we have $\mathcal{V}^+_{\mathrm{irr},1}(\lambda) \cong A(m, 1) \otimes L_1(\lambda)$ as $u(\mathfrak{g}_0)$-modules. Let $e_{\beta,k} = X^\beta D_k \in W(m, 1)$ and $e_{i,j} = X^{\epsilon_i} D_j \in \mathfrak{gl}_n(k)$. In [N1, Prop. 2.2.3–2.2.4], it was shown that the action of $W(m, 1)$ on $\mathcal{V}^+_{\mathrm{irr},1}(\lambda)$ is explicitly given by the following formula:

$$e_{\beta,k}.(X^\alpha \otimes v) = D_k(X^\alpha X^\beta) \otimes v + \sum_{j=1}^m X^\alpha D_j(X^\beta) \otimes e_{j,k}.v.$$

This idea is also employed by Shen in his "mixed product" construction.

Let $\lambda \in X_1(T)$. The weight λ is said to be a fundamental weight if λ is of the form $\lambda_0 = (0, 0, \ldots, 0)$ or $\lambda_t = \sum_{j=1}^t \epsilon_j$ where $1 \le t \le m$ for types W and S, and $1 \le t \le r$ for type H. For $X(m, 1)$ of type W, S and H, let the set of exceptional weights Λ be the set of fundamental weights. For type K, let $\sigma^\pm = -\sum_{i=1}^s \epsilon_i + (\pm(r+1-k) - r - 1)\epsilon_{r+1}$ for $0 \le s \le r$ where $m = 2r + 1$. The set of exceptional weights in this case is $\{\sigma^\pm : 0 \le s \le r\}$. The concrete realization of the action of $\mathfrak{g}$ on $\mathcal{V}^+_{\mathrm{irr},1}(\lambda)$ enables one to show that $\mathcal{V}^+_{\mathrm{irr},1}(\lambda)$ is a simple module when $\lambda \notin \Lambda$ [Sh, Prop. 1.2], [N1, Thm. 2.3.3], [Ho1, 1.2 Thm.].

Theorem. *Let $\mathfrak{g} = X(m, 1)$ be of type W, S, H or K.*

(a) *If $\lambda \notin \Lambda$, then $\mathcal{V}^+_{\mathrm{irr},1}(\lambda) \cong \mathcal{L}_1(\lambda)$.*

(b) *In this situation, $\dim_k \mathcal{L}_1(\lambda) = p^m \dim_k L_1(\lambda)$.*

3.3. The modules $\mathcal{V}^+_{irr,1}(\lambda)$ for $\lambda \in \Lambda$ turn out not to be simple. The $u(\mathfrak{g}_0)$-module $L_1(\lambda)$ for $\lambda \in \Lambda$ has a nice description. For type W and S, $L_1(\lambda_i) \cong \Lambda^i(V)$ for $i = 0, 1, \ldots, m$, where V is the natural representation. Let $\langle v_i : i = 1, 2, \ldots, m \rangle$ be a basis for V. A typical element of $\Lambda^i(V)$ will be a linear combination of elements of the form $\sigma = v_{j_1} \wedge v_{j_2} \wedge \cdots \wedge v_{j_i}$. There exists a natural $W(m, \mathbf{1})$-homomorphism δ_i from $\mathcal{V}^+_{irr,1}(\lambda_i)$ to $\mathcal{V}^+_{irr,1}(\lambda_{i+1})$ defined by

$$\delta_i(X^\alpha \otimes \sigma) = \sum_{i=1}^m D_i(X^\alpha) \otimes (\sigma \wedge v_i).$$

The reader might recognize this as being the "total derivative" when one identifies $\mathcal{V}^+_{irr,1}(\lambda_i)$ with the $A(m, \mathbf{1})$-module of differential forms in $\Omega(m)$ of degree i.

For type W, $\delta_i(\mathcal{V}^+_{irr,1}(\lambda_i))$ is an simple submodule of $\mathcal{V}^+_{irr,1}(\lambda_{i+1})$ (see [**Sh3**, Prop. 2.1] and [**N1**, Thm. 2.5.3]). An explicit basis for this module can be found in [**N1**, Prop. 2.5.1]. For types S and H the submodule structure of $\mathcal{V}^+_{irr,1}(\lambda_i)$ is given in [**Sh3**, §2]. These results are recorded in the following theorem.

Theorem [A]. *Let $\mathfrak{g} = X(m, \mathbf{1})$ where X is of type W, S or H.*
(a) $\dim_k \mathcal{L}_1(\lambda_0) = 1$.
(b) *For type W,* $\dim_k \mathcal{L}_1(\lambda_i) = \binom{m-1}{i}(p^m - 1)$, $i = 1, 2, \ldots, m$.
(c) *For type S,* $\dim_k \mathcal{L}_1(\lambda_i) = \binom{m-1}{i}(p^m - 1) - \delta_{i,1}$, $i = 1, 2, \ldots, m$.
(d) *For type H,* $\dim_k \mathcal{L}_1(\lambda_i) = \left[\binom{m-2}{i-1} - \binom{m-1}{i-1}\right](p^m - 1)$, $i = 1, 2, \ldots, r$.

Holmes [**Ho1–3**] was able to later compute the dimensions of the simple modules for type K corresponding to exceptional weights. His results are stated below.

Theorem [B]. *Let $\mathfrak{g} = K(m, \mathbf{1})$, $m = 2r + 1$.*
(a) $\dim \mathcal{L}_1(\sigma_0^+) = 1$.
(b) *For $0 < s \le r$,* $\dim_k \mathcal{L}_1(\sigma_s^+) = \left[\binom{m-1}{s-1} - \binom{m-1}{s-2}\right]p^m - \binom{m-1}{s-1}$.
(c) *For $0 \le s \le r$,* $\dim_k \mathcal{L}_1(\sigma_s^-) = \left[\binom{m-1}{s} - \binom{m-1}{s-1}\right]p^m - \binom{m-1}{s}$.

According to Theorem 3.2 and Theorem 3.3, the simple $u(\mathfrak{g})$-modules can be determined once the simple modules for reductive groups of types A and C are known. In fact, the converse is also true, so knowing the simple modules for $u(\mathfrak{g})$ is equivalent to knowing the simple modules for reductive groups of types A and C (see [**LN2**, Thm. 4.7] for more details).

3.4. The simple $D(G_r)u(\mathfrak{g})$ and $D((G_0)_r)$-modules are indexed by the set of p^r-restricted weights $X_r(T)$. Let $\mathcal{L}_r(\lambda)$ (resp. $L_r(\lambda)$) be the simple $D(G_r)u(\mathfrak{g})$ (resp. $D((G_0)_r)$)-module corresponding to the weight $\lambda \in X_r(T)$. The following result [**LN2**, Prop. 2.6–2.7] shows how to obtain the simple $D(G_r)u(\mathfrak{g})$-modules from simple $D(G_1)u(\mathfrak{g})$ and simple $D((G_0)_1)$-modules.

Theorem. *Let $\mathfrak{g} = X(m, \mathbf{1})^{[2]}$ where X is of type W, CS, S, CH, H or K.*

(a) *The simple $D(G_r)u(\mathfrak{g})$-module $\mathcal{L}_r(\lambda)$ lifts to a simple $D(G)u(\mathfrak{g})$-module.*

(b) *Let $\lambda = \lambda_0 + \lambda_1 p + \cdots + \lambda_s p^s \in X_r(T)$ where $\lambda_i \in X_1(T)$ for $i = 0, 1, \ldots, s$. Then there exists an isomorphism of $D(G_r)u(\mathfrak{g})$-modules*

$$\mathcal{L}_r(\lambda) \cong \mathcal{L}_1(\lambda_0) \otimes L_1(\lambda_1)^{(1)} \otimes L_1(\lambda_2)^{(2)} \otimes \cdots \otimes L_1(\lambda_s)^{(s)}.$$

We remark that $D(G_r)u(\mathfrak{g})/u(\mathfrak{g}) \cong D(G^{(1)})$ so $L_1(\lambda_i)^{(i)}$ becomes a $D(G_r)u(\mathfrak{g})$ by first extending the action to $D(G^{(1)})$ by letting $D(U^{(1)})$ act trivially, and then inflating the action to $D(G_r)u(\mathfrak{g})$.

4. Projective modules and the Cartan matrix

4.1. Let B be a finite-dimensional algebra over k. If S is a simple B-module then let $[M : S]$ be the number of times S appears as a composition factor in the finite-dimensional B-module M. Furthermore, let $[M]$ denote the formal sum of composition factors in the Grothendieck group of B. If X is a set which parametrizes the set of simple B-modules, then X also parametrizes the set of projective indecomposable modules. This correspondence is given by sending a projective indecomposable module to the simple module appearing in its head. In this case we say that this projective indecomposable module is the projective cover of the corresponding simple module. The computation of the composition factors of the projective covers of simple modules over a finite-dimensional algebra is often a difficult, yet central problem. This section will be devoted to the solution of this problem for $D(G_r)u(\mathfrak{g})$.

Let $A = D(G_r)u(\mathfrak{g})$ and $X = X_r(T)$. For $\lambda \in X$, let $\mathcal{P}_r(\lambda)$ be the projective cover of $\mathcal{L}_r(\lambda)$. Moreover, let B_0^- be a Borel subgroup of G_0 relative to the maximal torus T, B_0^+ be the opposite Borel subgroup, and $U_0^\pm$ be the unipotent radical of $B_0^\pm$. We will also let $U_{\mathfrak{g}}^+ = U_0^+ \ltimes U$. Now set $A_0 = D(T)$, $A^- = D((U_0^-)_r)u(\mathfrak{n}^-)$, and $A^+ = D((U_{\mathfrak{g}}^+)_r)$. If $\lambda \in X$, then set

$$\mathcal{Z}_r^-(\lambda) = \mathrm{ind}_{A_0 A^-}^A \lambda, \quad \mathcal{Z}_r^+(\lambda) = \mathrm{coind}_{A_0 A^-}^A \lambda.$$

The character $\lambda \in X$ is regarded as a one-dimensional module for $A_0 A^\pm$ by letting $A^\pm$ act trivially. The following theorem from [**HolN1**, Thm. 4.4–4.5] allows us to gain some information on the composition factors of $\mathcal{P}_r(\lambda)$.

Theorem [A]. *Let $\lambda \in X_r(T)$.*

(a) *The projective $D(G_r)u(\mathfrak{g})$-module $\mathcal{P}_r(\lambda)$ admits a filtration with factors of the form $\mathcal{Z}_r^-(\mu)$ where $\mu \in X_r(T)$.*

(b) *The number of times $\mathcal{Z}_r^-(\mu)$ appears as a factor in $\mathcal{P}_r(\lambda)$ is independent of the choice of such a filtration. This number will be denoted by $[\mathcal{P}_r(\lambda) : \mathcal{Z}_r^-(\mu)]$.*

(c) $[\mathcal{P}_r(\lambda) : \mathcal{Z}_r^-(\mu)] = [\mathcal{Z}_r^+(\mu) : \mathcal{L}(\lambda)].$

This theorem is motivated by Humphreys' results [**Hum**] for the restricted enveloping algebra of a classical Lie algebra and Jantzen's [**Jan1**] result for the hyperalgebras of a reductive group. In the classical situation the composition factors of the two types of induced modules coincide. However, because of the high degree of asymmetry in the graded decomposition of Cartan type Lie algebras, the two families of induced modules are not the same. The preceding result can also be proved in the context of $D(G_r T)u(\mathfrak{g})$-modules. This version can be found in [**LN2**, Prop. 4.1] and has also been proved in [**Chi3**, Thm. 5.1].

Let $\lambda \in X_r(T)$ and let $P_r(\lambda)$ be the projective cover of the simple $D((G_0)_r)$-module $L_r(\lambda)$. Recall that $\mathcal{V}^+_{\mathrm{irr},r}(\lambda) = \mathrm{coind}_{D(G_r)}^{D(G_r)u(\mathfrak{g})} L_r(\lambda)$.

Now set $\mathcal{V}^-_{\mathrm{proj},r}(\lambda) = \mathrm{ind}_{D((G_0)_r)u(\mathfrak{n}^-)}^{D(G_r)u(\mathfrak{g})} P_r(\lambda)$. There is another variation of Theorem 4.1a given below by using filtration of these generalized Verma modules [**HN1**] [**N1**, Thm. 1.3.5, 1.3.7].

Theorem [B]. *Let* $\lambda \in X_r(T)$.

(a) *The projective* $D(G_r)u(\mathfrak{g})$-*module* $\mathcal{P}_r(\lambda)$ *admits a filtration with factors of the form* $\mathcal{V}^-_{\mathrm{proj},r}(\mu)$ *where* $\mu \in X_r(T)$.

(b) *The number of times* $\mathcal{V}^-_{proj,r}(\mu)$ *appears as a factor in* $\mathcal{P}_r(\lambda)$ *is independent of the choice of such a filtration. This number will be denoted by* $[\mathcal{P}_r(\lambda) : \mathcal{V}^-_{\mathrm{proj},r}(\mu)]$.

(c) $[\mathcal{P}_r(\lambda) : \mathcal{V}^-_{\mathrm{proj},r}(\mu)] = [\mathcal{V}^+_{\mathrm{irr},r}(\mu) : \mathcal{L}(\lambda)]$.

Let $\lambda \in X_r(T)$ and $\lambda = \lambda_0 + \lambda_1 p + \cdots + \lambda_{r-1} p^{r-1}$ where $\lambda_i \in X_1(T)$. If λ_0 is not exceptional then $\mathcal{V}^+_{\mathrm{irr},r}(\lambda) \cong \mathcal{L}_r(\lambda)$. The reciprocity law in Theorem 4.1[B](c) shows that $\mathcal{P}_r(\lambda) = \mathcal{V}^-_{\mathrm{proj},r}(\lambda)$ when λ is not exceptional. Even though the dimensions of the projective covers of the simple modules with exceptional weight are known in general, the question of concretely realizing these modules still remains open.

4.2. Once again let B be a finite-dimensional algebra over k. The algebra B can be decomposed as $B \cong B_1 \oplus B_2 \oplus \cdots \oplus B_s$ where B_i are indecomposable two-sided ideals of B. The B_i are algebras and called the blocks of B. An indecomposable B-module M belongs to a block B_i if and only if $B_i.M \neq 0$. One can easily show that an indecomposable module belongs to precisely one block. The hyperalgebras for a reductive algebraic group have at least two blocks because the r th Steinberg module, $L_r((p^r - 1)\rho)$, is both simple and projective. Surprisingly enough, this is not the case for $D(G_r)u(\mathfrak{g})$.

Theorem. *The algebra* $D(G_r)u(\mathfrak{g})$ *has only one block.*

We will indicate the idea of the proof of Theorem 4.2, by considering the case $D(G_1)u(\mathfrak{g}) \cong u(\mathfrak{g})$ where $\mathfrak{g} = W(1, \mathbf{1})$. A basis for $\mathfrak{g}$ is given by $\{e_i : i = -1, 0, 1, \ldots, p-1\}$ where $e_i = T^{(i+1)}D_1$ for $i = -1, 0, \ldots, p-2$. Note that $\mathfrak{sl}(2) \cong \langle e_{-1}, e_0, e_1 \rangle$. Let $H^+ = u(\langle e_1 \rangle)$, and $\mathfrak{b}^+ = \langle e_j : j = 0, 1, \ldots, p-2 \rangle$. Then

$$[\mathrm{ind}_{A_0 A^-}^{u(\mathfrak{sl}(2))} \lambda] = [\mathrm{ind}_{A_0 H^+}^{u(\mathfrak{sl}(2))} \lambda']$$

for some $\lambda' \in X$. This observation allows us to rewrite $[Z_1^-(\lambda)]$ in terms of $[Z_1^+(\mu)]$ where $\mu \in X$:

$$[Z_1^-(\lambda)] = [\text{ind}_{A_0 A^-}^A \lambda]$$
$$= \left[\text{ind}_{u(\mathfrak{sl}(2))}^A [\text{ind}_{A_0 A^-}^{u(\mathfrak{sl}(2))} \lambda] \right]$$
$$= \left[\text{ind}_{u(\mathfrak{sl}(2))}^A [\text{ind}_{A_0 H^+}^{u(\mathfrak{sl}(2))} \lambda'] \right]$$
$$= [\text{ind}_{A_0 H^+}^A \lambda]$$
$$= \left[\text{ind}_{u(\mathfrak{b}^+)}^A [\text{ind}_{A_0 H^+}^{u(\mathfrak{b}^+)} \lambda'] \right]$$
$$= \sum_{\mu \in X} p^{p-4} [Z_1^+(\mu)].$$

For each $\lambda \in X$ the simple $u(\mathfrak{g})$-module $\mathcal{L}_1(\lambda)$ can be realized as the head of $Z_1^+(\lambda)$. The decomposition above shows that every simple $u(\mathfrak{g})$-module is a composition factor of $Z_1^-(\lambda)$ for a fixed λ. Therefore, by Theorem 4.1a, every projective indecomposable module has every simple $u(\mathfrak{g})$-module occuring as a composition factor, thus $u(W(1, \mathbf{1}))$ has one block.

For Lie algebras of type W (resp. K), one can generalize this argument by locating a copy of $\mathfrak{sl}(m+1)$ (resp. $\mathfrak{sp}(2(r+1))$) and applying the argument given above (see [N1, Prop. 3.2.3, 3.2.5]). Other techniques are necessary to prove the statement for the algebras of types S, CS, H and CH. Details can be found in [HN2] and [LN2].

4.3.   The Cartan matrix of $A = D(G_r)u(\mathfrak{g})$ is the decomposition matrix $C_r = ([\mathcal{P}_r(\lambda), \mathcal{L}_r(\mu)])$. The following result [LN2, Thm. 4.5] shows that this matrix can be factored into a product of five matrices.

Theorem. *Let $\mathfrak{g}$ be a restricted Lie algebra of Cartan type and $s = \frac{\dim_k A^+}{(\dim_k A_0)(\dim_k A^-)}$. Then*

$$C_r = (J_r K_r)^T A (J_r K_r).$$

where $J_r = ([Z_{0,r}^+(\lambda), L_r(\tau)])$, $K_r = ([\mathcal{V}_{\text{irr},r}^+(\tau) : \mathcal{L}_r(\mu)])$, and $A = (s)$ (all entries being s).

We shall illustrate how this theorem can be used by computing the Cartan invariants for $\mathfrak{g} = W(1, \mathbf{1})$.

Example.   Let $\mathfrak{g} = W(1, \mathbf{1})$ and $r = 1$. In this situation $\mathfrak{g}_0 \cong \mathfrak{t}$ where $\mathfrak{t}$ is a maximal torus. The representations over $u(\mathfrak{g}_0)$ are completely reducible, thus $J_1 = I$ where I is a $p \times p$ identity matrix. Therefore, $C_1 = K_1^T A K_1$. Let $X = \{0, 1, \ldots, p-1\}$. Recall that $\mathcal{L}_1(\lambda) \cong \mathcal{Z}_1^+(\lambda)$ for $\lambda \neq 0, 1$ (non-exceptional weights). The module $Z_1^+(0)$ has two composition factors $\mathcal{L}_1(0)$ and $\mathcal{L}_1(1)$ with $\mathcal{L}_1(0)$ appearing in the head of $Z_1^+(0)$. On the other hand, $Z_1^+(1)$ has the same two composition factors, but with $\mathcal{L}_1(1)$

appearing in the head. It follows that

$$
K_1 = \begin{pmatrix}
1 & 1 & 0 & \cdots & 0 \\
1 & 1 & 0 & \cdots & 0 \\
0 & 0 & 1 & \cdots & 0 \\
\vdots & \vdots & & \ddots & \vdots & \vdots \\
0 & 0 & 0 & \cdots & 1
\end{pmatrix}.
$$

We also have $A = (p^{p-4})$, because $\dim_k A^+ = p^{p-2}$, $\dim_k A^- = p$ and $\dim A_0 = p$. Hence,

$$
C_1 = K_1^T A K_1 = p^{p-4} \begin{pmatrix}
4 & 4 & 2 & \cdots & 2 \\
4 & 4 & 2 & \cdots & 2 \\
2 & 2 & 1 & \cdots & 1 \\
\vdots & \vdots & & \ddots & \vdots & \vdots \\
2 & 2 & 2 & \cdots & 1
\end{pmatrix}.
$$

This information also shows that $\dim_k \mathcal{P}_1(\lambda) = p^{p-2}$ for $\lambda \neq 0, 1$. Moreover, $\dim_k \mathcal{P}_1(\lambda) = 2p^{p-2}$ for $\lambda = 0, 1$.

For $r = 1$ and $\mathfrak{g} = W(2, \mathbf{1})$, the Cartan invariants and the dimensions of the projective covers are in [N1, Thm. 3.3.3], for $H(2, \mathbf{1}) = S(2, \mathbf{1})^{[2]}$ in [HN2, Thm. 4.6], and for $K(3, 1)$ (computed by Holmes) in [Ho2, Thm. 6.4]. When $r \geq 2$, these numbers have been calculated for $W(1, \mathbf{1})$ in [LN2, 4.6]. The characters for the simple $GL_2(k)$ and $CSP_4(k)$-modules are known. Therefore, in these cases the matrix J_r can be computed by using the preceding theorem. In fact, the computation of J_r is equivalent to computing the characters of simple $GL_n(k)$ and $CSP_{2r}(k)$-modules (see [LN2, 4.7]). As previously mentioned the solution to this problem is given by a conjecture due to Lusztig [Jan2, II 7.20]. This conjecture has been verified for Andersen, Jantzen and Soergel [AJS] for large primes, but with no effective lower bound on the prime. The matrix K_r is always computable by using the tensor product theorem and the decomposition of the induced modules for exceptional weights. Hence, the Cartan invariants are known up to verifying the Lusztig conjecture. We should also remark that a graded version of the Theorem 4.3 for projective $D(G_r T)$-modules can be found in [LN2, Thm. 4.2].

4.4. Let G_0 be an arbitrary reductive algebraic group scheme and $(G_0)_r$ be the kernel of the rth iteration of the Frobenius map. Moreover, let $X = X_r(T)$ where T is a maximal torus for G_0. In Section 3.4, the simple $(G_0)_r$-modules were denoted by $L_r(\lambda)$, $\lambda \in X$. Let $P_r(\lambda)$ be the projective cover of $L_r(\lambda)$. Ballard proved that any projective module over $(G_0)_r$ lifts to a module over G_0 provided that $p \geq 3(h - 1)$ where h is the Coxeter number associated to the root system of G_0. Jantzen later showed that this lifting property holds as long as $p \geq 2(h - 1)$. Moreover, if this lifting property holds then there is a twisted tensor product theorem which relates the projective modules for $(G_0)_r$ to the projective modules for $(G_0)_1$. Now let $G = \mathrm{Aut}(\mathfrak{g})$ and assume that G_0 is the maximal reductive subgroup of G. The following result is a lifting theorem for projective modules over $D(G_r)u(\mathfrak{g})$ [LN2, Thm. 3.4, 3.6].

Theorem. *Let* $p \geq 4m - 1$ *for* X *of type* W, CS *and* S, *and* $p \geq 6m - 1$ *for* X *of type* H, CH, *and* K.

(a) *For any* $\lambda \in X_r(T)$, *the* $D(G_r)u(\mathfrak{g})$-*module* $\mathcal{P}_r(\lambda)$ *has a unique* $D(G_0)u(\mathfrak{g})$-*structure.*

(b) *Let* $\lambda = \lambda_0 + \lambda_1 p + \lambda_2 p^2 + \cdots \lambda_{r-1} p^{r-1} \in X_r(T)$ *with* $\lambda_i \in X_1(T)$ *for* $i = 0, 1, \ldots, r-1$. *Then there exists the following isomorphism as* $D(G_r)u(\mathfrak{g})$-*modules:*

$$\mathcal{P}_r(\lambda) \cong [\text{ind}_{D((G_0)_r)u(\mathfrak{g})}^{D(G)u(\mathfrak{g})} \mathcal{P}_1(\lambda_0)] \otimes P_1(\lambda_1)^{(1)} \otimes \cdots \otimes P_1(\lambda_{r-1})^{(r-1)}.$$

One natural question that remains open is if any of the projective $D(G_r)u(\mathfrak{g})$-modules admit a $D(G)u(\mathfrak{g})$-structure.

5. Cohomology theory

5.1. We begin this section by describing two types of cohomology: ordinary Lie algebra cohomology and restricted Lie algebra cohomology. Let $U(\mathfrak{g})$ be the ordinary universal enveloping algebra of a Lie algebra $\mathfrak{g}$. The ordinary Lie algebra cohomology $H^j(\mathfrak{g}, M)$ for $j \geq 0$ with coefficients in a $U(\mathfrak{g})$-module M is defined to be $\text{Ext}_{U(\mathfrak{g})}^j(k, M)$ (see [**HS**]). For any $U(\mathfrak{g})$-module M, let M^* denote the dual of the module M. There exists a complex with terms of form $\Lambda^\bullet(\mathfrak{g}^*) \otimes M$ which allows one (in principle) to calculate these cohomology groups. The explicit formula for the differentials in this complex can be found in [**Jan2**, I 9.17]. For Lie algebras of Cartan type, there has been various work done on the computation of ordinary Lie algebra cohomology. The reader is referred to [**Dz1–5**], [**Chi1–2**].

The restricted Lie algebra cohomology for a given $u(\mathfrak{g})$-module M is

$$H^j(u(\mathfrak{g}), M) = \text{Ext}_{u(\mathfrak{g})}^i(k, M).$$

These cohomology groups tend to be more difficult to compute because there is no known explicit complex for calculational purposes and these groups can have infinitely many non-zero terms. There is a relationship between the ordinary Lie algebra cohomology and the restricted Lie algebra cohomology given by a spectral sequence. This spectral sequence is constructed by filtering the cobar resolution on the $u(\mathfrak{g})^*$ by powers of the augmentation ideal and has the form;

$$E_2^{2i,j} = S^i(\mathfrak{g}^*)^{(1)} \otimes H^j(\mathfrak{g}, M) \Rightarrow H^{2i+j}(u(\mathfrak{g}), M).$$

For Lie algebras of Cartan type, if M is a $D(G)u(\mathfrak{g})$-module, then the maps in this spectral sequence are all G-homomorphisms. This additional structure turns out to be useful for the purpose of calculating restricted Lie algebra cohomology.

5.2. The spectral sequence described in the preceding section has both computational as well as conceptual applications. In [**N3**], this spectral sequence was used along with the T-action to compute the cohomology groups $H^\bullet(u(W(1, \mathbf{1})), k)$ for $p = 5, 7$. The obstruction in generalizing this calculation was the computation of $H^\bullet(u(\mathfrak{b}^+), k)$. Dzhumadil'daev has indicated that this calculation will become increasingly difficult as

p gets large. Some of the same techniques also show that there exists a simple non-projective $u(W(1, \mathbf{1}))$-module S such that $H^\bullet(u(\mathfrak{g}), S) = 0$ [CNP, Ex. 4.2]. For finite groups there is no known example where the group cohomology with coefficients in simple non-projective module in the prinicipal block is zero. It would be interesting to see if one can generalize this example for other Lie algebras of Cartan type. Furthermore, it was shown in [CNP, Thm. 4.3], there exists a non-projective indecomposable module M in the principal block such that $H^\bullet(u(\mathfrak{g}), M) = 0$ for any Lie algebra of Cartan type $\mathfrak{g}$.

Some interest has focused on the computation of the first cohomology group with coefficients in a simple $u(\mathfrak{g})$-module. Chui and Shen [ChiS] have computed $H^1(u(\mathfrak{g}), \mathcal{L}_1(\lambda))$ for $\lambda \in X_1(T)$ and $\mathfrak{g} \cong W(1, \mathbf{1})$, $H(2, \mathbf{1})$ and $W(2, \mathbf{1})$. As one would expect, this computation relies heavily on knowing the simple modules for $u(\mathfrak{g}_0)$. Recently, Z. Lin and the author [LN4] have shown that $\mathrm{Ext}^1_{D(G_r)u(\mathfrak{g})}(S, S) = 0$ for any simple $D(G_r)u(\mathfrak{g})$-module S, as long as the prime is slightly larger than the Coxeter number for G_0. The reader should compare this to Andersen's earlier result [And] for reductive groups where he proves that self-extensions of simple modules over the rth Frobenius kernel split, if the characteristic is not 2 and the group does not contain a component of type C.

5.3. Let $\chi \in \mathfrak{g}^*$ and $u(\mathfrak{g}, \chi)$ be the reduced enveloping algebra. If M is a module for $u(\mathfrak{g}, \chi)$, then M^* is a module for $u(\mathfrak{g}, -\chi)$. Moreover, $\widehat{M} = M \otimes M^*$ is a module over $u(\mathfrak{g}, 0) = u(\mathfrak{g})$. The cohomology ring $R = H^{2\bullet}(u(\mathfrak{g}), k)$ is a Noetherian ring, and $\mathrm{Ext}^\bullet_{u(\mathfrak{g})}(\widehat{M}, \widehat{M})$ is a finitely generated module over R. Let J_M be the annihilator in R of $\mathrm{Ext}^\bullet_{u(\mathfrak{g})}(\widehat{M}, \widehat{M})$. Now let

$$|(\mathfrak{g}, \chi)|_M = \mathrm{Spec}(R/J_M).$$

The affine homogeneous variety, $|(\mathfrak{g}, \chi)|_M$, is called the *support variety* of M. The spectral sequence in 5.1 can be used to obtain a concrete realization of $|(\mathfrak{g}, \chi)|_M$. The edge homomorphism of the spectral sequence gives rise to a finite map of rings:

$$\Phi^\bullet : S^\bullet(\mathfrak{g}^*)^{(1)} \to H^{2\bullet}(u(\mathfrak{g}), k).$$

This map induces a map on varieties $\Phi : |(\mathfrak{g}, 0)|_k \to \mathfrak{g}^{(1)}$. Friedlander and Parshall [FP1–2] have shown that

$$\Phi(|(\mathfrak{g}, \chi)|_M) = \{x \in \mathfrak{g} : x^{[p]} = 0, \text{ and } M_{\langle x \rangle} \text{ is not free}\} \cup \{0\}$$

where $\langle x \rangle$ denotes the cyclic subalgebra in $u(\mathfrak{g}, \chi)$ generated by x. The *complexity* of a module $c_{\mathfrak{g}}(M)$ is the vector space rate of growth of the minimal projective resolution of the module. The complexity turns out to be a non-negative integer for any finitely generated $u(\mathfrak{g}, \chi)$-module M and equals the dimension of $|(\mathfrak{g}, \chi)|_M$.

In [LN1], the support varieties for modules over Lie algebras of Cartan type were investigated. A projectivity criterion, similar to Cline, Parshall and Scott's result [CPS] for infinitesimal kernels of reductive algebraic groups, was shown to hold for $D(T)u(\mathfrak{g})$-modules. The complexity and support varieties for all simple $u(W(1, \mathbf{1}))$-modules were also computed [LN1, Thm. 5.4]. This result is stated below.

Theorem. *Let* $\mathfrak{g} = W(1, 1)$. *The complexity and support varieties for the simple* $u(\mathfrak{g})$-*modules are given by*

$$c_\mathfrak{g}(\mathcal{L}_1(\lambda)) = \begin{cases} p - 1 & for \quad \lambda = 0, 1 \\ p - 2 & for \quad \lambda \neq 0, 1, \frac{p-1}{2} \\ p - 3 & for \quad \lambda = \frac{p-1}{2}, \end{cases}$$

$$\Phi(|\ (\mathfrak{g}, 0)\ |_{\mathcal{L}_1(\lambda)}) = \begin{cases} \overline{G \cdot e_{-1}} & for \quad \lambda = 0, 1 \\ ke_1 \oplus ke_2 \oplus \cdots \oplus ke_{p-2} & for \quad \lambda \neq 0, 1, \frac{p-1}{2} \\ ke_2 \oplus ke_3 \oplus \cdots \oplus ke_{p-2} & for \quad \lambda = \frac{p-1}{2}. \end{cases}$$

Farnsteiner [**Fa**, Thm. 5.1] has some results pertaining to the support varieties for non-restricted simple $u(W(1, 1), \chi)$-modules. He uses this information to determine the representation type of $u(W(1, 1)), \chi$ for arbitrary χ.

Let $\mathfrak{g}^\chi = \{x \in \mathfrak{g} : \chi([x, \mathfrak{g}]) = 0\}$. Gottman [**Got**] conjectured that

$$\mathfrak{g}^\chi \cap \Phi(|(\mathfrak{g}, 0)|_k) = \cup_M \Phi(|(\mathfrak{g}, \chi)|_M)$$

where M ranges over all simple $u(\mathfrak{g}, \chi)$-modules. This conjecture has been verified in several cases (see [**Got**, §4]). Recently, Premet has formulated a weaker version of this conjecture for classical Lie algebras [**Pr3**, 1.5] and has shown that this conjecture has some interesting implications.

5.4.   Throughout this paper our focus has been on the representation and cohomology theory of the finite-dimensional Hopf algebras $D(G_r)u(\mathfrak{g})$. A natural question the reader might ask is what can be said about the infinite-dimensional Hopf algebra $D(G)u(\mathfrak{g})$? The simple $D(G)u(\mathfrak{g})$-modules are indexed by the set of dominant integral weights $X(T)_+$ and labeled by $\mathcal{L}(\lambda)$. If $\lambda \in X_r(T)$ then $\mathcal{L}(\lambda) \cong \mathcal{L}_r(\lambda)$ as $D(G_r)u(\mathfrak{g})$-modules. Moreover, if $\lambda = \lambda_0 + \lambda_1 p + \cdots + \lambda^s p^s$, then

$$\mathcal{L}(\lambda) \cong \mathcal{L}_1(\lambda_0)^{(1)} \otimes \mathcal{L}_1(\lambda_1)^{(1)} \otimes \cdots \otimes \mathcal{L}(\lambda_s)^{(s)}.$$

We refer the reader to [**LN2**] for more details.

In [**LN3**], we defined "induced" and "Weyl" modules (objects) for $D(G)u(\mathfrak{g})$. The induced modules, unlike the classical situation, are infinite-dimensional and labeled $\mathcal{H}^0(\lambda)$, $\lambda \in X(T)_+$. The Weyl objects are finite-dimensional $D(G)u(\mathfrak{g})$-modules and are labeled $\mathcal{V}(\lambda)$, $\lambda \in X(T)_+$. With this setup one can define good filtrations for $D(G)u(\mathfrak{g})$-modules. Roughly speaking, a $D(G)u(\mathfrak{g})$-module admits a good filtration if it has a filtration with factors isomorphic to induced modules. Moreover, a necessary and sufficient criterion (motivated from the classical case) for a $D(G)u(\mathfrak{g})$-module to admit a good filtration is given:

Theorem. *Let* $A = D(G)u(\mathfrak{g})$ *and* M *be an* A-*module. Moreover, assume that for all* $\lambda \in X(T)_+$, $\dim_k \mathrm{Hom}_A(\mathcal{V}(\lambda), M) < \infty$ *and the set* $\Lambda(M)$ *with respect to the partial order in* $X(T)_+$ *satisfies the minimal condition (i.e., any nonempty subset has a minimal element).*

(a) *The module M admits a good filtration if and only if* $\mathrm{Ext}^1_A(\mathcal{V}(\lambda), M) = 0$ *for all* $\lambda \in X(T)_+$.

(b) *If M admits a good filtration then* $[M : \mathcal{H}^0(\lambda)] = \dim_k \mathrm{Hom}_A(\mathcal{V}(\lambda), M)$.

An immediate consequence of this theorem is if $\mathcal{Q}(\lambda)$ is the injective hull of $\mathcal{L}(\lambda)$, then $\mathcal{Q}(\lambda)$ admits a good filtration. Furthermore, $[\mathcal{Q}(\lambda) : \mathcal{H}^0(\mu)] = [\mathcal{V}(\mu) : \mathcal{L}(\lambda)]$.

At this point we hope that the reader is convinced the approach presented in this paper opens up various avenues for further work in the subject. We conclude with a few open questions.

(5.4.1) For a classical Lie algebra, the cohomology ring $H^{2\bullet}(u(\mathfrak{g}), k)$ is isomorphic to the coordinate algebra of the nullcone as long as the prime is larger than the Coxeter number. One can also see that $H^{2\bullet}(u(\mathfrak{g}), k)$ has a good filtration relative to classical induced modules $H^0(\lambda)$. Can the same be said of the cohomology ring for the restricted enveloping algebra of Lie algebra of Cartan type?

(5.4.2) Jantzen has introduced a translation principle for comparing blocks of modules over reductive groups and their Frobenius kernels. Is there a suitable translation functor which will induce a self-equivalence of the single block of $D(G_r)u(\mathfrak{g})$?

(5.4.3) In the classical theory there is a theory of tilting modules. Is there an appropriate notion of tilting modules for the category of $D(G)u(\mathfrak{g})$-modules?

(5.4.4) Let G_0 be a reductive group of type A or C. There is a polynomial representation theory for $\mathrm{GL}_n(k)$ and one recently developed by Doty for $CSP_{2r}(k)$ **[D]**. Does a polynomial representation theory exist for a certain subcategory of $D(G)u(\mathfrak{g})$-modules? If so, can one construct finite-dimensional algebras, like Schur algebras, which reflect this polynomial representation theory?

Bibliography

[And] H. H. Andersen, Extensions of modules for algebraic groups, Amer. J. Math. **106** (1984), 489–504.

[AJS] H. H. Andersen, J. C. Jantzen, W. Soergel, Representations of quantum groups at a p th root of unity and of semisimple groups in characteristic p: independence of p, Astérisque, 220, (1994).

[BO] G. M. Benkart, J. M. Osborne, Representations of rank one Lie algebras of characteristic p, in: Lie algebras and related topics, D. Winter (Ed.), Lecture Notes in Math. **933**, Springer-Verlag, 1982, 1–37.

[Bl] R. E. Block, On the Mills–Seligman axioms for Lie algebras of Cartan type, Trans. Amer. Math. Soc. **121** (1966), 378–392.

[CNP] J. F. Carlson, D. K. Nakano, K. M. Peters, On the vanishing of extensions of modules over reduced enveloping algebras, Math Ann. **302** (1995), 541–560.

[Ch] H. J. Chang, Über Wittsche Lieringe, Abh. Math. Sem. Univ. Hamburg **14** (1941), 151–184.

[Chi1] S. Chui, Central extensions and $H^1(L, L^*)$ of the graded Lie algebras of Cartan type, J. Algebra **149** (1992), 46–67.

[Chi2] S. Chui, The cohomology of modular Lie algebras with coefficients in a restricted Verma module, Chin. Ann. Math. Ser. B **14** (1993), 77–84.

[Chi3] S. Chui, Principal indecomposable representations for restricted Lie algebras of Cartan type, J. Algebra **155** (1993), 142–160.

[ChiS] S. Chui, G. Y. Shen, Cohomology of graded Lie algebras of Cartan type of characteristic p, Abh. Math. Sem. Univ. Hamburg **57** (1987), 139–156.

[CPS]   E. Cline, B. J. Parshall, L. L. Scott, On injective modules for infinitesimal algebraic groups, J. London Math. Soc. **31** (1985), 277–291.

[D]   S. R. Doty, Polynomial representations, algebraic monoids, and Schur algebras of classical type, J. Pure Appl. Algebra **123** (1998), 165–199.

[Dz1]   A. S. Dzhumadil'daev, Deformations of general Lie algebras of Cartan type, Dok. Akad. Nauk SSSR **251** (1980), 1289–1292.

[Dz2]   A. S. Dzhumadil'daev, Relative cohomology and deformations of Lie algebras of Cartan type, Dokl. Akad. Nauk SSSR **257** (1981), 1044–1048.

[Dz3]   A. S. Dzhumadil'daev, Cohomology of modular Lie algebras, Math. USSR-Sb. **47** (1984), 127–143.

[Dz4]   A. S. Dzhumadil'daev, Irreducible representations of strongly solvable Lie algebras over a field of positive characterstic, Mat. Sb. **123** (165) (1984), 212–229; Math. USSR-Sb. **51** (1985), 207–223.

[Dz5]   A. S. Dzhumadil'daev, Central extensions and invariant forms of Cartan type Lie algebras of postive characteristic, Funct. Anal. Appl. **18** (1984), 331–332.

[DzK]   A. S. Dzhumadil'daev, A. I. Kostrikin, Deformations of the Lie algebra $W_1(m)$, Trudy Mat. Inst. Steklov **148** (1978), 144–155.

[Er1]   Y. B. Ermolaev, Computation of a central element of the universal enveloping algebra of the Witt algebra, Soviet Math. (Iz. VUZ) **19** (5), (1975).

[Er2]   Y. B. Ermolaev, A formula of exponentiation of a product of two elements of an associative ring, Soviet Math. (Iz. VUZ) **20** (8), (1976).

[Er3]   Y. B. Ermolaev, The minimal polynomial of a central element of the universal enveloping algebra of the Witt algebra, Soviet Math. (Iz. VUZ) **20** (10), (1976), 32–41.

[Er4]   Y. B. Ermolaev, On the central elements of the universal enveloping algebra of the Zassenhaus algebra, Soviet Math. (Iz. VUZ) **22** (6), (1978), 53–65.

[Er5]   Y. B. Ermolaev, On the structure of the center of the enveloping algebra of the Zassenhaus algebra, Soviet Math. (Iz. VUZ) **22** (12), (1978), 29–39.

[Er6]   Y. B. Ermolaev, On irreducible modules with filtration $p > 0$, Soviet Math. (Iz. VUZ) **25** (7) (1981), 93–98.

[Fa]   R. Farnsteiner, Representations of blocks associated to induced modules of restricted Lie algebras, Math. Nachr. **179** (1996), 57–88.

[FP1]   E. M. Friedlander, B .J. Parshall, Support varieties for restricted Lie algebras, Invent. Math. **86** (1986), 553–562.

[FP2]   E. M. Friedlander, B. J. Parshall, Modular representation theory of Lie algebras, Amer. J. Math. **110** (1988), 1055–1094.

[Got]   J. E. Gottman, Characters on restricted Lie algebras Northwestern Univ. Dissertation, (1993).

[Gr1]   A. N. Grishkov, Irreducible representations of completely solvable Lie p-algebras, Algebra and Logic **16** (1977), 186–198.

[Gr2]   A. N. Grishkov, Irreducible representations of completely solvable Lie algebras, Math. Notes **30** (1981), 496–499.

[HS]   P. J. Hilton, U. Stammbach, A course in homological algebra, Grad. Text in Math. **4**, Springer-Verlag, Berlin, Heidelberg, New York 1971.

[Ho1]   R. R. Holmes, Simple restricted modules for the restricted contact Lie algebra, Proc. Amer. Math. Soc. **116** (1992), 329–337.

[Ho2]   R. R. Holmes, Cartan invariants for the restricted toral rank two contact Lie algebra, Indag. Math. (N.S.) **5** (1994), 1–16.

[Ho3]   R. R. Holmes, Dimensions of the simple restricted modules for the restricted contact Lie algebra, J. Algebra **170** (1994), 504–525.

[HN1]   R. R. Holmes, D. K. Nakano, Brauer-type reciprocity for a class of graded associative algebras, J. Algebra **144** (1991), 117–125.

[HN2]   R. R. Holmes, D. K. Nakano, Block degeneracy and Cartan invariants for graded Lie algebras of Cartan type, J. Algebra **161** (1993), 155–170.

[Hu1]   N. Hu, The graded modules for the graded contact Cartan algebra, Comm. Algebra **22** (1994), 4475–4497.

[Hu2]   N. Hu, Irreducible constituents of graded modules for graded contact Lie algebras of Cartan type, Comm. Algebra **22** (1994), 5951–5970.

[Hum]   J. E. Humphreys, Modular representations of classical Lie algebras and semisimple groups, J. Algebra **19** (1971), 51–79.

[Jan1]   J. C. Jantzen, Über Darstellungen höherer Frobenius-Kerne halbeinfacher algebraischer Gruppen, Math Z. **164** (1979), 271–292.

[Jan2]   J. C. Jantzen, Representations of Algebraic Groups, Pure Appl. Math. **131**, Academic Press, Boston 1987.

[KW]   V. Kac, B. Weisfeiler, Coadjoint action of a semisimple algebraic group and the center of the enveloping algebra in characteristic p, Proc. Kon. Nederl. Akad. Wetensch. **79A** (1976), 137–151.

[Kor1]   N. A. Koreshkov, On the irreducible representations of a certain Lie algebra, Soviet Math. (Iz. VUZ) **22** (9) (1978), 38–43.

[Kor2]   N. A. Koreshkov, On the irreducible representations of the Hamiltonian algebra of dimension $p^2 - 2$, Soviet Math. (Iz. VUZ) **22** (10) (1978), 28–34.

[Kor3]   N. A. Koreshkov, On the irreducible representations of a Lie p-algebra W_2, Soviet Math. (Iz. VUZ) **24** (4) (1980), 44–52.

[Kor4]   N. A. Koreshkov, On the irreducible representations of Lie p-algebras W_n, Soviet Math. (Iz. VUZ) **27** (3) (1983), 16.

[Kry1]   Y. S. Krylyuk, Maximum dimension of irreducible representations of simple Lie p-algebras of Cartan series S and H, Math. Sb. **123** (165) (1984), 108–119.

[Kry2]   Y. S. Krylyuk, The index of Cartan type Lie algebras over a field of arbitrary characteristic, Izv. Akad. Nauk SSSR Ser. Mat. **50** (1986), 393–412.

[LN1]   Z. Lin, D. K. Nakano, Algebraic group techniques in the representation and cohomology theory of Lie algebras of Cartan type, J. Algebra **179** (1996), 852–888.

[LN2]   Z. Lin, D. K. Nakano, Representations of Hopf algebras arising from Lie algebras of Cartan type, J. Algebra **189** (1997), 529–567.

[LN3]   Z. Lin, D. K. Nakano, Good filtrations for Hopf algebras arising from Lie algebras of Cartan type, to appear in J. Pure Appl. Algebra.

[LN4]   Z. Lin, D. K. Nakano, Extensions of modules over Hopf algebras arising from Lie algebras of Cartan type, preprint, (1996).

[Mil1] A. A. Milner, The irreducible representations of modular Lie algebras, Math. USSR-Izv. **9** (1975), 1169–1187.

[Mil2] A. A. Milner, The irreducible representations of the Zassenhaus algebra, Uspekhi Mat. Nauk **30** (6) (1975), 178.

[Mil3] A. A. Milner, On maximal degree of the irreducible representations of a Lie algebra of positive characteristic, Funct. Anal. Appl. **14** (1980), 136–137.

[N1]   D. K. Nakano, Projective modules over Lie algebras of Cartan type, Mem. Amer. Math. Soc. **470**, Amer. Math. Soc., Providence 1992.

[N2]   D. K. Nakano, Filtrations for periodic modules over restricted Lie algebras, Indag. Math. **3** (1992), 59–68.

[N3]   D. K. Nakano, On the cohomology of the Witt algebra $W(1, \mathbf{1})$, Proc. II International Conference on non-associative algebra and its applications (ed. S. Gonzales), Kluwer, (1994).

[Pan1] V. V. Panyukov, Centers of the universal enveloping algebras of some Lie algebras, Moscow Univ. Math. Bull. **37** (1982), 21–26.

[Pan2] V. V. Panyukov, The centers of the universal enveloping algebras of nilpotent Lie algebras in characteristic $p > 0$, Moscow Univ. Math. Bull. **36** (1981) (3) 35–41.

[Pan3] V. V. Panyukov, On representations of Lie algebras in positive characteristic, Moscow Univ. Math. Bull. **38** (2) (1983), 64–70.

[Pan4] V. V. Panyukov, Generators and relations of the center of the universal enveloping algebras of certain nilpotent Lie algebras, Moscow Univ. Math. Bull. **39** (1984), 47–53.

[Pr1]   A. A. Premet, Algebraic groups associated with Lie p-algebras of Cartan type, Math. USSR-Sb. **50** (1985), 85–97.

[Pr2]   A. A. Premet, A theorem on the restriction of invariants and nilpotent elements in W_n, Math. USSR-Sb. **73** (1992), 135–159.

[Pr3]   A. A. Premet, Support varieties of non-restricted modules over Lie algebras of reductive groups, preprint, 1996.

[Rud]   A. N. Rudakov, On the representations of classical semisimple Lie algebras in characteristic p, Math. USSR-Izv. **4** (1970), 741–742.

[RS]   A. N. Rudakov, I. R. Shafarevich, On the irreducible representations of a simple three-dimensional Lie algebra over a field of finite characteristic, Math. Notes **2** (1967), 439–454.

[Sch1] J. Schue, Representations of solvable Lie p-algebras, J. Algebra **38** (1976), 253–267.

[Sch2] J. Schue, Representations of Lie p-algebras, in: Lie algebras and related topics, D. Winter (Ed.), Lecture Notes in Math. **933**, Springer-Verlag, 1982, 191–202.

[Sh1]   G. Shen, Graded modules of graded Lie algebras of Cartan type I, Scientia Sinica **29** (1986), 570–581.

[Sh2]   G. Shen, Graded modules of graded Lie algebras of Cartan type II, Scientia Sinica **29** (1986), 1009–1019.

[Sh3]   G. Shen, Graded modules of graded Lie algebras of Cartan type III, Chinese Ann. Math. Ser. B **9** (1988), 404–417.

[St1]   H. Strade, Representations of the Witt algebra, J. Algebra **49** (1977), 595–605.

[St2]   H. Strade, Darstellungen auflösbarer Lie p-Algebren, Math. Ann. **232** (1978), 15–32.

[St3] H. Strade, Zur Darstellungtheorie von Lie-Algebren, Abh. Math. Sem. Hamburg **52** (1982), 66–82.

[SF]   H. Strade, R. Farnsteiner, Modular Lie algebras and their representations, Monographs Textbooks Pure Appl. Math. **116**, Marcel Dekker, New York 1988.

[WK]   B. Weisfeiler, V. G. Kac, The irreducible representations of a Lie p-algebra, Funct. Anal. Appl. **6** (1972), 111–117.

[W1]   R. L. Wilson, Classification of generalized Witt algebras over algebraically closed fields, Trans. Amer. Math. Soc. **153**, (1971), 191–210.

[W2]   R. L. Wilson, Automorphisms of graded Lie algebras of Cartan type, Comm. Algebra **3** (1975), 591–613.

[Yak] N. I. Yakovlev, The center of the enveloping algebra of the Witt algebra, Funct. Anal. Appl. **6** (1972), 171–172.

[Zas1] H. Zassenhaus, Uber Lie'sche Ringe mit Primzahl Characteristik, Abh. Math. Sem. Univ. Hamburg **13** (1939), 1–100.

[Zas2] H. Zassenhaus, Darstellungstheorie nilpotenter Lie-Ringe bei Characteristik $p > 0$, J. Reine Angew. Math. **182** (1940), 150–155.

[Zas3] H. Zassenhaus, The representations of Lie algebras of prime characteristic, Proc. Glasgow Math. Assoc. **2** (1954), 1–36.

Department of Mathematics and Statistics
Utah St. University
Logan, UT 84322-3900, U.S.A.
E-mail: nakano@sunfs.math.usu.edu